# OPERATIVE TRADITIONS VOL. I

*A Book Where Ernst Jünger & Julius Evola Meet at Last*

MIGUEL ANGEL FERNANDEZ

Operative Traditions Vol. I
Miguel Angel Fernandez

© Manticore Press, 2017

All rights reserved, no section of this book may be utilized without permission, including electronic reproductions without the permission of the author and publisher. Published in Australia.

BIC Classification:
HPX (Popular Philosophy), HPS (Political Philosophy), HBLL (Modern History), HBLW (20th Century History), HRAB (Philosophy of Religion)

978-0-9945958-6-7

MANTICORE PRESS
WWW.HADEAN.BLACK

*Can one thus measure oneself against what is most advanced in contemporary thought and lifestyle, while remaining inwardly determined and governed by a completely different spirit?*
Julius Evola. *Ride the Tiger*

*It was necessary to deepen the roots, piercing a very arid soil, in order to reach those sources where is placed that magical fusion of blood and spirit which makes the word irresistible.*
Ernst Jünger[1]

*Our life is not what matters; what matters is to make possible again in the world a return to a way of living with great style and with great criteria. To this, we contribute if we are much more demanding with our requirements.*
Ernst Jünger[2]

*Never had I discovered in physical action anything resembling the chilling terrifying satisfaction afforded by intellectual adventure. Nor had I ever experienced in intellectual adventure the selfless heat, the hot darkness of physical action. Somewhere, the two must be connected. Where, though?*
Yukio Mishima[3]

---

[1] "Tief mußten die Wurzeln durch dürren Boden hinabgetrieben werden, um die Quellen zu erreichen, in die die zauberische Einheit von Blut und Geist gebettet ist, die das Wort unwiderstehlich macht" Ernst Jünger. *Der Arbeiter*. Ernst Klett, Stuttgart 1981. Klett-Cotta, pg 5.

[2] "Es kommt nicht darauf an, daß wir leben, sondern daß überhaupt auf der Welt wieder die Führung eines Lebens im großen Stile und nach großen Maßstäben möglich ist. Man trägt dazu bei, indem man die eigenen Ansprüche schärft." Ernst Jünger. *Der Arbeiter*. Ernst Klett, Stuttgart 1981. Klett-Cotta, pg 100.

[3] Yukio Mishima. John Bester. *Sun and Steel*. Kodansha USA (2003), pg 86.

## CONTENTS

Introduction — 6

### PART I

The Missing Link of Tradition — 35
Aiming to an Operative Tradition — 45
The Meaning of the Earth — 48
The Decline of the West — 55
Cosmic Attractors — 59
The Growth of Complexity — 91

### PART II

Theory & Phenomenology of the Absolute Individual (TAI-PAI) — 124
Main structure of TAI-PAI — 129
The Concept of Value and the Empirical State of Being — 186
Phenomenology of the Absolute Individual — 204
IT: The Spiritual "Flash," the Lightning of Zeus — 231
Conclusion — 252

## PART III

| | |
|---|---:|
| *Social and Political Implications of TAI-PAI* | 264 |
| *Der Arbeiter/The Worker:* | 268 |
| The Modern Operative Figure Foreseen by Ernst Jünger | |
| *Technique* | 284 |
| *Technique as a Civilization Force that Transcends Scientific and Economic Discourse* | 308 |
| *Caste as an Operative Factor* | 319 |
| *Introduction to Ernst Jünger's Notion of Type & Individual* | 335 |
| *The Operator:* | 352 |
| A Figure that Transcends Nations | |

# INTRODUCTION

The aim of *Operative Traditions* is to present to the readers a perspective on Tradition that is not new, but has been mysteriously veiled during the last few centuries while other aspects of Tradition have been much more emphasized. We are referring to the *operative*[4] aspect of Tradition.

The intention of *Operative Traditions* is to provide a new framework for all those individuals in the modern era who feel deeply akin to the world-views present in the Traditionalist domain, but who are also challenged with the task of applying these ideas in contemporary times. This application doesn't entail any direct socio-political expression, but rather is a new form of interaction with the material conditions of our world that to a large extent are determined by the specific techno-scientific developments taking place amidst modern societies.

When facing this challenging task of reconstruction it can be easily verified that Traditionalism is derived from the mystical, exoteric/esoteric, and speculative domains, while the strictly operative aspects of Tradition became more and more relegated by the cultural domain. In the West, some of the most influential inheritors of the operative modes of Tradition were in Historical

---

[4] "Operative" is related to the aim of inner transformation or *metanoia* by means of *opus*. Opus, which is a term that is recovered from medieval accounts, corresponds to the idea of exerting towards perfection any form of activity, as a discipline that was good in itself, that had a *magical* character (opus magnum, opus magicum) and that is equivalent to the Hindu ideal of the *karmayoga*. In this context *opus* aims for achieving metanoia which entails a change of inner perception of *physis*. Peter Senge. *La Quinta Disciplina*. Barcelona: Granica. 1995.

times the Operative Freemasons, who constituted the main initiates, artists, and architects responsible for developing the Gothic medieval *praxis* and world-view, which afterwards impregnated very diverse cultural domains. During medieval times, the Operative Freemasons were entitled to play an important role in the entire cultural development; Tradition was then fully *alive* and fully *integral*. A temporary "bridge" had been established between the metaphysical domain and the physical domain, spreading on a wide scale through the European continent, affecting the most elementary tools, utensils, arts, artefacts, and conceptual structures. The Operative Freemasons constituted a way of relating to the world that was purely initiatory and artistic. Initiation was only possible through artistic disciplines, and the main purpose of the artistic development was that of facilitating spiritual initiation. By restoring some core Pre-Christian views on the cosmos, these Master architects and artists conceived that the developments taking place in nature *(physis)* always correspond to symbolic expressions of metaphysical powers that are in active in the world, and the whole idea of freedom, dominion and responsibility cherished by these Masters was that of disciplining the spirit of the individual in order to allow convergence with such cosmic development, through the discipline of art as the main foundation. Operative Freemasons had very perceptive vision…they could perceive the determinisms and aims existent within the developments taking place in the material domain, and the symbolising and conceptualising of this phenomena was particularised in scientific knowledge, which became a science of sacred character that required active participation in the initiatory process in order to properly assimilate its contents.

However, even though the symbols of Tradition are eternal, the capacity of such transcendent power to magically direct the artistic developments taking place at a material or human level can eventually become exhausted, and the metaphysical tension can ultimately be extinguished. One of the main consequences of the weakening of the operative thread was that it abruptly affected the human vision of the cosmos, and this mutation of perception was expressed in the genesis of modern science, where any question in regard to the developments, aims and synthetic aspects of the most diverse scientific domains not only become harder to answer, but were no longer asked or posited by a huge percentage of the specialists and experts who developed scientific knowledge itself.

Obviously, in such a context, the Operative Freemasons were not required to assist with cultural developments, and the initiates progressively disappeared from most Historical accounts. Yet this does not mean that the Tradition they transmitted vanished, or that their presence disappeared…it only means that their core individuality became less and less perceptible through the paradigms and ideological components embedded in our modern cultural frameworks and political or social institutions. And whenever any of these highly rationalistic and secular approaches to our past (ideologies and paradigms) become imprecise in order to capture the personalities, then highly irrational myths and legends of mysterious initiates, alchemists, and Rosicrucians emerge as the inevitable counterpart of the approaches.

If any specific date can be presented as the moment in History when the initiatory message of the Operative Freemasons was officially removed from the dominant culture of the West, it was during the constitution of Speculative Freemasonry in London, 1717. This event can be considered as one which represented a subtle, yet extremely determining, mutation in the way culture, politics, economics, and sciences started to be conceived of in the West, and a new faith was presented in the hypothetical capacity of men to drive and determine society, science, politics, and the material domain in general through the development and promotion of ideologies and highly rationalistic conceptual frameworks. Firstly, this new conception of culture and politics transmitted to the peoples was inevitably lacking the firm initiatory foundations that could allow a individual to relate qualitatively to the developments taking place, and secondly, we encounter an absolute reference point for discernment among all cultural productions of a progressively higher speculative character. Before the latter mutation had taken place, such qualitative synthesis had been formerly expressed in the symbolic and alchemical world-views that were cherished by the Operative Freemasons, who aimed to produce in the individual's consciousness a spiritual state that could bridge the metaphysical powers with the immanent developments that were taking place at all levels. But eventually all this wisdom became discarded by the philosophical currents that emerged during the Enlightenment, which began to focus on speculation rather than on the characteristic modes of mundane action; so ultimately the deep comprehension and transference of Traditional symbols into the cultural domains had also been affected and compromised.

One of the key missions of *Operative Traditions* is to recover and describe the distinction between the *operative*, *mystical* and *speculative* derivations of Tradition. Strictly speaking, we can't actually refer to a *living/integral Tradition* unless these three domains (the operative, the mystical and the speculative) are correctly integrated and actual. In short, the *operative* domain refers to those domains of activity where a higher metaphysical principle integrates the physical actions and interactions of a given individual with the environment/territory. This operative domain implies the establishment of territorial links that serve as the foundation for ulterior developments at the level of society, institutions, etc.

\* \* \*

The fundamental ideas of Tradition were recovered by eminent figures (René Guénon, Julius Evola, Titus Burckhardt, etc.) during the last century, but it is crucial to take into account that Western men had also inherited specific spiritual predispositions at an operative level which were highly influenced by their intergenerational and progressive subjection to modern technique, science, propaganda, and reinforced customs inherited through progressive individual exposure to the technical conditions present amidst typically urban-industrial environments. These predispositions caused a considerable gap between the frameworks provided by the aforementioned Traditionalists and the instinctive/subconscious operative predispositions of modern men.

This gap has also widened due to the speculative approaches to Tradition which arrived during the last century from the East, embedded in conceptual and paradigmatic structures that are alien to Western conceptual mindsets and to the original character present in the cultural roots of the West. The inevitable consequence of such incompatibility entailed the apparition of "spiritualism," theosophy and "New Age" tendencies, where speculation, evasion of reality, the private focus on pleasurable happiness for the self, and surrendering to hedonism or to irrational subconscious drives, were the main driving factors of such movements.

The operative character of Tradition is present in Zen Buddhism, where most views on the cosmos are highly practical, non-linear and cyclical. And yet...could Zen Buddhism have fulfilled the operative gap existent already in the Traditions of the West since the times of the Enlightenment? Very unlikely. As it is pointed out in *Operative Traditions*, the techno-industrial development taking place after the Industrial Revolution corresponds to a unique historical operative process, qualitatively different than any of the former technological developments taking place in other civilizations, both in the West as in the East. As History has shown, the last Traditions existent in the East and Middle-East have also been incapable of directing the technological process towards their own aims, thus failing to integrate the material/industrial developments into their own traditions. And yet in *Operative Traditions*, in parallel to a strictly didactic exposition, we shall also follow the very important accounts of German philosopher, Eugen Herrigel (1884-1955), who in his most important book, *Zen and the Art of Archery*, provides a deep and extremely appealing account of the experiences he went through when being in contact with the Zen Tradition that was still alive during the 1930s in Japan. Even though the operative elements that Herrigel had to learn and handle are very different from the objects, systems and devices we use today on a daily basis, the reader of *Operative Traditions* will eventually gain a deep insight into the important analogies that exist—from the highest Imperial viewpoint—between the Operative Traditions of the West and the East.

It is not difficult to sense that the process of hypertrophic techno-industrial development and hyper-multiplication of human needs has endangered the short-term future of the global techno-economic civilization and the growth is clearly related to many planetary dysfunctional processes, such as abrupt climate change, economic instability and severe resource depletion. It is unlikely that today's human population, at a large scale, shall be capable of overcoming this process. Yet Tradition always exists...Tradition is eternal, supra-temporal, and always shows its invisible yet indestructible power whenever some humans aim for embracing evolutionary processes of absolute destruction-construction. If we want to aim for a creative purpose in order to overcome this process of rampant global destruction, it is worth recalling that such vertiginous development was initially triggered in the West, during the times of the French Revolution and the Industrial Revolution. As formerly pointed out, these important Historical events also coincided with the progressive extinction of the Operative Freemasons as cultural Masters and representatives, in addition to the constitution of Speculative Masonry in London in 1717. At a progressively faster pace, the cultural ideals that became "brainstormed" by this influential group started to flow through communicating vessels of an institutional kind within almost all political, economic, cultural, and social structures of the West.

1717 was a tipping point of the History of the West when everything at an operative level began to run amok, like a Promethean Frankenstein or Golem.

## THE STRUCTURE OF OPERATIVE TRADITIONS

A very important aspect of this book, which is directly linked to the aim of recovering the principles required for establishing Tradition in operative terms, is that of fusing, synthesizing, and integrating as briefly and didactic as possible the most unknown—yet very crucial—aspects of the works of Italian philosopher and esotericist Baron Julius Evola (1898-1974), and the German writer and World War I hero, Ernst Jünger (1895-1998).

Julius Evola is highly regarded in the Traditionalist cultural domain mostly because of his views on politics, culture, sex and esotericism, which all embraced standpoints very much in opposition to the subversive movements taking place at a political and State level in the West and East. Yet, surprisingly enough, there still remains a facet of the Italian thinker which is generally dismissed. Julius Evola developed a whole body of philosophical thought, mostly during the 1920s-1930s, where he succeeded in reintroducing the operative elements existent in Tradition within a philosophical framework that is strictly adapted to the Western conceptual mindset. This philosophical endeavour was initially intended to be a reaction against the philosophies or "transcendental idealisms" present in Western thought (Hegel, Kant, Schelling). The Italian author initially named this philosophical development as "Magical Idealism," in a publication[5] still tinged with esoteric components drawn from other non-Western Traditions. In his following works, Evola developed the *"Theory of the Absolute Individual"* and the *"Phenomenology of the Absolute Individual,"* two volumes which set the basis for a new empirical/practical process for the dominion of reality.

Neither of these important texts by Evola have been published in English. Such works however exhaust all the questions and contradictions arising from the developments of Western philosophy until the 20th century, revealing a method for attaining absolute *certitude*, and contrarily to almost all the speculative and idealistic approaches to the key philosophical and epistemological

---

[5] Julius Evola, *Saggi Sull'idealismo magico*, Todi-Roma, Atanòr, 1925.

issue of certitude, Evola no longer intends to establish a rigorous system of categories of thought, but rather a method and *praxis* to progressively approach the experience of reality. As with any culture that provides a specific set of frameworks, rites and techniques for developing the multiple state of being in the individual, it was precisely a reaction like Evola's that should have been expected from Western culture, in order to provide the foundations for attaining a state of absolute certitude by the individual, and rendering such certitude afterwards into a directing force for politics and society in general. Yet during the existentialist and deeply nihilistic times of the last century, Evola's work in this field remained absolutely unnoticed by all main cultural representatives. Even today, one can only find a couple of reviews of his philosophical endeavours, published by Italian publisher Ediciones Mediterranee.

This state of absolute certitude is directly linked to what Evola constantly refers to as dominion, persuasion, freedom and responsibility. This state is, however, an ideal, and in this sense Evola's position still demands an idealistic *first step*, yet no longer in the sense of adopting an abstract philosophical standpoint, but rather in the sense that confidence and trust in the entire process constitutes the most fundamental requirement, even though the spiritual "fruits" of the process cannot be initially demonstrated dialectically or empirically.

In terms of restoring an *integral* and *living* Tradition, the gap between theory and practice—or between speculation and operation—must no longer be widened, since this only entails the ultimate loss of any integral character in terms of Tradition. Strict operative approaches are always present whenever a craft or artistic process is still fully alive, and also whenever the productions relate to the actual territorial conditions of existence. In our times, most remaining traditional arts and crafts—though still requiring purely operative processes, disciplines, and methods—have lost contact with the higher determining forces of our times (industry, technology, etc.) and are remaining in a subsidiary position in regard to the forces. In the same way, most of today's Martial Arts traditions eventually discipline individuals to attain skills and combat techniques that, by themselves at a physical level, have no chance to gain dominion whenever facing most of the global industrial weapons of our times.

The former description on the issues that arise when intending to apply traditional knowledge in a completely new set of circumstances encourages *Operative Traditions* to base its didactic structure in the "centrifugal-centripetal" process of teaching stated by Confucius more than two millennia ago in a text called *The Great Teaching*:

*"The ancients who wished to illustrate illustrious virtue throughout the kingdom first ordered well their own states. Wishing to order well their states, they first regulated their families. Wishing to regulate their families, they first cultivated their persons. Wishing to cultivate their persons, they first rectified their hearts. Wishing to rectify their hearts, they first sought to be sincere in their thoughts. Wishing to be sincere in their thoughts, they first extended to the utmost their knowledge. Such extension of knowledge lay in the investigation of things.*

*Things being investigated, knowledge became complete. Their knowledge being complete, their thoughts were sincere. Their thoughts being sincere, their hearts were then rectified. Their hearts being rectified, their persons were cultivated. Their persons being cultivated, their families were regulated. Their families being regulated, their states were rightly governed. Their states being rightly governed, the whole kingdom was made tranquil and happy."*[6]

\* \* \*

---

[6] Confucius. *The Great Learning,* http://classics.mit.edu/Confucius/learning.html

This first volume of *Operative Traditions* is therefore divided into three parts: the first part approaches the immanent conditions of existence in our times. The second part presents the philosophical approach capable of assisting men for rooting their core being with the cosmos, and the third part exposes the important conclusions drawn from the former parts, in order to re-dignify the value of operative activities.

In the first part of *Operative Traditions* is described the progressive mutation of the operative material conditions in the West, which were culturally conceived as a process of *decline* by many philosophers and historians. This approach towards decline in operative terms requires a study of the empirical conditions of existence, showing to a large extent how the modes of interaction with reality significantly changed over the last few centuries.

In order to accomplish this task it is necessary to describe the crisis experienced in the West—and afterwards in the East—when it comes to the issue of relating to means of power and when intending to gain operative mastery of the means. We shall see how this crisis is clearly related to the incapacity on the part of the Western cultural framework for grasping the actual and effective idea of the State—which Oswald Spengler referred to as *"a purely Western problem that simply does not arise in the intellects of other Cultures"*[7]—not only at a political level but also in terms of integral dominion exerted on the part of the individual in relation to the means he has at reach. Hence it shall be shown how the crisis affects both the macroscopic and microscopic levels. It shall also be shown that whenever powerlessness is manifest in the case of an individual, group, corporation or nation that intends to gain mastery over means, and "pilot" them in a very specific direction, such individual, group, corporation or nation eventually becomes dependent (mostly through debt mechanisms) on the determinations of such development, entering irreversibly into a vicious cycle where impotency is demonstrated by the fact that despite all ideological, economic, and political colourings, nobody is yet capable of answering

---

[7] Oswald Spengler. *The Decline of the West*. Alfred A Knopf. 1926. New York, pg 176.

the question in regards to what is the aim of the entire development of means and that of the economic resources.

Because of the lack of firm reference points, we shall also see the limitations of several scientific paradigms which relate to the aim of the development of the material world. This aspect is crucial to point out since, mostly due to Judaeo-Christian influences, one of the dangerous prejudices that we have inherited from modern culture is the belief that we have no need to establish any connection with the development of the cosmos and that we can master the development exclusively through techno-science. The latter assumption is completely naïve from the spiritual viewpoint embraced by any Tradition since the incapacity to relate to the fatal determinisms existent in the material domain cannot provide any basis for establishing a stable connection or *rooting* to the cosmos. The idea of *rooting* is very much connected to the ancient pagan *ethos* which is at the core of developments in civilization which took place in the past, such as the Sacred Roman Empire. However, the modern concept of paganism or neo-paganism, which to some extent inspires many people to reconnect with folklore or their national heritage, to reject abstract/dualist thinking, and to feel in harmony with nature, is still powerless when it comes to rooting the individual in the actual interplay of forces existent in our times; forces that act upon ourselves in the most unexpected ways. One of the main strategies for overcoming this romanticism is by defining with rigor, clarity, analytically, and synthetically, the basic configurations of material power existent in our times which have been capable of overcoming, integrating and encroaching countless rural and virgin "pagan" natural landscapes all around the globe. One of the consequences of this approach is to realise that we can no longer resort to the romantic concept of nature expressed by Jean Rousseau centuries ago where the distinction between *nature* and *civilization* was commonplace. At this moment we are living in a completely novel cosmos where the local environments are affected by new functional laws and determinations that overcome those of the natural ecosystems, communities, flora and fauna.

The incapacity on the part of our modern scientific paradigms for allowing us to relate to the aim of material development is of huge importance, since the attainment of the aim constitutes the required condition for establishing a set of ethical or moral disciplines that in an operative Tradition are embedded with an instrumental character, or as Spengler wrote:

> "Moral. It is, wherever true and strong, a relation that has the full import of ritual act and practice; it is (to use Loyola's phrase) "exercitium spirituale" performed before the deity."[8]

As modern science is incapable of providing an aim of a metaphysical nature, any set of ethics or morals that are not legitimized by the constitution of aims can be easily eroded by individualistic interest, social conformism or existentialist critique. These nihilistic developments clearly took place during the last decades, corresponding to symptoms of the decline of the West.

The metaphysical, moral and ethical crises that are all related to the incapacity of men to establish new values based on completely new conditions of existence, was surprisingly overcame and "healed" by the apparition of a completely new mode of relating to the world, where men started to progressively remain "out of the equation." As the aims of the material development were no longer graspable beyond ideological or economic discourse, or beyond that which Ernst Jünger referred to as the *"artificial game of words,"*[9] a new set of methods, techniques, and paradigms allowed men from the 1920s to no longer experience the anxiety caused by any metaphysical tension or the need to establish aims for gaining a sense of responsibility. This was accomplished by the development of a new structure of power of a *cybernetic* nature that "feeds"—so to speak— on the deep nihilism of those very active men who, though unable to define aims, are given the chance to establish temporary control over the varied means or contents. As we shall see, this structure is

---

[8] Oswald Spengler. *The Decline of the West*. Alfred A Knopf. 1926. New York, pg 272.

[9] "Das kunstvolle Spiel seiner Begriffe" Ernst Jünger. *Der Arbeiter*. Ernst Klett, Stuttgart 1981. Klett-Cotta, pg 9.

characterized by very specific properties that are intrinsically alien to any metaphysical and autonomous regulation or responsibility, and also characterized by properties that allow it to be considered as a *system* where human actions become integrated, but where human actions cannot determine its development, which is autonomous in regard to human morals, ethics and economics.

The systemic aspect of the new territorial configuration of the planet can be proven mostly by resorting to Ludwig von Bertalanffy's analysis on systems, as well as by reintroducing the concept of *technological operation* and *technological phenomena* inaugurated by French philosopher Jacques Ellul. As it is shown in *Operative Traditions Vol. II* the object of study for comprehending local determinations is no longer the local environment, but the planetary object constituted by the Techno-System, which is a term masterly described by Ellul in his book "*The Technological System.*" This system corresponds to a particular mode of power, in the sense that it enhances the proliferation and activation of specific human techniques that determine the very specific character of the material configurations at a planetary level, which are no longer defined nor mastered by the State-nations governmental powers. Such material configurations correspond to an extremely complex technological ensemble, which, contrary to all other forms of technology that have appeared in the History of the West, has the special attribute of being systemic and cybernetic, but not *homeostatic*, a term that shall be approached in all its descriptive power in the first section of *Operative Traditions*.

The awareness and capacity on the part of the individual for identifying this *system of techniques* or *Techno-System* corresponds to a vital part for introducing both Julius Evola's theory/phenomenology and Ernst Jünger's perspectives on *opus*. If dominion and mastery on the part of the individual are to be accomplished in our times, such a quest is to be faced by employing the system as the main operative challenger, since the more the individual begins to "use" the advantages for control and exploitation provided by the system, the more he becomes controlled by the overall system as a fatal consequence. Due to the key characteristics of this "system of techniques" or *Techno-System* presented in *Operative Traditions*, it appears obvious to the reader that such a system contributes to that what Evola refers to as "the path of the other," *privation*, or the path of "alienation," in contrast to the path of the *Absolute Individual* or that of *persuasion*, which corresponds with Jünger's idea of *dominion*. The core idea in this synthetic approach is that of pursuing the creation of *opus*, a synthetic mode of creative activity that justifies and integrates all former developments, thus transmuting and producing a meaning amidst the functional and horizontal nonsense encroached by the Techno-System. In effect, if any aim can be attributed to such a planetary impersonal and highly systemic contender, it is that of testing the potency for autonomy and absoluteness on the part of any individual, group, or community, since the more a given individual, group, or community relies on such a system for gaining control over economic means or resources, the more they inevitably lose their capacity for autonomous action and absolute freedom.

The latter phenomenon has not been addressed by modern science, and yet Evola's phenomenology relies much more on the issue of the modes of experience with the empiric domain (technique) than that of science, which has constituted the main object of study in many of the phenomenological approaches to reality. Contrary to Evola's standpoint, all other epistemological studies have limited their phenomenology exclusively to the study of science, and it has been historically shown that the incapacity to find an absolute point of reference was covered by misty clouds of relativist and existentialist conclusions. In the case of *Operative Traditions*, technique is the determining factor for structuring consciousness in a very specific mode, a mode that is then transferred to the paradigms that instinctively determine not only the individual's scientific conceptions, but also thinking processes, to the extent that the operative disciplines can even alter psychosomatic reactions, gestures, and subconscious predispositions of the individual, and the predispositions, once developed and cultivated, become hereditary. We are entering into a domain that has evolutionary undertones, in the sense that the establishment of new relations to the actual territorial conditions imply a profound change in the deepest recesses of the individual, in connection with the global challenges presented in our times.

By aiming to provide a basic framework for the reader, the important mutations taking place historically in the domain of technique and its important consequences shall be described, in order to prepare the reader to progressively approach the key issue of technique as an operative value. In this approach, one key element to consider is that different techniques are characterized by different levels of cosmic integration, which basically means that the techniques put forth in a given society and civilization always depend on the nature of the civilization. In contrast to the particular focus on the specific operative traits of past civilizations, most studies on civilization and the diverse modes of State have focused on the myths, religions, economic systems and ideologies portrayed by cultural configurations of power, yet in operative terms, most modern State-nations have clearly surrendered to the use of equivalent techniques, and this surrendering is a key factor—as we shall see—for allowing the development of techno-industrial infrastructures that are no longer circumscribed by differences in ideologies, religions, race, or secular traditions.

* * *

We live in times when vast sectors of our modern societies and institutions give credit to science as a form of knowledge, since to a large extent science is assumed as the form of knowledge that provides us with technology, and consequently with comfort and well-being. Even though this assumption of the relations between science and technology shall be constantly questioned in this book, it is important to recall that science, as a form of knowledge that is intended to provide us with a coherent and objective view of the universe, has actually been in crisis for almost a century. This crisis refers to the incapacity on the part of the main scientific paradigms to provide a view of the cosmos that empowers and assists men to relate *directly* to the cosmos, in practical and qualitative terms. The image of the scientist who impresses us with his accurate descriptions of quantum physics, human biology and physiology, yet who cannot handle his dog or social relationships comes inevitably here to our minds.[10] Modern science, not in terms of its function as an intellectual resource for technological development, but as a world-view that gives justification to our lives in the world, *here and now*, has inevitably surrendered to the same dead-end as many other ideologies, philosophies, religions and spiritualisms; as a world-view, modern science is dangerously embedded with the same character for consolation and evasion, like any other spectacular content, and thus it has become degraded to the level of a popular *faith;* a faith still very much demonstrated whenever in order to provide popular credentials to any journalistic publication it is enough to state it was elaborated by "experts." Yet this corresponds merely to the realm of journalism, where science—like many other spectacular and evasive contents—has become easily integrated and objectified.

Besides journalism, in terms of the application of scientific knowledge based on modern paradigms, *application* corresponds to an *operative* domain. Hence in order to understand the potentials for the application of science, it first requires a study of *operations*, of *actions,* and of *direct experience*. This operative realm corresponds precisely to that of *technique*. Yet this study of technique, if intended to be accomplished integrally and constructively, requires an approach that is absolutely alien to our common understanding of what "studying" or "research" really means. An integral study of technique requires the cultivation of an intense discipline between the individual and the surrounding objects and/or means of power available; it requires the progressive elimination of all the

---

[10] "If one believes in science, one must accept the possibility—even the probability—that the great era of scientific discovery is over. By science I mean not applied science, but science at its purest and grandest, the primordial human quest to understand the universe and our place in it. Further research may yield no more great revelations or revolutions, but only incremental, diminishing returns"

J. Horgan, The End of Science: Facing the Limits of Knowledge in the *Twilight of the Scientific Age* (London: Little, Brown and Company, 1996), pg 6.

subjective interpretations of phenomena and requires focus on the actions and reactions established between the spirit that impels the individual/subject and the spirit that impels the objects; it requires the development of a deep and symbolic comprehension of the aims that the configurations follow, to the point that no artificial separation between object and subject is necessary, and to the point that object becomes subject and subject becomes object. The latter corresponds, briefly, to an *operative* discipline, such as that taught by the Operative Freemasons, or that which Ernst Jünger experienced in warfare situations, whereby learning to relate to the processes and developments of the material development, one learns to relate to oneself. Once this spiritual state of convergence I-"non-I" is established, a deep inner criteria or value is then timelessly set which allows to configure and classify all the elements provided by science; a configuration no longer valid for everyone—which is the main assumption of modern science and Immanuel Kant's "practical reason"—but only valid and vivified as a set of conceptual and symbolic frameworks or reference points available for those who have gained an analogous and qualitative experience of reality.

This analysis of the complex character of our operative conditions of existence corresponds to the main contribution of the first part of *Operative Traditions*, where it shall be outlined how the progressive dispersion of human responsibility at a strictly operative level is mainly due, is mainly due to the emergence of a completely new set of interactive and technical conditions.

\* \* \*

The former context is, therefore, the main pre-requisite for exposing and presenting Julius Evola's philosophy as an adequate framework in order to approach the operative issues presented by our age. Evola's "*Theory and Phenomenology of the Absolute Individual*" gives us the chance to start developing a new relation with cosmic forces in a completely primordial and original way. Once described, the main characteristics of both the empirical domain and our relations to it are well established in order to present to English readers Evola's philosophical work. Science requires a material domain in order to prove its hypothesis, and the realm of technique will constitute the "matter" required for framing Evola's theory. In conjunction with Evola's philosophy the second part of *Operative Traditions* will also deal with the operative challenges faced by Eugen Herrigel when aiming to develop himself via the Zen art of archery. Herrigel's account is one of the few texts available that recovers a spiritual practice which Zen Masters such as Awa Kenzo still incarnated—mostly before 1945 in Japan—a very specific and non-physical mode of energy that completely altered the character of the body's movements, and that corresponds to the empiric proof of the "bridge," "thunderbolt"—or what is referred to in *Operative Traditions* as a "flash"—between the transcendent and the immanent.

Though Evola presents us with an understanding of the process that grants a "point of departure" which is the same for everyone, this does not imply that everyone will have the qualifications and virtues required to further progress through the process, which is embedded with ascetic traits that resemble those of the Buddhist disciplines in the Pali Canon.

Evola's theory—as with all theories—has to be verified in practice and empirically: in *gesture and body movement*. The specific mode of energy released in the peak achievements of the spiritual practices can be cherished by the practitioner as a solitary yet sacred proof of his empirical connection with the sacred or the transcendent. If the social conditions are adequate and receptive, a Master might even intend to channel the transcendent energies into societal forms. In the latter case, the elaboration of a set of normative arrangements is required which ritually "dose" the suggestive power of the energy into the consciousness of those initiates who have received a proper operative preparation. Afterwards these initiates can also become sources of irradiation, which allow a whole imperial configuration of efforts and convergences at a wide scale, resembling analogically the socio-political configurations which constituted the creative core of Solar Empires such as the Japanese or even the ecumenical configuration of power prevalent in the European Gothic world-view.

Evola's framework and Herrigel's extraordinary experience challenge us to present in *Operative Traditions* one of the key factors of our times: *technique*. At this point it then becomes necessary to reconnect Evola's philosophical work with the writing of Ernst Jünger, which is the aim of the third part of this volume.

* * *

The issue of *technique* is one of the greatest mysteries of Western thought, having being studied in depth during the last century, by Lewis Mumford, Oswald Spengler, Ernst Jünger, Marshall McLuhan, Friedrich George Jünger, Gilbert Simondon, Jacques Ellul, Neil Postman, and Jerry Mander. Among all these authors it was definitely Ernst Jünger who was the first to perceive the revolutionary power of technique, and this perception clearly influenced his brother Friedrich George. It is plausible to consider that these authors had a powerful influence on other authors like French philosopher Jacques Ellul—who during his life was devoted to studying technique in a more detailed, sociological, and theological way than Ernst Jünger or Julius Evola. When referring to the study of technique, it is convenient to point out that we are not referring exclusively to engineering or to the study of technology. As we shall show in this book, in the English language the term *technique* is very often equated with *technology,* and yet in French, German and Spanish there is still a conceptual distinction between both terms. It can't be a coincidence that the progressive marginalization of the historical development of technique as an object of study ran in parallel to the progressive disappearance of operative disciplines, traditional arts, and crafts.

Ernst Jünger's first writings arose from the experiences he gained in the battlefield during World War I, where the development and power of the industrial machinery of warfare was rendering human individuality into something extremely fragile, and when new international infrastructures of power completely altered the previous modes of activity within the military, eventually spreading into civil societies as well. The idea that Jünger transmitted to those individuals who "*are not to be found in those sectors of the front that are to be defended, but in those where one acts*"[11] was that all the ideological views on men and their place in the cosmos—diffused to the peoples of the West to a great extent after the fatidic date of 1717—fell like a house of cards whenever ideals of the Enlightenment were challenged by raw material powers that did not follow the laws of human designed superstructures. The *total mobilization*[12] of all material factors intensively rebelled at progressively higher paces against all the bourgeoisie ideals of security that were based mostly on a speculative view on the human condition and society. Jünger not only gained a crucial experience of this development but was also capable of describing it with a sense of mastery. Ernst Jünger profoundly captured how a new mode of organic development was taking place at a strictly *operative* level; the entire conflict of powers between nations, banks, corporations and many other power-structures was taking place in this operative "battlefield" where the efficiency of machine conglomerates determined what political or economic entity would be victorious. In this scheme of things, the ensemble formed by ideologies, political slogans and ideals were rendered into propagandistic contents driven by technical means that were characterised by a very clear, rational, analytical and mechanistic nature. In other words, the new relations of power and the new State configurations were being released from all individualistic or social subjections and becoming linked to technical operations that were increasing at all levels. "*Ideas,*" considered here in an abstract and discursive sense, were also less crucial on the battlefield since another mode of configuring the material domain appeared that was transcending all others: *technique*.

In *Der Arbeiter (1932)*, Ernst Jünger referred constantly to the latter idea and the crucial importance of focusing on the intrinsic

---

[11] "Man darf nicht an den Teilen der Front angetroffen werden, die zu verteidigen sind, sondern an denen, wo angegriffen wird" Ernst Jünger. *Der Arbeiter.* Ernst Klett, Stuttgart 1981. Klett-Cotta, pg 46. This quote is rephrased as well by Julius Evola in his text "Orientations."

[12] Ernst Junger, *Die Totale Mobilmachung* (Berlin: Verlag der Zeitkritik, 1931); English translation: "Total Mobilization," trans. Joel Golb and Richard Wolin, in Richard Wolin, ed., *The Heidegger Controversy: A Critical Reader* (New York: Columbia University Press, 1991).

power embedded in operation and technique. This book, though subtly recovering core conceptions that go deep into the history of the West, has not yet been assimilated adequately by academia, and whenever approached, academia has tended to an ideological interpretation of Jünger's ideas. Consequently, this essay, among some other writings of Ernst Jünger written before the 1950s, is a testimony to a new way of seeing and perceiving the world. Hence, Ernst Jünger's intent was to express in words a world he could only conceive as a symbolic conflict of diverse figures aiming for dominion. It was clearly a metaphysical struggle that every time it was projected into language became degraded in regard to its pristine significance. The title of Ernst Jünger's masterpiece ("*Der Arbeiter*"/English: *The Worker*) is extremely paradigmatic in this regard, since Jünger did not intend to describe a way of relating to the world conceived in terms of *work*, but rather a way of relating to the world which has to be now recovered as *opus*.[13] Hence in Operative Traditions our intent is to approach Jünger's main figure of dominion with a new name: the *Operator*, so any confusing interpretations of his essay can, at last, be considerably minimised.

During the same decade, Jünger published *Der Arbeiter*, Julius Evola—as aforementioned—had also produced a philosophical framework potentially capable of sheltering and protecting the views of the German author from all the cultural distortions of those times, distortions that were already embedded in the structure of language. It is known that both authors wrote to each other during their lifetimes, yet it is not totally clear if they ever met in person.[14] It can be hypothesized that such an encounter would have

---

[13] Opus, a term that is recovered from medieval accounts, corresponds to the idea of exerting towards perfection any form of activity, as a discipline that was good in itself, that had a magical character (opus magnum, opus magicum) and that is equivalent to the Hindu ideal of the karmayoga. In this context opus aims for achieving metanoia which entails a change of inner perception of physis. Peter Senge. *La quinta disciplina*. Barcelona: Granica. 1995.

[14] Though there is only one letter documented to be written by Julius Evola to Ernst Jünger in 1953, the German author later stated in Il Secolo d'Italia during 1986 that both had maintained "a long epistolary contact." This opens an interesting mystery about both authors. Gianfranco de Turris considers five important questions: *"1) Is it possible that Jünger had a "false recalling" with such an accuracy in 1986? 2) Or was it rather a translation error, and the German writer might have met Evola in his house in Rome after 1953? 3) Or also, instead, it was all an invention, but what would be the point of it? 4) And if there*

definitely constituted the triggering of a very dangerous alchemical synthesis, that is, the reconstitution of the symbolic "sword" of Tradition that had been broken since medieval times into two pieces.[15] Such a "Sword" could have been finally wielded together, and the revivification of a Traditional world-view which had been buried for centuries could have been resurrected and transmitted with the power of a thunderbolt to the entire world. And yet this synthesis apparently did not take place. Evola's philosophical work, characterised by a grandeur only comparable to that of a magnificent cathedral or temple, is still awaiting to become illuminated by the new dawn of an integral Tradition, awaiting new guests who aim to be sheltered by this architecture of thought. The fact that this work has remained in "underground" culture for about a century is clearly due to the fact that such a brilliant accomplishment can very easily trigger fearful nightmares in any subsidised scientist, technician, politician, scholar or nihilistic personality who, though feeling very self-satisfied in our times, are completely unaware of the raw power that drives their lives at a subconscious and operative level. In *Operative Traditions* we shall show that, due to the very specific development of powers emerged since the 1920s-1950s, the potential synthesis we are referring to between the works of both authors was not completely accomplished since both authors were capable of expressing symbolically in their writings the strong determinations or fatalisms existent in their times (Evola by resorting to myth and Jünger to figures such as the Operator, the Anarch, etc.). As we shall see, the powers were still at a very initial stage of development since both authors lived as *men among the ruins*.[16] So the hypothetical synthesis

---

*was really a meeting: but in this case one has to imagine Evola traveling in a paralyzed state to Germany after 1953, something not impossible (during those years the philosopher was however in good physical conditions and often traveled to Bologna, Modena, and other places for visits) yet highly unlikely, since there is no proof of such encounter. 5) And such "long contacting through letters" exists or does not exist? And if it exists, which was the aim? All this questions remain unanswered, and they won't be resolved until conclusive documents appear that can reveal the whole issue."* L'Operaio nel Pensiero de Ernst Jünger. Ediciones Mediterranee, pg 10. [Excerpt translated by Miguel A. Fernandez].

[15] That is, the divorce between speculation and operation or between spirit and matter.

[16] Julius Evola. *Men Among the Ruins: Post-War Reflections of a Radical Traditionalist* is a book by Julius Evola. Inner Traditions.

between both authors was very much challenged by the global development of power that was in total opposition to the spirit of the synthesis. As we shall point out in this book, any such synthesis Evola referred to as *immanent transcendence* could not be feasible until the immanent developments could reach a stationary phase that could facilitate a fixed relation to the powers, and permit the establishment of a method, discipline or doctrine that could "galvanise" the whole process. If intending to express this with an image, such a task would be as challenging as planting the seed of a powerful tree (a being) in a soil (becoming) that is constantly in motion, and constantly changing its material conditions, configurations and properties. When faced with this key issue, both authors decided—especially after the 1950s— to adopt more interior positions and to become human testimonies of a way of understanding life and destiny which itself is characterized by a transcendent importance. The parallelisms between both authors among the ruins of the Western civilization are present in Evola's books like "*Ride the Tiger,*"[17] and in the case of Ernst Jünger in literary figures such as the Anarch,[18] just to provide a couple of examples.

The heritage both authors provided to us is of extraordinary value. In the third part of this volume the main focus shall be redefining and revaluing a domain of activity that in our times has powerful operative elements, and that grants the chance of powerfully developing consciousness in convergence with the metaphysical powers *acting* in the world. We are referring here to the domain of *work*.

---

[17] Julius Evola. *Ride the Tiger. A Survival Manual for the Aristocrats of the Soul.* Inner Traditions.

[18] The heroic inner attitude in correspondence with modern constrictions is well expressed in Ernst Jünger's Anarch (*Eumeswill* 1977): "The Anarch is the positive counterpart of the anarchist. I am an anarch—not because I despise authority, but because I need it. Likewise, I am not a nonbeliever, but a man who demands something worth believing in. The anarch sticks to facts, not ideas. He suffers not for facts but because of them, and usually through his own fault, as in a traffic accident. Certainly, there are unforeseeable things—maltreatments. However, I believe I have attained a certain degree of self-distancing that allows me to regard this as an accident. As I have said, I have nothing to do with the partisans. I wish to defy society not in order to improve it, but to hold it at bay no matter what. I suspend my achievements – but also my demands. Although I am an anarch, I am not anti-authoritarian. Quite the opposite: I need authority, although I do not believe in it. My critical faculties are sharpened by the absence of the credibility that I ask for. As a historian, I know what can be offered. The Anarch is to the anarchist, what the monarch is to the monarchist."

\* \* \*

One of the aspects where both Julius Evola and Ernst Jünger shared a common view was in the need to overcome economic determinism, or in other words, to overcome the surrendering of individuals to liberal, neo-liberal or Marxist views on life. Accomplishing such liberation allows the individual to relate to the active modes of power existent in the world, and as Evola outlines in his theory, it allows for establishing *empirical conditions of existence*, so the relationship with operative powers becomes actual and effective, acting as a prerequisite for absolute freedom and dominion in the domain. Hence, *work*—at all levels—has in *Operative Traditions* the chance of gaining a new dignity, not as a means for material production and satisfaction of needs (as is mainly considered by socialists and liberals), but as an activity characterised by a operative character, especially in our techno-scientific times. Depending on the specific sectors, work environments always provide a greater or lesser margin for developing activity according to a style Jünger described as corresponding to the figure of the *Operator* while keeping in mind the *aim* Evola described in his *Theory of the Absolute Individual*. When an individual approaches work from this perspective, activity eventually becomes the most crucial chance to relate to the effective powers of our times; work becomes a means for self-liberation and dominion, allowing an individual to attain progressively higher and more integrated realms of operative/professional competence, to the point where such an individual can finally reach *sacrum otium*—the sacred redemption from all economic determinations and a new focus in establishing power from this centre. This is exactly what Ernst Jünger referred to as *dominion* and the constitution for new modes of State—new castes—at the exact moment when all liberal, socialist and Marxist world-views exhaust their capacity to ideologically drive social, political, and economic developments. There is no other way to conceive this path but as a strictly *magical* path, redeemed from all modern occultism, traditionalism, and esotericism; that is, a path made effective, actual, and empirical.

This is also an ascetic path, not in the sense of physical mortification and repression, on the contrary: in the affirmation of all bodily potentials, yet always intending to aim a higher and synthetic goal.

One necessary requisite in this third part of *Operative Traditions Volume I* is to emphasize how the technical factors develop the economical, so before acting on economics it is required to act on the operative aspect of the configurations of power. As in the case of all Traditional States, the economy constitutes an organ or sub-function that determines the domain of production, capital, resources and workforce, yet above this domain a sacred hierarchy and political elite has to embody and symbolize *absolutely* and *homeostatically* the form/figure/gestalt or—in scientific terms—"attractor" or *autotélos*[19] which all economic, cultural, and ritual practices aspire to. In our times the fact that the economic sub-function is so overwhelming and intense doesn't necessarily imply that the whole process is less determined by the higher drives of a given structure or system. It is precisely the unawareness of the system—and to selectively use its means for our strictly autonomous purposes—which necessarily entails economic alienation (the need to believe in economic determinism, or that "money makes the world go around") as a form of what Evola refers to as a mode of privation. Evola's *Theory and Phenomenology* allows us to be released from this state of privation and attain a state of dominion, where the individual integrates within himself the diverse powers at reach, yet no longer is determined by such powers. This individual state of dominion corresponds to the *Absolute Freedom* and the individual who mysteriously reaches it is the *Absolute Individual*. Once attained a given power structure can no longer affect the individual who has acquired such freedom. In practical terms, this is equivalent to Evola's idea of "riding the tiger" which eventually is affirmed in the emergence of a new Order of men characterized by very particular biogenetic and spiritual traits.

Before delving into the domain of *work*—which shall be further developed in*Operative Traditions Volume II*—the progressive

---

[19] A thing which is autotelic is described as "having a purpose in and not apart from itself." The word comes from the Greek αὐτοτελής autoteles from αὐτός autos, "self" and τέλος telos, "goal." *Merriam-Webster Dictionary*.

outlining of Ernst Jünger's ideas in *Der Arbeiter* allows us to address the key issue of technique as a core operative factor. Evola's Theory and Phenomenology also enables the reader to perceive technique as a configuring force present in *physis,* which develops independently of all individualistic traits of the self.

This approach allows us to unlock all linguistic association between *technique* and *machines*, *technique* and *technology*, or *technique* and *science*. Technique, as conceived originally by Ernst Jünger, is overall, an operative magnitude that transcends any human or anthropomorphic conception, so ultimately it is absolutely unrealistic and naïve to believe that technical development can be determined by morals, or by a strictly economic perception of the world. The only way technique can be mastered is when all individualistic "skins" of the human ego are removed, and a new human type emerges who no longer projects on the actual conditions of modernity an illusory world-view of human affairs where ethics, good intentions, ideologies, spiritualisms, utopias, or ideals of happiness and the "good for humanity" are supposedly intended to govern cosmic forces characterised by a volcanic and titanic character. There are macrocosmic forces present in the cosmos that transcend the very thin layer of all our human intentions, and the only way to not be attacked by the telluric and titanic forces present in the underground and our psyche is by experiencing a relationship with these forces in terms of strict power relations. This is what Jünger constantly refers to as an active nihilism developed by a deep sense of *heroic realism*. By integrating the spirit of technique, technique shall no longer constitute a force intended to be artificially used in order to strengthen a weak and ephemeral sense of the individual self, but rather shall serve as a force that strongly empowers dominion of means in relation to the power configurations present in the world.

This corresponds to the transition Ernst Jünger refers to as from the "individual" to the "type." This transition—which can be highly traumatic for individuals anchored in a romantic or bohemian view of life—also entails that all sociological references to "class" (the social position of an individual based on a higher or lower attachment to material production or political predominance in the domain of 20th, 21st century State-nations) lose all practical and substantial validity. Ludwig Von Bertalanffy once wrote that *"never had been seen before the individual so tied, so dominated, so directed in the field of his most intimate activities by impersonal social forces, frequently inhuman forces."*[20] Jünger's concept of "type" resembles in practice the traditional idea of "caste," or what Frithjof Schuon referred to as *"the operative principle behind the phenomena of social distinction,"*[21] a hierarchical organization of a community not based on the amount of one's possessions, but rather in how one has produced what one possesses, to the extent that the higher members of a caste (emperors, kings, etc.), don't even produce anything material, but rather produce specific energy modes through ritual technique (magic). This "how" or "know how" typical of a given caste refers to the "form" of a given activity, which is ultimately always of a technical nature. The more an individual thinks in terms of "class," the more he is determined by technique and economics (technical/economic alienation); the more an individual becomes a "type," the more he can master techniques, developing a novel mode of economic relations which surpasses any liberal or Marxist conception of the economic domain.

The concept of "caste" is thus also, by extension, of an operative magnitude, and has very little to do with the biological conception of "race." The key issue present in our times is that the operative aspects that are naturally implicit in the old castes, aristocracies and nobilities became less and less applicable to the specific conditions of our highly urban-industrial and networked times, mostly due to the fact that modern technique is substantially different from the previous prevalent technical modes embraced by the castes, thus causing some specific caste attributes to become what Jünger would refer to as belonging to the typically "museum" world, as opposed

---

[20] Ludwig Von Bertalanffy. *Towards a New Image of Men.* Clark University (Worcester. Massachusetts) January 1966.

[21] Frithjof Schuon, Seyyed Hossein Nasr, *The Essential Frithjof Schuon* (Library of Perennial Philosophy), World Wisdom (2005) pg 196.

to the actual domain of machinery, warfare, technique, cybernetics, etc. where two types prevail, the *technician* and the *technocrat*, who become types capable of undertaking feasible operative decisions at a planetary scale, thus rendering impotent all State-nation political structures based on the conditions of existence previous to the 20th century.

The power of technique as an operative factor is mainly expressed in the economic domain when projected in a very specific activity: *work*. An insight into the nature of technique shall also allow us to present all the mystifications embedded in the term "work" as exclusively related to economic necessity. Ernst Jünger's key figure, the Worker/Operator, is liberated from economic necessity to the same extent he is liberated from a purely individualistic sense of identity, and this causes the figure to no longer constitute an "object" mobilised by vast technical processes, but rather to constitute an ultimate justification and source of power that attracts centripetally all means and experiences required in order to develop *opus*, that is, a form of purely artistic activity that aims to integrate all techniques, as an "unmoved mover" which affirms a new state of freedom and responsibility. This purpose is thus linked to the aim of attaining *sacrum otium*, no longer focused in consumerist leisure, but rather political and artistic production in the highest sense. Hence, one of the inevitable corollaries of the former is to realise that without an adequate focus on the operative/ascetic aspects of work, then domains such as modern technique, modern science, modern politics and modern ideologies can constitute the most dangerous counter-traditional forces in the sense that they are *counter-operative* and thus that they inevitably promote all the falsifications spread after the French Revolution and after the constitution of the Speculative Masonry. We could even admit that in this regard they also favour *counter-initiation* practices, in the context used by René Guénon. In terms of Evola's theory, all these forces favour individual privation, hence compensating with insubstantial surrogates for the lack of dominion of reality on the part of the individual.

# PART I

# THE MISSING LINK OF TRADITION

*"What is necessary is not to talk about tradition, but to create it."*
Ernst Jünger[22]

At the current time, the concept of Tradition is awakening more and more interest around the world. This interest is demonstrated by the high level of publications referring to the subject, and by the presence of studies of Tradition existent in many cultural domains. During the last decades, the Internet has facilitated enormously the diffusion of information in this regard, and all the various traditional symbols, writing, scriptures, artistic works and studies on Tradition have been made available to people around the world.

This interest in Tradition could be due to many different reasons that are impractical to list here. And yet, as a first approach, when we talk about "Tradition," what are we actually referring to? In *A Handbook of Traditional Living*[23] it can be read:

> *"Etymologically, 'Tradition' derives from the Latin tradere, a verb formed from trans (= 'beyond') + dare (= 'to hand over'); hence, it indicates the act of passing something over and should be understood as 'that which is transmitted'. Tradition consists not in the conservation or consolidation of exterior appearances or of things the meaning of which is no longer understood: Tradition*

---

[22] "es nicht von Tradition zu reden, sondern Tradition zu schaffen gilt." Ernst Jünger. *Der Arbeiter*. Ernst Klett, Stuttgart 1981. Klett-Cotta, pg 103.

[23] Raido. *A Handbook of Traditional Living*. Arktos 2010.

*rather indicates the direct and effective transmission of a heritage that is non-human and essentially spiritual in origin."*

It is precisely when it comes to considering what this "heritage" might actually be, that we can end up with diverse interpretations. Interpretations can range from the transmission of a set of beliefs, artistic canons, rituals, customs, lifestyles, folklore, techniques, religious ideas, myths, etc. to many other approaches where the word "tradition" is commonly associated with conservatism in order to maintain some stability of thoughts, morals, policies, and values. Accordingly, the spectrum of views on what Tradition might be is extremely vast.

It appears to be, that if as a first approach we associate the term "tradition" with that of handing over to our lives the elements that exist Historically in the past, then we should have to necessarily consider that the elements of the past are currently more present than ever, as a vast number of individuals worldwide are gaining access to sources of information online. All of the latter is in addition to a progressively more abundant amount of specialized books and magazines in regard to History and Civilization published worldwide, in correspondence with the production of music and films that reproduce the socio-political and artistic conditions of our past and cultural roots.

Despite the availability of this information which relates to the entertainment and tourist economic sectors, it is undeniable that whoever is deeply interested in gaining access to the sources of our cultural heritage has more access than ever to the contents of such heritage. We only have to imagine ourselves in the plight of an English peasant of the 17th century, and his particular conditions of existence, in order to realize how much things have changed in terms of access to the contents of our cultural heritage, and that today such contents are more available than ever by a simple click on the mouse, or by just pushing buttons.

So it is undeniable that if we approach Tradition as the particular "handing over" of information and contents from our historical past, it is inevitable to deduce that the interest in Tradition is apparently more remarkable than ever. And yet, even by considering what elements and contents might be transmitted by Tradition, if we dare to release our eyes from the TV screen, books, and films on History, we shall realize that—in terms of our actual modern lifestyles, activities, modes of relationship, and social structures—there appears a progressively widening gap between the lifestyles of our ancestors and that of our particular lifestyles. And it is not even necessary to resort to historically documented evidence for the realization, as any of us can easily verify the enormous gap existent between the current generation and that of our grandparents.

The interest in the traditions of the past by today's public is taking place in parallel with the highest demolition of all intergenerational links with the lifestyles embraced by our ancestors. The younger generations are well aware that the disciplines, ethics, crafts, lifestyles and beliefs that are typical of their ancestors are less and less applicable to the immediate conditions of a highly modified environment, which is mostly of an urban/industrial kind. Due to the developments taking place in the technological and educational environment, very young individuals are now already attaining high levels of social prestige and personal influence in society whenever participating in activities (like software engineering, for instance) that our grandparents can't relate to at all…thus, by observing the socio-economical advantages of release from any actual Tradition. Why would the new generations be actually interested in recovering them?

If we start approaching the idea of Tradition in terms of the intergenerational transmission of lifestyles, skills, ethics, and crafts, we can clearly conclude that this transmission is vastly impeded due to practical reasons. In our times, such lifestyles, skills, ethics and crafts are treated with the same condescension and even disdain as the values of today's elders, who are inevitably the embodiments of the existential traits of the past. Ethnologist Konrad Lorenz (1903-1989), when considering the processes that consist in the transmission of a set of traits that allow the human species to adapt and to relate harmonically to new environments, viewed progressive demolition of the intergenerational links as one of the *8 Civilized Man's Deadly Sins.*[24] Thus it is clear that when examining the term "tradition" based on the former approaches, some important contradictions can be easily pointed out, contradictions that refer to the apparently growing contrast between the *interest* in traditions by the public and the actual *practice* of the traditions on a daily basis.

Friedrich Nietzsche already foresaw this progressive gap, and he did not perceive it as a useful trait for the future of the human species, and to some extent, the German philosopher even thought that it corresponded to one of the key signs of human decadence. In *Thus Spoke Zarathustra* it is written: *"This, however, is the other danger, and mine other sympathy:—he who is of the populace, his thoughts go back to his grandfather,—with his grandfather, however, doth time cease."*[25] In the context of the book, Nietzsche uses the term "populace" or "plebs" to refer to those individuals who are disrespectful of their traditions, who have no regard for their cultural heritage, thus corresponding with the hierarchical conception of many ancient civilizations where nobility was linked to having had "divine ancestors." Friedrich Nietzsche even considered that the greatness of a personality is provided by the depth of his *roots*. When referring to Napoleon, for instance, the German philosopher writes: *"Napoleon was different, the heir of a stronger, older, more ancient civilization than the one which was then perishing in France."*[26]

One of the issues that appear in our times in the case of the men forged by Western culture, especially in the case of the men highly devoted to action—or what Oswald Spengler denominated

---

[24] These are some thoughts by Konrad Lorenz in this regard. "All that is transmitted by tradition over long periods of time, ends up adopting the characteristics of a "superstition" or a "doctrine" (...) The attitude that much of the current generation of young people has for the generation his parents has a generous measure of arrogant contempt, there is nothing good about it. The revolution of today's youth is driven by bitterness (...) The pace of change that is imposed on the current culture by technological change implies that what a generation still considers as good legacy for tradition is considered obsolete for a very important part of the critical youth (...) At a somewhat later age can be seen, especially in children, the disturbance caused by the loss of the father image. Except in the peasant environments, a child rarely sees his father working, let alone they have the opportunity to help in their work and thus to perceive the superiority of a mature man" K. Lorenz. *The 8 Civilized Man's Deadly Sins.*

[25] Friedrich Nietzsche. *Thus Spoke Zarathustra.* Old And New Tables.

[26] Friedrich Nietzsche. *The Portable Nietzsche.* Walter Kaufmann. Penguin books 1977. pg 547.

as "*faustian* spirit"—is that of how to select traditional forms that can be applied to today's conditions. A perfect expression of this challenge was once written by Gilbert Keith Chesterton when affirming in *The Mystagogue* that:

> *"If the idea does not seek to be the word, chances are that it is an evil idea. If the word is not made flesh it is a bad word."*

This raises another issue, which is related to the extent to which the concept of an "idea" has become banal, as is employed in varied and contradictory contexts. By assuming in classical terms that ideas are primordial to words and language, such a banal character adopted by the concept of "idea" derives from the progressive acceptance of language as a form of communication equally banal, that is, as a potential means for *communion* among individuals that is deprived of any capacity for *meaning* in convergence with the meaning of our lives.

In many domains, ideas have been reduced to ideologies or theories, which both constitute intellectual constructs that provide us with a schematic world-view that *justifies* the integration of ourselves within given conditions of existence. And yet as History has shown very clearly during the last century, all these discursive constructs can fall very easily, whenever our eye becomes capable of perceiving new facts that can not be explained by the conceptual frameworks or ideologies. This process of erosion has been more obvious in the case of modern scientific theories and work hypotheses, which are applicable in those domains where entropic dissipation is low and where time and space maintain their coherence (highly mechanistic domains, for instance), yet are forced to rely on a statistical approach where the empirical approach enters into highly entropic domains (weather forecasts, etc.) or into atomic and intergalactic dimensions. As in scientific theories, we could consider a parallel process taking place in many of the most common modern ideologies (Marxism, liberalism, equality, etc.) which have all been overthrown by facts that once again contradict their basic thesis.

Whoever in our times maintains a sense of honesty in regard to the aforementioned contradictions seems to have only two extreme options to resort to, as a sort of existential consolation. Ernst Jünger wrote in 1977:

> *"No living values are left. The historical material has consumed itself in passion. Ideas have become untrustworthy, and the sacrifices made for them are disconcerting."*[27]

In this historical context, one option is that of surrendering to a nihilistic or existentialist perspective of life where nothing makes sense and everything is pure chaos; the other option is that of avoiding the pain and vertigo caused by facing the nihilistic chaos or abyss by surrendering to rigid dogma. During the last century, the adoption of a dogmatic existential approach to such contradictions was dressed-up in the form of religious, ideological, consumerist, and political[28] masks. In terms of religion, we can quote both Spengler's expression "second religiosity," and that of René Guénon when referring to "spiritualism," where basically both traits coined by the German and French authors can be summarised in one single existential attitude: *evasion*. It is an inevitable trait of human beings to need to relate to a justification of life, no matter how irrational such justification might appear whenever it is confronted with effective reality. Yet whenever we loose the skills and attitudes required in order to grip reality in all its powerful dimensions, then evasion of reality becomes the most tempting option. So by returning for a moment to the initial issue of Tradition, it appears to be that if there is an extremely valuable element that has to be recovered from the vast array of traditions, it is precisely that of identifying what individual *skills* and *attitudes* are required in order to grasp the multidimensional reality we are living in, and ultimately to be one with it. In other words, recovering the practices or *praxis* that can allow us to *root* ourselves in the cosmos, *here and now*.

---

[27] Ernst Jünger. *Eumeswil*. The Eridanos Library. 1980.

[28] We refer here to politics as reduced to the political contents present in the public opinion, and not politics in terms of activities in relation to the power of a given State.

Some might be tempted to believe in the option of philosophical activity or speculation as an alternative for surpassing the formerly described dead-ends. This book itself can be considered as a philosophical approach, and yet what is absolutely fundamental to understand, is that the roots from which the pages of *Operative Traditions* have emerged belong to a domain that is all but philosophical; the roots can easily be compared to what some philosophers or scholars would consider—quite disdainfully sometimes—as "profane" or "mundane" domains of reality. Due to very particular interpretations of History spread by modern culture, we still have present in our minds the image of the classical philosopher living a contemplative life in a "Ivory tower" or university, alienated from the world…and yet it wouldn't be completely fruitless to consider that this particular image presented to us as just one of those many ideals transmitted by the cultural establishments of the West, which as any ideal, law or simplification, hides elements very much worth considering. In the same way that the cultural elites once transmitted to the public a rather speculative and contemplative image of the philosopher, constituting in many cases an individual who aims for the moral "good" of humanity, but also in popular culture wolves have been presented as "bad." There may be hidden elements to this moral interpretation. Even a key philosopher of modern times, Friedrich Nietzsche, became extremely sceptical of philosophy and started to grant more importance to the forms of art, and to the reconnection to the individual senses and the body. The German philosopher writes:

*"I agree more with the artists than with any philosopher hitherto: they have not lost the scent of life, they have loved the things of "this world"—they have loved their senses. To strive for "desensualization": that seems to me a misunderstanding or an illness or a cure, where it is not merely hypocrisy or self-deception."*[29]

---

[29] Friedrich Nietzsche, Walter Arnold Kaufmann, R. J. Hollingdale *The Will to Power,* 1968, pg 434.

In this book, we shall eventually deal with the impotence of strictly philosophical or speculative approaches used to recover the missing link of Tradition. This "missing link" constitutes the subtle link which *bridges* the immanent conditions of existence with the transcendent, allowing the material productions to constitute symbolic expressions of metaphysical forces. And in this entire scheme of things, men—as initiates—constitute the main agents capable of developing the configuration based on the conditions of power and determinisms of a given place and age.

This leads us to the key issue of *initiation* amidst the specific contemporary conditions. To be as brief as possible: the act of initiation implies the deep modification of one's core being, causing the individual to develop activity in a integral domain where all rational and irrational elements of the inner and outer world converge harmonically. This is initiation *in theory*, but in practice, things become problematic, especially in our times. How can we verify that a regular initiatory procedure or discipline has had a successful effect on the individual, and modified his relation to the cosmos? In other words: How do we know if the initiatory procedure actually *works* beyond all self-denominations and labelling? As we shall show in *Operative Traditions*, the fact that our modern urban-industrial societies are conceived to be divided into economic classes, impedes us to verify the connections an individual has with the technical conditions of production, since what socio-economically defines the rank of the individual in our times is the higher or lesser attachment to industrial production in general or the level acquired of socio-political prestige. And yet this contemporary context prevents us from seeing the relation an individual has with the effective powers of the world beyond any sociological or economical consideration.

When societies were organised in castes, the connections of a given individual to the conditions of production were clearer and more direct, since ultimately it was the modes of relation with the material conditions that defined the belonging or not to a caste, and the level of spiritual initiation. The latter modes of relation constituted the *empirical proof* of the individual serving a transcendent/metaphysical factor of development. So ultimately the capacity of an individual to impregnate spontaneously any activity (art, craft, politics, etc.) with a qualitative character caused *action* itself to be highly regarded as full of symbolic signification; hence, the way an individual *operated* upon the cosmos corresponded to an honest language of the spirit. What one produced defined work, *how* one produced it defined *opus*. Hence, the diverse modes of *opus* correspond to the diverse modes of individual initiation.

The focus on *opus*, which implies focusing on the *operative aspects of one's activity* (technique), is what allows one to effectively bridge theory and practice, and ultimately acquire absolute certainty in regard to the effectiveness or not of a given spiritual discipline or initiatory procedure. This corresponded to the key knowledge cherished by the Operative Freemasons, that is, key responsible figures of the medieval European Gothic grandeur, who nonetheless—as René Guénon points out—suffered from a process of spiritual degeneration that necessarily affected not only the substantial/magical character of their operations, but also their own denomination, being eventually named from the 1700s as "Speculative Masonry." As the French author points out:

> *"its name [Speculative Masonry]indicates quite clearly that it is confined to pure and simple speculation, that is, theory without conducting (…) not being, from many points of view, rather than a degeneration of the Operative Masonry (…) The latter, indeed, was truly complete, possessing both the theory and relevant practice, and their designation may in this respect be understood as an allusion to the "operations" of the "sacred art" of which construction according to traditional rules was one of the applications."*[30]

And yet historically, not only the substantial mutation of Masonry but also practically all castes and aristocracies, were progressively substituted by the merchant, technocratic and proletarian types,

---

[30] *Studies in Freemasonry and the Compagnonnage*. René Guénon. Hillsdale, NY: Sophia Perennis, 2004.

causing the entire issue of addressing *opus* to become complicated, due to the extreme modification of the worldly power configurations. This itself compromises the chance for any "verification" of validity in any serious spiritual discipline.

Though progressively marginalised by big industry, today there are still remaining traditional arts, crafts and disciplines in many domains. Yet the key question still remains: To what extent are these old operative traditions capable of serving as an artistic initiatory framework, as art was conceived originally by the Operative Freemasons? Unfortunately, the initiatory potentials of such traditional modes of art are also highly doubtful, since they are disconnected from the operative powers that act in the specific conditions of our times. Not only is the outcome of these arts fulfilment of aesthetic or emotional delight, this "disconnection" to the worldly powers also impedes any bridging between the immanent and the transcendent forces, which not only effects one at an individual level but also at a macroscopic one. Hence by aiming to recover the core message of the war hero Ernst Jünger, a new chance now emerges to potentially recover the power of Tradition, no longer in its mystical or speculative aspects, but rather in terms of those that contribute to the "root" of key traditional doctrines amidst today's immanent conditions of existence.

Hence, *Operative Traditions* provides to individuals the "tools" required to detect the direction of an absolute reference point, like using a compass. In effect, whenever we use a compass, we know the direction; we have no idea, however, of what we shall encounter along the way. These reference points shall serve firstly to reconnect to an idea that transcends the purely intellectual categories of scientific theories or ideologies; secondly, to legitimate a *praxis* that can *root* ourselves to the contradictory conditions of existence present in our times, and thirdly, to implement a language capable of recovering a meaning in correspondence with such an idea.

Before undertaking this endeavour, however, let's recall the important account of a modern philosopher's struggle when aiming to become rooted in Tradition.

# AIMING TO AN OPERATIVE TRADITION

*...we still have it, the whole need of spirit and the whole tension of its bow! And perhaps the arrow too, the task, and—who knows? The goal...*

Friedrich Nietzsche. *Beyond Good and Evil*

Eugen Herrigel (1884-1955) was a German philosopher who had studied Protestant Theology and Neo-Kantian Philosophy at Heidelberg University from 1908 to 1913. And yet in Herrigel's quest for knowledge, pure philosophical speculation never managed to satisfy his hunger for something much more substantial. The following are his own words in regard to such deep dissatisfaction:

*"Even as a student I had, as though driven by a secret urge, been preoccupied with mysticism, despite the mood of the times, which had little use for such interests. For all my exertions, however, I became increasingly aware that I could only approach these esoteric writings from the outside; and though I knew how to circle around what one may call the primordial mystic phenomenon, I was unable to leap over the line which surrounded the mystery like a high wall. Nor could I find exactly what I sought in the extensive literature of mysticism, and, disappointed and discouraged, I gradually came to realize that only the truly detached can understand what is meant by "detachment," and that only the contemplative, who is completely empty and rid of the self, is ready*

*to "become one" with the "transcendent Deity." I had realized, therefore, that there is and can be no other way to mysticism than the way of personal experience and suffering."*

As he tells us in his most important book, *Zen and the Art of Archery*, Herrigel was asked in 1924 to teach philosophy at Tohoku Imperial University in Sendai, Japan, a chance he didn't reject as it would allow him to establish a practical and methodical link to the core spirit of Buddhism and the Tradition of Zen, during a time—before the Second World War—when such teachings in Japan were still very consistent and alive.

Initially it wasn't easy for Herrigel to have access to the millenarian teachings of the "Great Doctrine," since Kenzo Awa, the highly respected Zen Master he had addressed for such purpose, had gone through bad experiences when he previously taught the doctrine to foreigners. However, the Great Master finally accepted Eugen Herrigel's proposition.

One of the most surprising aspects discovered by Herrigel, who had been influenced by modern tendencies in Western knowledge and philosophy, was realizing how the actual learning of the "Great Doctrine" corresponded to a task that required the *learning of an art*. In effect, Master Kenzo Awa was not only the highest representative of the "Great Doctrine" at that time in Japan, but also a Master of a specific art, *kyūdō*, or traditional Japanese archery. According to the doctrine, the *bridge* to be eventually built between the pupil and the ultimate realization of the doctrine required the practice of what was considered to be an "artless art," such as traditional Japanese archery, which in its spirit and aims is diametrically opposed to most of the competitive sports of our times. It's convenient to recall that the Tradition of Zen, having its roots in Buddhism, clearly impregnates the formation of many of the Japanese activities and arts (painting, architecture, tea ceremony, flower arrangement, weaponry production, martial arts, etc.) and corresponded to a key doctrinal element in the forging of the Samurai warrior castes. This traditional domain is characterized by a foundational focus in *habit* and *repetition*, and is highly embedded with ritual elements that don't have the same character as most modern social ceremonies, since in this context, *ritual* is conceived overall as a *technique;* a technique in its highest sense, related to the ancient form of *techné/poiesis,* upon which we shall delve many times in this book.

The whole issue of this Traditional teaching process deals with a progressive relation to the forms of identification typical of the "I," a term—the "I"—we shall examine in regard to its dependencies, and determinations. In this specific Traditional context, the ideal of liberation is conceived of as a progressive liberation from the constrictions of the "I," and yet this liberation cannot be equated with an *evasion* of the material domain and its constrictions, but rather as the attainment of higher and more integrated forms of dominion, power and freedom through specific modes of interaction with the developments of objects and symbols.

The accounts of Herrigel on his six years of experience—which we shall refer to very often in this book—offer us an adequate memory of the key elements required in order to transcend the limitations of pure philosophical thought, whenever intending to attain the highest levels of spiritual development. As shown by the reconnection to a integral Tradition, which still existed in Japan during those times, Herrigel explains how the experience of the "I" can not be separated from the actions taking place in the world, and to a large extent, the actions exerted on the world or material domain which "mirror" the state of the "I."[31]

---

[31] L'azione, in generale, è individuazione: è essa che ora crea il corpo della persona (v.d. ciò che vive come ταυτον dentro una libertà riflessa). Il nostro mondo è lo specchio della nostra persona o, meglio, la nostra persona è la coscienza, lo specchio di quel suo atto, che è il nostro mondo: si è quali il mondo è, epperò quali ci si fa, agendo nel mondo o facendo divenire un mondo. Noi stessi determiniamo le condizioni e le forme in cui la realtà, la verità e, in un certo senso, anche la libertà dovranno apparirci [Action, in general, is individuation: it is this what creates therefore the body of the person (namely, that which lives as ταυτον amidst a "reflective" freedom). Our world is the mirror of our person or, better said, our person is the awareness, the mirror of the latter upon the act, which is our world: if such is the world, as well as what it does, acting on the world of doing becomes itself a world. We ourselves determine the conditions and the forms where reality, truth and, to some extent, also freedom must emerge] Julius Evola. *Fenomenologia Dell'individuo Assoluto Iii* ed. corretta: Edizioni Mediterranee, Roma 2007.

# THE MEANING OF THE EARTH

*Continents will be staked, India, China, South Africa, Russia, Islam called out, new techniques and tactics played and counterplayed. The great cosmopolitan foci of power will dispose at their pleasure of smaller states - their territory, their economy and their men alike—all that is now merely province, passive object, means to end, and its destinies are without importance to the great march of things.*

Oswald Spengler[32]

Nietzsche's Zarathustra states: "*Remain true to the earth, my brethren, with the power of your virtue! Let your bestowing love and your knowledge be devoted to be the meaning of the earth! Thus do I pray and conjure you.*"[33]

The account of the experiences of Eugen Herrigel presents us with a discipline in which the material domain gains a new sense of dignity as the most perfect "mirror" of consciousness. Deep down, both in the former sentence of Zarathustra as in the learning process of Herrigel there is the recovery of a need to root one's being with the multiple states of being present in the cosmic order. In the West, it was firstly Christianity and secondly the emergence of modern ideologies what progressively alienated men from this meaning of the earth. By "meaning of the earth" we can conceive here at first the convergence in the development of the material conditions with the development of man. And yet as insightfully Robert M. Pirsig writes in *Zen and the Art of Motorcycle Maintenance*: "in

---

[32] Oswald Spengler. *The Decline of the West*. Alfred A Knopf. 1926. New York, pg 429.

[33] Friedrich Nietzsche. *Thus Spoke Zarathustra*. The Bestowing Virtue.

*that strange separation of what man is from what man does we may have some clues as to what the hell has gone wrong in this twentieth century."*[34] Let's recall that if there is something that characterizes pagan cultures in comparison to modern cultures, is that reality is conceived symbolically as an expression of transcendent meanings. In relation to this, Alain de Benoist writes: *"the fact is that in Ancient Europe the sacred was not conceived as opposed to the profane but rather encompassed the profane and gave it meaning."*[35] In pagan culture, it is by relating to the symbols present in the cosmos that the individual can transcend the limitations and determinations linked to necessity, and afterwards attain states of being that can be conceived as purely divine. This pagan vision of the cosmos is clearly an attack on the typical dualism that appeared later in Christianity and all its branches (Protestantism, etc.) where the human domain and the divine domain become split, and the initiatory bridge between both domains is finally broken, arriving to a point where the Western eyes became amputated in regard to a very specific organ, not that organ capable of perceiving the objects of reality, but that organ capable of perceiving the *aim of such object's development*. To some extent all modern ideologies on progress and all utopias intend to answer this key question, yet the attempts necessarily foster nihilism, that is, the incapacity of apprehending any aim for the complex of surrounding objects, the incapability to perceive values *here and now*; in *this* world. Once an individual is incapable of rooting to the values existent *here and now*, as a necessary surrogate, all forms of outlandish speculations on the world emerge as pure justifications of defeat. Yet Nietzsche, even though still today very misunderstood in this regard, clearly provides important advice for us to ground virtue into the here and now, by means of an *active nihilism*... Zarathustra states in regard to virtue:

---

[34] Robert M. Pirsig. *Zen and the Art of Motorcycle Maintenance: An Inquiry into Values.* 1974 (William Morrow & Company).

[35] Alain de Benoist. *On Being a Pagan.* Ultra, pg 17.

*"Let it not fly away from the earthly and beat against eternal walls with its wings! Ah, there hath always been so much flown-away virtue!"*[36]

And as Jünger writes, by following Nietzsche's *weltanschauung* [world-view]:

*"If a person of strength and good will who draws his nourishment from the past isn't able to find firm ground under his feet in the present, he is doomed to impotence. If he strives for the impossible, he must destroy himself."*[37]

However, having briefly referred to the experience of Eugen Herrigel in order to initially present in this book the factors required to reconnect to the "missing link" of an integral Tradition, it is inevitable to affirm, as a corollary of the previous assumptions, that if the configuration of the material domain eventually changes, then one can reasonably hypothesize as well that the specific disciplines of the "Great Doctrine" present in Tradition have to become necessarily inapplicable, as they originally referred to human interaction with specific objects. It is key to recall as well that the golden thread of the Hermetic Tradition has always been linked to doctrines that were imperative when referring to the configurations that exist in the material domain. For instance, one of the maxims of the Emerald Tablet of Hermes Trismegistus states: *"That which is Below corresponds to that which is Above, and that which is Above corresponds to that which is Below, to accomplish the miracle of the One Thing."* But what might be the consequences of the modifications taken place in the conditions of "below"?... By grasping the significance of the former Hermetic maxim, the latter hypothesis we presented that refers to the inapplicability of specific disciplines when related to such doctrine is not completely misleading, as in Herrigel's accounts. Master Kenzo Awa provided the German teacher with practices and disciplines that to a large

---

[36] Friedrich Nietzsche. *Thus Spoke Zarathustra.* XXII The Bestowing Virtue.

[37] Ernst Jünger. *The Glass Bees.* Straus and Giroux. New York, pg 72.

extent were referred to the very specific objects he handled, that is, to the arc, the bow and to his own body. As operative, artistic and strictly Traditional objects, the arc and bow succeeded in timelessly overcoming the determinations of time for many generations. Yet in contrast to such objects appears the body of the student, which is highly determined by time.

By conceiving of Tradition from this integral perspective where a synthetic transcendent domain is bridged into very specific configurations in the material world, being men—as initiates in the arts—are required to implement the transformation, and we can easily recognize that the abrupt transformation of power infrastructures taking place in the world during the times of WW I-II had to compromise the traditional "thread" that could link the domain of the supernatural with that of the natural. Since the 1920s the configuration of objects and infrastructures changed for the whole planet; all techniques and technologies developed at exponential paces, and all cultures and natural environments that had once been molded by the principles of Traditions suffered from the progressive dominion of means, economic interests, and networks that compromised the practice of old customs and habits. New modes of military, economic, and industrial power subverted the old orders, especially in Europe, and the development of science and technique configured new rationalistic languages for men in order to relate to the developments and participate actively in them.

The German writer Ernst Jünger, in addition to Georges Friedmann[38] was one of the first to perceive the new character of power that emerged during the 1930s-1940s. He became a pioneer who had realized the new form of dominion established over new technical means, new territories, especially within the domain of what Erich Ludendorff[39] referred to as "Total War," where the

---

[38] See Georges Friedmann. *Problèmes humains du machinisme industriel*, Gallimard, 1946 & Georges Friedmann. Sept études sur l'homme et la technique. Le pourquoi et le pour quoi de notre civilisation technicienne, Gonthier, 1966.

[39] General Erich *Ludendorff in Total War* (1936) already eliminated the distinction between soldier and civil person. These are Winston Churchill words: "There is another more obvious difference from 1914. The whole of the warring nations are engaged, not only soldiers, but the entire population, men, women and children. The fronts are everywhere to be seen. The trenches are dug in the towns and

limits between civil society and the military castes were more diffuse, because society was progressively employing means that had formerly been developed in the military domain. The fact that Ernst Jünger described this mutation taking place in the power configurations does not imply that he was the only individual affected by such change, since the *spirit of the times* or *Zeitgeist*[40]—a term Ernst Jünger often resorts to in his writings—determines everything, it is *total*.

Hegel was the first to articulate the spirit of the times, as a Historical force that effects cultural, political, artistic, religious, social, and economic forms. Modern approaches to Historical development have followed Hegel's framework. In this book the Zeitgeist shall be regarded in terms of the specific modes of power that prevail when dominating the specific material and operative conditions of reality. Cultural, political, artistic, religious, social, and economic forms shall be considered as secondary to the ever-encroaching force of the Zeitgeist. Therefore, it can be postulated that the cosmic Zeitgeist is a transcendent form of creative-destructive power or *subtle neguentropic force* that determines the activation and deactivation of specific human faculties and techniques in order to act creatively-destructively on the world according to very specific patterns. This idea has correspondences in many cultures. We can find its equivalent in the Hindu term *akasha*,[41] the *baraka* of the Sufis,

---

streets. Every village is fortified. Every road is barred. The front line runs through the factories. The workmen are soldiers with different weapons but the same courage." Winston Churchill on the radio, June 18 ; and House of Commons 20 August 1940. http://en.wikipedia.org/wiki/Total_war#cite_note-13

[40] Zeitgeist is a German term composed of both words spirit (Geist) and time (Zeit). The term is often attributed to the philosopher Georg Hegel, but he never actually used the word. In his works such as Lectures on the Philosophy of History, he uses the phrase der Geist seiner Zeit (the spirit of his time)—for example, "no man can surpass his own time, for the spirit of his time is also his own spirit." Glenn Alexander Magee (2011), "Zeitgeist," *The Hegel Dictionary*, Continuum International Publishing Group, pg 262.

[41] The Sanskrit word is derived from a root kāś meaning "to be visible." It appears as a masculine noun in Vedic Sanskrit with a generic meaning of "open space, vacuity." In Classical Sanskrit, the noun acquires the neuter gender and may express the concept of "sky; atmosphere" (Manusmrti, Shatapathabrahmana). In classical Vedantic Hindu philosophy, the word acquires its technical meaning of "an ethereal fluid imagined as pervading the cosmos." In many modern Indo-Aryan languages, the corresponding word (often rendered Akash) retains a generic meaning of "sky." *Dictionary of World Philosophy* by A. Pablo Iannone, Taylor & Francis, 2001, pg 30.

the *prana* by the ancient yogis, the *mana* of the Polynesians, the Vis Medicatrix Naturae (Healing Force of Nature) of Hippocrates, the *huaca* of the ancient Peruvians, and the *sila* of the Eskimos, etc. All these ancient terms are applied to the particular conditions of the respective cultures, yet the term Zeitgeist has a global character. In spite of the fact that the Zeitgeist cannot be directly perceived, it is existent in the productions of a given age and the configuration of the power structures in general. Its "footprints" can be followed without surrendering to its power. The aim is to integrate such power, to *master* it, in an analogous way, just as an electric motor dominates and channels the invisible currents of electricity.

In truth, the Zeitgeist of a given age affects all developments; it is *total*. Gilbert Merlio defines it as "*the Form of forms, what 'informs' reality in the manner of the Aristotelian entelechy; it is the morphological unity that one perceives beneath the diversity of historical reality, the formative idea (or Urpflanze!) that gives it coherence and direction.*"[42] It is a power meant to prevail over all other powers due to its cosmic, transcendent character. By resorting here to a quick analogy, the *Spirit of the Times* is as decisive in terms of creative processes as the seasons of the year; when winter arrives, all beings perceive it consciously or unconsciously through their senses and take action or re-action to face it; some beings might detect it first and others a little later, yet eventually winter shall be there to stay for some months. Those beings who have gained practical knowledge of how to survive in winter shall prevail and gain power within the transformed environmental conditions, and those who do not shall eventually become subservient or die. Even in our times, it is easy to verify that a considerable amount of individuals with high techno-scientific training or education in humanities are still unaware of the true character and implications of the transformations taking place in our surrounding environment, and the mutual conditionings established when developing an activity amidst it, were pointed out by the psychologist Carl Jung almost a

---

[42] Gilbert Merlio. *Les images du guerrier chez Ernst Junger.*

century ago.[43] This does not have an impact on their importance as a determining force in our lives.

Let's start by examining the characteristics of the spiritual "winter" of our times, in order to comprehend how this affected all integral and living Traditions.

---

[43] "Every Roman was surrounded by slaves. The slave and his psychology flooded ancient Italy, and every Roman became inwardly, and of course unwittingly, a slave. Because living constantly in the atmosphere of slaves, he became infected through the unconscious with their psychology. No one can shield himself from such an influence." Carl Jung. *Contributions to Analytical Psychology*, London, 1928.

# THE DECLINE OF THE WEST

> "Perfection of means and confusion of goals seem to characterize our age."
> Albert Einstein[44]

> "We note the psychological change in those social classes that had hitherto been creators of culture. Their power and creative energy are dried; men lose interest in the activity of creating and cease to value it. They are disillusioned; their effort no longer tends to a creative ideal for the benefit of humanity; their minds are not concerned with anything other than business or indifferent material ideals to life."
> Michael Rostovtzeff. The Social and Economic History of the Roman Empire

It was Oswald Spengler who first put forth the idea of the *decline of the West,* and apart from the German philosopher, there have been several other authors who delved into the issue of what appears to be the irreversible decline of all civilizations, which—according to Spengler—follow cyclical laws similar to those of any living being. When it comes to identifying the element that declines, there have been many hypotheses set forth. By transcending the approaches that consider the decline of a civilization as exclusively related to a political decline or a racial miscegenation of the elites—as hypothesized by De Gobineau and H. S. Chamberlain—authors such as Oswald Spengler, Arnold Toynbee,[45] Alfred Louis Kroeber,[46]

---

[44] Cited in Reid, D., 1995. *Sustainable Development: An Introductory Guide.* London, UK: Earthscan Publications Ltd, pg 169.

[45] Toynbee, Arnold J. (1962). *A Study of History (twelve volumes).* Oxford University Press, Oxford.

[46] Alfred Louis Kroeber. *Style and Civilizations.* Cornell University Press, Ithaca. 1957.

Rushton Coulborn[47] and Charles Edward Gray[48] considered that the decline was very much related to the transformation of the cultural aspects of a given civilization.

We can now approach this issue with new tools and paradigms that were not available for the aforementioned authors. As we shall see, these conceptual tools arise mainly from fields such as those developed by biologist and philosopher Ludwig Von Bertalanffy and Nobel laureate Ilya Prigogine, among others, who intended to overcome the limits of reductionist, Newtonian, or cybernetic scientific paradigms, when intending to grasp the characteristics of the processes that determine life. The work developed by these authors has been quite exceptional during the last century, as most of the research in science has surrendered to the requirements of technical application and economic profitability, and has thus focused in very specialized domains that have inevitably lost contact with the determining forces of the cosmos that transcend those of individual interest or technical determination.

In order to visualize as accurately as possible the idea of cultural splendor and decline, we shall start with the key concept of *homeostasis*. Before approaching this concept it is important to point out that this term has been widely applied in cybernetic technologies and servo-systems. However, the notion of homeostasis surpasses the limits established by the paradigm of cybernetics, and this is why we have considered it more satisfactory to introduce the term *homeostasis* before introducing the scientific paradigm of *cybernetics*.

\* \* \*

*"There is no difference between the physical and chemical processes that are developed in a living organism from those that take place in a corpse. They all obey the same physical and chemical laws,*

---

[47] Coulborn, Rushton (1954). *The Rise and Fall of Civilizations*. Ethics 64: 205-16.

[48] Gray, Charles Edward (1958). *The Epicyclical Evolution of Graeco-Roman Civilization*. American Anthropologist 60: 13-31.

*and that is all that can be said. But for the biologist and the doctor, there is a deep difference between what is organized to conserve a system, and what disorganizes it to destroy it. Which are the principles of the ordering and organization? What does 'health' and 'norm' mean in contrast to 'disease' and 'pathology'?"*

*- Ludwig Von Bertalanffy*[49]

\* \* \*

*Homeostasis*[50] is a term from Greek: [ὅμοιος homœos, "similar" and στάσις stasis, "standing still"] which corresponds to the property of a system where variables are regulated so that the internal conditions remain stable and relatively constant, within conditions of dynamic balance.

In living beings, a correct homeostasis is directly related to the concept of *health*. A living being that has all its diverse metabolic functions in adequate homeostasis corresponds thus to a healthy being. Optimal health corresponds to optimal reactions in situations of stress and anxiety. In the case of optimal health, any external perturbation can become easily counterbalanced through internal feedback mechanisms, the so-called *negative feedback responses*. The only way these negative feedback responses can be effective is if the organization or system *already has information about the attributes of the diverse perturbations*, and thus can compensate for the perturbations with opposite responses (negative feedback). For

---

[49] Ludwig Von Bertalanffy. *Towards a New "Natural Philosophy*. Clark University (Worcester. Massachusetts) January 1966.

[50] The concept was described by French physiologist Claude Bernard in 1865 and the word was coined by Walter Bradford Cannon in 1926. Although the term was originally used to refer to processes within living organisms, it is frequently applied to automatic control systems such as thermostats. Homeostasis requires a sensor to detect changes in the condition to be regulated, an effector mechanism that can vary that condition; and a negative feedback connection between the two. Cannon, W. B. (1926). "Physiological regulation of normal states: some tentative postulates concerning biological homeostatics." In A. Pettit(ed.). A Charles Richet : ses amis, ses collègues, ses élèves (in French). Paris: Les Éditions Médicales. pg 91.

instance, when someone is vaccinated against smallpox, the body is provided with antigenic material in order to slightly perturb the immune system, allowing the development of adaptive immunity to the pathogen, that is, by improving adequate responses to the perturbation (negative feedback), and therefore increasing the levels of health or homeostasis.

Whenever there is an optimal *state* of health, the whole organism can respond very well to perturbations, since there is clearly present a "stable balance" behaviour and this "stable balance" corresponds to the *state* of health. In living beings, the *state* of health, like a *vector*, points to an *attractor* or *gestalt* entity that emerges mysteriously in states of high entropic dissipation and non-linear physical conditions. The state determines the character of the phenomena existent within the domains it encompasses. Heidegger referred to this idea when employing the term *Dasein* (German: da "there"; sein "being") considering it *"that entity which in its Being has this very Being as an issue."*[51] Nietzsche referred to the same idea when stating in his Zarathustra: *"Spirit is life which itself cutteth into life: by its own torture doth it increase its own knowledge—did ye know that before?"*[52]

---

[51] Heidegger, M. (1962). *Being and Time:* Translated by John Macquarrie & Edward Robinson. London: S.C.M. Press, pg 68.

[52] Friedrich Nietzsche. *Thus Spoke Zarathustra*. The Famous Wise Ones.

# COSMIC ATTRACTORS

*We see the filings, but we do not see the magnetic field that determines its order*
Ernst Jünger[53]

*What can be seen is not by chance the final order, but the modification of disorder under which a law can be divined.*
Ernst Jünger[54]

The mathematical concept of an *attractor* was put forth initially by French mathematician René Thom (1923-2002) in physics and mathematics during the decades of the 1970s-1990s as a hypothetical mathematical structure that discovered the stable configurations that appear in conditions of turbulence. Since then many attempts have appeared afterwards in order to approach the concept mathematically. Yet all mathematical approaches to attractors, such as those simulated in computers—like the Lorenz attractor—never end up as gestalts or figures having a qualitative integrating character, but rather end up as merely chaotic figures that in some cases are characterized by fractal geometry. This itself shows us the mysterious character of attractors, which can't be defined in strictly mathematical terms. Beyond computer simulations and mathematical topologies, attractors in nature are characterised by their *symbolic* aspect, a symbol characterised by a qualitative value that transcends and integrates the limitations

---

[53] "Wir sehen die Feilspäne, aber wir sehen nicht das magnetische Feld, dessen Wirklichkeit. Ernst Jünger. *Der Arbeiter*. Ernst Klett, Stuttgart 1981. Klett-Cotta pg 42.

[54] "Was gesehen werden kann, ist nicht etwa die endgültige Ordnung, sondern die Veränderung der Unordnung, unter der ein großes Gesetz zu erraten ist." Ernst Jünger. *Der Arbeiter*. Ernst Klett, Stuttgart 1981. Klett-Cotta pg 45.

of language, and even transcends the many topological forms of today's mathematical structures. The attractor of a given organism is its irreversible destiny. Whenever we are in the presence of an organism that dissipates energy, and the organism is capable of self-organization, an ultimate *attractor* or State is always present. The fact that we might not perceive this *attractor* or state of being with humans senses or by the filtering of nature by technological devices, doesn't mean they don't exist, it's just that in order to perceive them we don't have what medieval scholastics referred to as *intellectual intuition*, known in Hinduism as *vidyâ* and in Buddhism as *prâjna* or *bohdi*. In the cosmos, attractors are of a metaphysical nature, that is, they can not be approached exclusively through dialectics; dialectics are rather the "shadow" of the path towards an attractor or state in terms of language and concepts. As Ernst Jünger wittily writes:

> *"language does not live by its own laws. If this were the case, the world would be dominated by grammarians."*[55]

Heidegger saw the key influence of technique in language development itself when writing: *"it is the technical possibilities of the machine what prescribes how language can and should still be language."*[56]

And yet attractors determine the nature of the interactions of a given system, they are its *idea*, they are the "centre" they revolve around. In order to provide an appealing image that can help inner visualisation, it can be expressed that *attractors are to organisms what gravitational fields are to Newtonian physics*. If we want to have an adequate first approach to Julius Evola's idea on power or *potency*, or Jünger's idea on *Macht*, it is precisely in this concept of attractor or gestalt where find the most satisfactory analogies, which the Italian author constantly related to myth and symbol.

---

[55] "Die Sprache lebt nicht aus eigenen Gesetzen, denn sonst beherrschten Grammatiker die Welt." Ernst Jünger. *Der Waldgang*. Klett-Cotta. Stuttgart 1980, pg 95.

[56] Martin Heidegger. *Langue de tradition et langue technique*, Lebeer-Hossmann, Bruselas, 1990, pp. 39 y 41 (1962 text).

By now we can identify the entity as the *figure* of the organism or organization. For instance, just for didactic purposes, we can consider a flower as the *figure* of a given plant; the flower corresponds to the core symbol that impels centripetally all organic functions towards a specific centre, amidst conditions of countless perturbations and influences coming from the environment. During this "upstream" stage, material production (economy) aims to ultimately *synthesis the figure* (a process called in biology as *autopoiesis*). Once this qualitative centre is attained and the symbols inherent to the *figure* are fully expressed in relation to a given environment, then the attractor loses its directing capacity, and the organic processes characterised by dissipation of energy no longer pull "upstream" the subtle fluxes of entropy towards the *figure*. It is precisely at this point when corruption and anarchy start to take place, which means that all the material and biological elements of the plant or living being progressively start to lack a common directing aim whenever facing external perturbations. This implies a loss of regulation capacity for the whole organism and more presence of *mechanization* and *mobilization* factors. In this case there is no state of health anymore, no "centre of gravity" of all productions; the *attractor* or *figure*, having constituted formerly an effective and operative transcendent power, becomes then just a memory, a powerless shadow, a reflection or image that can then only become dialectic, rationalized, and systematized in order to direct the last remnants of the organism, until all the separate and mechanistic elements finally die and become inert, ultimately waiting for a new *figure* to mobilize all elements towards a new direction and purpose…

The previous description can be applied quite easily to beings at both a microscopic and macroscopic level. The *figure* corresponds to an attractor that "pulls" all material phenomena into a specific synthetic direction, and only expresses its directing power whenever the phenomena take place in organisms, that is, in dissipative systems far from equilibrium (systems that generate entropy), characterised by attractors. Life itself corresponds to a complex set of dynamics where the dissipative structures emerge at all levels, generating an organic hierarchy from the level of the cells, to the organs, to the whole body, to the whole environment. The diverse integration of *figures* or *multiple states of Being* within the cosmos cannot be manipulated by human means or human attributes since the activation of given human means or actions is determined by such overall powers.

Humberto Maturana and Francisco Varela[57] concluded that it is not possible to *direct* those states of attraction or attractors by technical or scientific procedures, but only to *perturb* the elements the attractors mobilize.[58] This is equivalent to the fact that in Newtonian physics if an object is relatively small we can interfere temporarily in its momentum, yet the gravitational fields that determine it won't be significantly affected. Humans can easily change the conditions of the soil or water in order to impede the correct flourishing of a plant, yet this act cannot change the being of the plant, that is, the *figure* or *essence* expressed in the flower, which can easily appear elsewhere given the correct conditions. Figures and attractors are therefore beyond the constrictions of time and space, in a similar way as how today's golden eagle (*aquila chyrsaetos*) is the same golden eagle of ancient time. Of course, genetic manipulation or the production of hybrids can be accomplished with biological intrusion techniques, yet these selections are all qualitatively inferior to the original archetypal flower, which is not at all being affected.

Whenever homeostasis is intense and powerful in a given organism, qualitative organic aspects become enhanced at all levels, and whenever the organisms face perturbations, the capacity for regulation is also much more effective. This implies the creation of a hierarchy of functions and specializations (a concept conceived

---

[57] *Autopoiesis and Cognition: The Realization of the Living* (Boston Studies in the Philosophy and History of Science) Maturana, H. R., Varela, F. J. Published by Springer

[58] "The classical demonstration of this phenomenon was in Hans Driesch's experiments on sea-urchin embryos. When one of the cells of a very young embryo at the two-celled stage was killed, the remaining cell gave rise not to half a sea urchin but to a small but complete sea urchin. Similarly, small but complete organisms developed after the destruction of any one, two, or three cells of embryos at the four-celled stage. Conversely, the fusion of two young sea-urchin embryos resulted in the development of one giant sea urchin" Rupert Sheldrake. Morphic Resonance. The Nature of Formative Causation. Park Street Press. H. Driesch, *Science and Philosophy of the Organism* (London: A. & C. Black, 1908).

by Ernst Jünger as *types*), where each organ relates to the cosmos in particularized and symbolic ways. Under these circumstances, no organ can attain health or homeostasis unless it is powerfully directed by the regulatory capacity of the whole Being.

So whenever a given civilization reaches its *maximum symbol*—as Oswald Spengler coined it—then the process of decline and mobilization of factors inevitably take place, and there is no way it can be stopped by human means. The civilization might adopt rigid, tyrannical, and dictatorial forms that correspond to what Spengler called *caesarism,* but in any case, the figures and symbols present in the culture lose their impelling and agglutinating power. When a civilization reaches its splendorous peak, chaos and order synthesize in symbolic configurations at all levels and the State doesn't need to rely on the mathematical tool of statistics, that is, a powerful yet myopic approach on sociological phenomena that can only trace exclusively quantitative aspects on human behavior. In the former situation of splendor and potency, entropy generation—in terms of waste, energy and chaos—flows "upwards," to the production of the cultural symbols. And yet when a civilization starts to decline, the civilization is nonetheless still *alive,* which means that there is still a generation of entropy in the form of waste, chaos and heat, to the point that even scientist Tim Garrett argues modern civilization is a "heat engine."[59] Yet the waste and entropy become less and less "recyclable" by the civilization itself due to the diminishing returns of all efforts. Russian-American sociologist Pitirim Sorokin[60] considers that whenever the decline of a given civilization takes place, there is also the potential emergence of new cultural forms. These new cultural forms and symbols express relations of power with a reality that has been profoundly modified, and this itself generates a different form of civilization characterised by symbols qualitatively different than those of the civilizations in decline, which all become integrated as archetypes. Along the same lines, Ernst Jünger points

---

[59] http://www.sciencedaily.com/releases/2009/11/091123083704.htm

[60] Sorokin, Pitirim A. *Social Philosophies of an Age of Crisis.* Beacon Press. Boston. (1950) *Social and Cultural Dynamics.* Porter Sargent, Boston (1957).

out: *"nowhere, neither in the mechanical world nor in the organic world, neither in nature nor in history, we see one force reduced to dust without bringing another to overcome it."*[61]

Oswald Spengler made the division between *civilization* (*Zivilisation*) and *culture* (*Kultur*) by considering that the initial blooming phase corresponds to culture, and the inevitable phase of decline corresponds to civilization. Based on Spengler's distinction, culture implies a qualitative ordering of human groups in a hierarchy of castes that derive from the primordial nobility and priestly castes which embraces the symbols of culture; civilisation, on the contrary, implies for Spengler the disempowerment of the castes and the emergence of individualism and economic determinism, both associated with the apparition of the masses. By applying the concepts formerly explained, and by referring to Spengler's distinction, we can assert that a culture becomes a civilization to the extent it loses its autonomy and capacity for self-regulation, and to the extent new mechanistic forms of control are established in order to minimise chaos and anarchy. In the case of decline, the capacity for self-regulation no longer is a responsibility of the politics of the culture itself, instead responsibility is outsourced to another political structure that bases its power in a materialistic approach to all domains of the cosmos. Francis Parker Yockey expressed a very accurate approach to these dynamics in what he coined as the *Law of Constancy of Inter-Organismic Power*. By explicitly defining such law he writes: *"In any age, the amount of power in a State-system is constant, and if one organic unit is diminished in power, another unit, or other units are increased in power by the same amount."*[62]

---

[61] "Nichts ist naheliegender als dies, denn nirgends, weder in der mechanischen noch in der organischen Welt, weder in der Natur noch in der Geschichte beobachten wir eine Kraft, die sich ohne Ablösung zerstäubt" Ernst Jünger. *Der Arbeiter.* Ernst Klett, Stuttgart 1981. Klett-Cotta, pg 107.

[62] In this regard Yockey also referred to the The Law of Totality by stating that it "affects individuals by embracing them existentially in the life of the organism. Politics places the life of every man within the political unit in the balance. It demands, by its very existence, the readiness of all individuals in the service of its fulfillment to risk their lives (…) Like all organic laws it is existential: if any inner force can challenge it, the organism is sick; if the challenge is attended with success, the organism is in severe crisis and may be annihilated. (…) it is not the rulers who are sovereign within the meaning of this law. Their powers in fact are derived from their symbolic-representative position. If a stratum represents and acts in the Spirit of the Age, revolution against it is impossible. An organism true to itself cannot

One of the key features that clearly expressed the decline of the West in the 19th century was the fact that the State-nations progressively lost their strict autonomous dominion over the powerful development of all industrial, economic, technical, scientific and commercial activities that transcended their selfsame national frontiers. New configurations appeared that challenged all former frontiers; new forms. How then can we perceive the State as *"the inward form of a nation, its "form" in the athletic sense,"* as Oswald Spengler defines it?[63] In order to tackle this key issue and provide consistency, legality and control over the developments, an international financial system had to be necessarily developed in parallel to the emergence of corporations and networks that gained progressively international jurisdiction. Formerly it had been the bourgeoisie classes who had handled the political and government issues of the State-nations, yet the concept of "State" was becoming less and less substantial, in the sense that it could no longer direct national developments to one single aim. This is strictly a key characteristic of the overall decline of configuring power in the West that necessarily was expressed in the most diverse dialectical approaches that appeared in those times to the idea of the State. For instance, in the so-called *positive school* (that is, the States constituted according to Marxism and Bolshevism) the concept of law, and those of political order and the obligations of the State no longer serve to create a bridge between the State and human personality, conceived in its qualitative terms. In the liberal-democratic case, the State is conceived also in a highly abstract way, where conceptions such as "objective person," "Rule of Law," Hobbes' Leviathan and the "State of Wealth" emerge. In the case of liberalism, there is no *interest* in approaching the idea of State, and the State becomes rather the main antagonist, to the extent that Alexander Dugin even writes: *"liberalism did everything possible to ensure the collapse of politics."*[64] Other definitions and approaches to the modern State emerge from

---

be sick or in crisis" Francis Parker Yockey. *Imperium*. The Philosophy of History and Politics.

[63] Oswald Spengler. *The Decline of the West*. Alfred A Knopf. 1926. New York, pg 137.

[64] Alexander Dugin. *The Fourth Political Theory*. Arktos 2012.

the *historic-sociologic school*, where the ideological basis is still highly abstract, and it is defined as the "people's conscience," the "spirit of the nation" and the "immanent life of history" based on the philosophical thought developed by Hegel. These latter conceptions were also mixed with those of the *idealist-immanent school*, where the concept of State is completely volatilized into misty clouds with terms such as "transcendental I," "absolute spirit," "pure act."

All these attempts of approaching dialectically the State were influenced by the transcendental philosophy existent during those times. In the second part of this book we shall delve briefly on this particular modern philosophical approach, but for now it is useful for the reader to question if the State, or any homeostatic property, can be approached exclusively in a dialectical way.

The dialectical approach to the concept of State assumes that all processes taking place in reality can be deduced by static categories of thought and immutable laws. However, this very idealistic and dialectic approach to the cosmos or *physis* was already showing its important limitations in the most immediate material domain. The decadence that appeared when approaching the concept of State took place in parallel to the erosion of the foundations of knowledge or gnosiology, when these disciplines intended to establish a commonly accepted and systematic approach to the reorganization of the most diverse areas of science. Hence, modern history has shown how the diverse branches of science and all scientific productions of knowledge developed quite anarchically, since even by accepting that scientific discourse develops rationally, the ways that scientific problems are posited,[65] or the premises of the scientific approaches by the scientist are highly irrational, or determined by cultural ideology or myth.[66] So the decadence of modern science

---

[65] Non vi è concetto scientifico che in correlazione all'esperienza come materia del dato intrisa di qualità e di soggettività personale [there is no scientific concept that correlated to experience imbues it with quality and personal subjectivity] Julius Evola. *Fenomenologia Dell'individuo Assoluto Iii ed. corretta*: Edizioni Mediterranee, Roma 2007.

[66] The case of one of the most renown scientists of modern times, Charles Darwin, is very paradigmatic in this regard. Darwin particularly projected in his work the "endless fight for existence" proposed already by Thomas Malthus, as well as the ideas of Arthur Schopenhauer regarding the struggle for self-preservation, the human intellect as the perfect means for such fight, and sexual

was not so much in terms of the production of massive amounts of analytical scientific knowledge, but rather in terms of the lack of capacity for developing a synthesis. As Lord Northbourne remarks:

*"Modern science is well equipped to provide certain kinds of information, but the possibility of interpreting that information is denied; therefore, such task is left to the interplay of opinion, individual or collective, informed or ignorant. Consequently, its cardinal error lies in its assertion that it is science itself, the only science possible, the only science that exists."*[67]

The materialistic vision of the universe that comforted the idea of immutable laws within the material domain was the world-view that emerged basically from the Newtonian vision of nature. Hence, one of the main assumptions in modern science was the existence of immutable laws in nature independent of the observer. And not only the latter, but also modern science dismissed any teleological direction in the processes of nature. As Robert M. Pirsig writes in regard to modern science:

*"It's good for seeing where you've been. It's good for testing the truth of what you think you know, but it can't tell you where you ought to go, unless where you ought to go is a continuation of where you were going in the past. Creativity, originality, inventiveness, intuition, imagination..."unstuckness," in other words...are completely outside its domain."*[68]

Thus nature was no longer supposed to be governed by gods, but by scientific laws; objects in nature were supposed to surrender passively to the external laws, as objects and matter were therefore considered

---

love as an "unconscious selection that favours the species." Already a century before Charles Darwin, Erasmus Darwin had proposed the idea of adaptation, inheritance, struggle and self-protection as the principles of evolution. In short, the interpretation of Darwin that affirms that men act according to the same predispositions of animals was merely a psychological suggestion caused by the English authentic Zeitgeist (spirit of the times..) of those times.

[67] Lord Northbourne. Hombre y Naturaleza. La crisis espiritual del Hombre Moderno. Kier.

[68] Robert M. Pirsig. *Zen and the Art of Motorcycle Maintenance: An Inquiry into Values*. 1974 (William Morrow & Company).

as lifeless, passive, strictly "material" in the most inert sense, and with no capacity for self-determination. Any form of autonomy and self-determination in nature thus constituted an important menace for the reductionist scientific framework. And in order to technically facilitate the separation of local phenomena from the whole, laboratories were developed as the most adequate environments for the separation and control of phenomena. Scientific knowledge became a by-product of laboratories and technical equipment, which then became an important intellectual assistance for developing even more sophisticated technical equipment. And yet in order to understand life based on mechanical laws, life had to necessarily become static, mechanic, "cold" and vivisected in the laboratory. Once set forth the historical scientific project of modernity aiming to discover all the laws of the mechanical universe, the entire universe was eventually supposed to be predictable, and at last any existential fear in regard to the unpredictable behaviors of nature could also then be frozen and redeemed. This optimistic vision on the prediction of the becoming of the world by immutable laws was expressed by Laplace when stating:

> *"We may regard the present state of the universe as the effect of its past and the cause of its future. An intellect which at a certain moment would know all forces that set nature in motion, and all positions of all items of which nature is composed, if this intellect were also vast enough to submit these data to analysis, it would embrace in a single formula the movements of the greatest bodies of the universe and those of the tiniest atom; for such an intellect nothing would be uncertain and the future just like the past would be present before its eyes."*[69]

This ideal view of the universe had been also expressed mathematically in the field of physics, where the first law of thermodynamics—established and stated mathematically by Rudolf Clausius (1822-1888) in 1850—defined the exact energy

---

[69] Laplace, *A Philosophical Essay*, New York, 1902, pg 4.

correspondences between heat, work and the internal energy of a given system. The formulation of the law satisfied a vision of the processes taking place in nature where the fluxes of energy are supposed to not be lost or destroyed, and where the relations defined the law are ultimately *conserved*, timelessly. Hence, the flux of energy between heat and mechanical work could go in one direction or the other; that is, all heat could be transferred into mechanical work in the same way all mechanical work could be transferred into heat. All entropy into energy and all energy into entropy. The formulation of this law satisfied all those scientific mindsets prone to think in terms of strict mechanical causality and exactitude of all phenomena, that is, all those mindsets who find appeal in whatever laws are considered to be static, abstract, and where all magnitudes are perfectly acknowledged. The first law of thermodynamics could easily justify the view of a strictly mechanical universe where every single magnitude becomes acknowledgeable.

Yet the discovery of the principle of entropy and its formulation in the 1860-1870s by Rudolf Clausius introduced the idea of the inevitable irreversibility of all spontaneous processes taking place in nature, mostly due to the energy losses caused by friction. And friction in actual nature is inevitable. This idea was expressed mathematically in the second law of thermodynamics, where the core idea that can be deduced from the second law when applied to nature, is that without any external intervention from outside a given system, the system shall necessary degrade its energy levels into entropy production. The thermodynamic death of the universe is then assumed as inevitable and irreversible. The former mechanistic and reductionist view of nature in modern science that was inherited from the times of Newton and Laplace, and founded on supposedly immutable laws independent of time and space, could only be applied in those systems where nature doesn't generate entropy in those systems where all developments are predictable and reversible, thus allowing even the variable of time itself to have negative values. Yet when entropy "intrudes" in a process, the certainty of all scientific laws become "eroded" to the same extent as such natural processes experiment friction. In relation to the turbulent medium in which such phenomena take place, the following assertion is attributed to Albert Einstein:

*"I am going to ask God two questions: the cause of relativity and the cause of turbulence. I'm optimistic in finding the answer to the first question..."*

And yet as formerly explained, anything that is alive generates entropy, which allows participation in long-range non-linear patterns, following a *destiny* that transcends any form of mechanical causal links. This destiny is never expressed in the form of scientific mathematical laws but is rather expressed in symbols. But with the irruption of the second law of thermodynamics (the law of entropy) the fixed character of natural laws became highly compromised and all sorts of chaotic phenomena and dispersion of control started to prevail when intending to analyze the phenomena of the universe. Not even quantum mechanics can adequately approach this mysterious domain, and the proof of such impotence is that when it comes to predicting molecular structures, it is an extremely powerless tool that relies on statistics. As the editor of *Nature*, John Maddox expressed in 1988:

*"One of the continuing scandals in the physical sciences is that it remains impossible to predict the structure of even the simplest crystalline solids from their chemical composition."*[70]

\* \* \*

So what we can realize at this point is that if there has been t a decline in the civilization of the West that has affected the macroscopic and microscopic forces, and that also has affected every single State-nation and individual, it is that whenever relating to the material domain and to what Nietzsche referred to as "the meaning of the

---
[70] J. Maddox, "Crystals from First Principles," *Nature* 335 (1988): 201.

earth," all dialectical approaches and all categories of thought become more and more powerless in order to "capture" the aim of the material developments, their *attractor*, their *idea*. Whenever in the political arena the term "progress" was used ideologically in order to answer the question of the ultimate aims, the "progress" referred basically to the progressive surrendering of humanity to the developments taking place in the material conditions, and this surrendering is inevitably linked to a necessary surrendering to the economic conditions. Yet the economic field is referred to as the domain of means and not to the domain of aims.

Politics, by constituting traditionally and classically the domain in charge of defining aims for society, progressively surrendered to ideological propagation in order to satisfy the needs of the masses in regard to their activism and progressive lack of autonomy and self-organization, that is, by the progressive transformation of organically rooted peoples into masses. Politics was no longer an issue that *"implies (and often politics ignores its own implications) a certain idea of men, and even an opinion on the fate of the entire species, a whole metaphysics that sprouts from the most brutal sensationalism to the boldest forms of mystique"* as considered by Paul Valéry,[71] but rather when the transformation of peoples into masses finally takes place, any ideology that makes use of the term *democracy* becomes just treacherous and misleading propaganda, since even the Ancient Greeks who developed the democratic ideal considered that the power of the *demos* or the political participation of the people only made sense if the State was a well-organized and actual governing force, not just an intellectual abstraction. To think that today's demo-liberal societies don't resort to propaganda is extremely naive. Noam Chomsky writes in this regard:

*"It should be expected that it's in the democracies that these ideas [of propaganda] would develop. Because in a democracy, you have to control people's minds. You can't control them by force. There's a limited capacity to control them by force, and since they have*

---

[71] Elementos. *Revista de Metapolítica para una Civilización Europea*. Jesús J. Sebastián.

*to be controlled and marginalized, be 'spectators of action', not 'participants,'...you have to resort to propaganda."*[72]

It can be hypothesized that the problem of defining aims during the decline of the West might have arisen due to the powerlessness of the existent language and procedures available in order to capture the aim, or it might have arisen because the industrial, economic, scientific and social phenomena that emerged were extremely novel. All these fields extensively developed new concepts and branches of science, yet when it came to discerning any teleology, any direction, or any meaning in the entire process, everything became too "misty"; techniques for relating with orderly structure and mechanistic environments were developed in countless areas, whereas techniques and methods to discipline the spirits in approaching chaos were completely lacking in the new configuration of educative forces of the West. As Robert M. Pirsig writes:

> *"What you've got here, really, are two realities, one of immediate artistic appearance and one of underlying scientific explanation, and they don't match and they don't fit and they don't really have much of anything to do with one another. That's quite a situation. You might say there's a little problem here."*[73]

As we can already observe here, the progressive lack of methods and disciplines for facing chaos in the West was in opposition to the core teachings of Zen that Eugen Herrigel had to learn from Master Awa Kenzo, where both the chaos of the outer domain, as the chaos in the self is faced, to the point that in such teachings chaos only exists to the extent that one has not attained yet the condition of Mastery in the Art.

As modern science was more and more in crisis due to its incapacity to define the ultimate aim of the phenomena taken place

---

[72] Barsamian, D., N. Chomsky, N., *Propaganda and the Public Mind: Conversations with Noam Chomsky.* 2001. Cambridge, MA: South End Press, pg 252.

[73] Robert M. Pirsig. *Zen and the Art of Motorcycle Maintenance: An Inquiry into Values.* 1974 (William Morrow & Company).

in the material domain, all forms of previous ethical and moral forms became powerless *in terms of application*, that is, in *technical* terms. The typical morals and sense of ethics promoted by the bourgeois classes in the West had arisen some centuries before due to the moralization of religious contents (mostly Christian) that served to strongly support the political constitutions created by merchant classes within the State-nations. But when the State-nations became highly disempowered as when were forced to deal with power structures that transcended their frontiers, the moral forms had to be inevitably compromised, accessible and eroded by dialectical critique, such as in the case of the critique of morals by Karl Marx, who considered them a cynical and inapplicable superstructure having no direct influence in the economic forces prevailing at a wide transnational scale during those times.

Nietzsche, during the same years had already noticed that the development of modern science was deeply embedded by the causal mechanistic patterns derived from the "punishing and rewarding" character of secular or moralistic Christian forms applied to human actions, or in other words, that modern science had idealistically assumed that the same causes rigorously imply the same effects. In a similar way as it was idealistically conceived in many moralistic forms of Christianity that given "good" moral actions implied "heavenly or social compensation." Yet when it comes to actions that are exerted on the actual conditions of the world, Nietzsche also showed that such actions are never characterized by a rigorous moral character. Moral attributes are an *a posteriori* projection of human interpretation in extremely reductionist terms. For instance, in technical applications, we might be capable of interpreting a "good" use of a technique based on a moral conception of "good" as "the increase of comfort," for instance, but then the technical application can indirectly cause unknown opposite effects. Nietzsche also showed the frequently morbid character of compassionate actions where the individual facilitates the well-being and happiness of the faceless crowds.

All this makes patent that the establishment of a given human ethics or morals makes no sense if it is not firstly approached by metaphysics as the core reference point of value. For instance, in the case of Zen, morals and ethics are characterized by a strictly technical character, they are only means for self-disciplining in order to attain spiritual convergence with the metaphysical. Once these aims are attained, the initiate is "beyond good and evil," that is, beyond the determinisms that require in the case of most individuals a moralistic interpretation of their surrendering to the determinisms.[74] In this specific context, the ritual is not ceremony, but *technique*. The potential symbols that emerge in situations of chaos or danger of the self can only be apprehended by a constant inner spiritual work where specific morals and ethics are "sign points" or "pointers" that allow self-guidance along the entire process, that is, the moral disciplines are not aims in themselves; they rather constitute temporary "crutches" with the metaphysical domain, that is, with the ultimate aim of all material developments.

Traditionally, it is conceived that the material domain and its development constitutes the "mirror" of consciousness and its development. Thus, based on this doctrinal principle, we have the chance of looking at our deepest selves by seeing our reflection in the material configurations of the cosmos we relate to. Yet the irruption of entropy in science, the emergence of a statistical approach to chaos, and the realization of the contingency of all scientific laws of the universe, necessarily compromised not only any capacity for self-recognition and any moral or ethical development based on the new conditions of existence, but also compromised the construction of social contexts that could favor an integral development of the modern individual.

José Ortega y Gasset, in his work "*The Rebellion of the Masses*" when dealing with the issue of the decline of the West in the first chapters, points out the fact that it might appear rather delusional to talk about "decline" when during the first decades of the 20th

---

[74] Julius Evola writes: "it is not a question of "values" but of "instruments," instruments of a virtus, not in the moralistic sense but in the ancient sense of virile energy. Here we have the well-known parable of the raft: a man, wishing to cross a dangerous river and having built a raft for this purpose, would indeed be a fool if, when he had crossed, he were to put the raft on his shoulders and take it with him on his journey. This must be the attitude-Buddhism teaches-to all that is labelled by ethical views as good or evil, just or unjust." Julius Evola. *The Doctrine of Awakening*. Inner Traditions.

century societies were expressing powerful imagery of mobilization, dynamics, and initiative. The Spanish philosopher exposes how much greater the chances of individual participation in society, economy and politics were. "How can we then be talking of "decline"?"—he posits. This is indeed a very challenging question.

If, as formerly hypothesized, the "fish starts to get rotten by its head," and the cultural impotence of a State is what provides the first sign of its decline, how is it that we can refer to the decline of the West as the decline of its States if the confidence in politics and State administration on the part of the individual apparently was greater than ever? The empowerment of the individual and vast population expansion had been channeled by the intense development of cities, industries, and a sophisticated banking and commercial conglomerate. Social participation in the developments increased in parallel to more opportunities for individual expansion. Hence, it was not difficult to be seduced by the mobilization factors, feeling instinctively that the issue in regard to the aims of development was being handled adequately by the State in clear directions, thus constituting the most diverse States as the guarantee of freedom for the individual. But was this actually the case?

In the case of the US, the Bill of Rights establishes the concept of "freedom" in terms that were similarly developed by Montesquieu and Rousseau during the French Revolution. In this context "freedom" is conceived of in terms of "release," that is, in terms of release from all former ethical, moral, and political "straightjackets," thus causing a surrendering to instinct, to the thrill of emotion and intuition.[75] In many countries such as France, the UK, and especially in the US, this freedom was guaranteed to the citizen by the demo-liberal constitutions. Hence, it was no longer necessary to maintain any inner disciple in regard to former modes of action, ethics or morals. One was allowed release from internal tension, thus allowing

---

[75] In the US the removal of the roots has always been perceived as the essential condition for the growth of freedom. The key symbols of American life, the "border" and the melting-pot, have contributed, among other factors, to develop the idea that only the uprooted can reach a true intellectual and political freedom. Lasch, Christopher. "Mass Culture Reconsidered," in *Democracy*, 1, 4, October 1981 (pp. 7-22).

one's life to be determined by external economic, industrial, and social determinisms. In this regard, Alexander Dugin writes:

> *"Liberty implies freedom from something. It is from here that the name liberalism is derived. Liberals fight for this freedom and insist on it. As for 'freedom to'—that is, the meaning and goal of freedom—here liberals fall silent, reckoning that each individual can himself find a way to apply his freedom."*[76]

Inevitably, this progressive erosion of inner self-discipline and its "outsourcing" necessarily empowered social conformism as the main active morals. And we use the word "active," as even though there were a considerable percentage of individuals who were still conditioned by Puritan ethics that repress the instinctive and unconscious realms, in terms of all those human actions that established new forms of dominion and power in civil life (such as in the industry, science, economy, etc.) based on specific ethical values were becoming less and less operative. In this situation the individual becomes less and less responsible for his actions; hence at a deep existential level, he owes his life's development to external sources which he inevitably has to grant a mystical or religious power. And all these dynamics even had their particular justification in the philosophical school of vitalism, as an intellectual escape into highly emotional and sentimental approaches to life; as an evasion from all former moral and ethical constrictions of the societies of the West, that were interpreted more and more as romantic residues, and the evasion was developed *as fast as possible*. Lewis Mumford wrote during those times: *"the going becomes the goal."* When addressing this specific character of a highly technical civilization, Jünger wrote: *"the greater the speed we move, the less we reach the goal."*[77]

In a scientific-technical context, if we have a container having a gas at high pressure, and we release the gas slowly into the atmosphere

---

[76] Alexander Dugin. *The Fourth Political Theory*. Arktos 2012, pg 145.

[77] "Je schneller man sich zu bewegen vermag, desto weniger kommt man zum Ziel." Ernst Jünger. *Der Arbeiter*. Ernst Klett, Stuttgart 1981. Klett-Cotta, pg 89.

we shall generate friction and entropy in the whole system, causing the levels of chaos to increase. However, if I connect a turbine to the container, then the pressure shall be channeled through a mechanism that produces useful work. The action of release, by having a specific configuration, articulates the energy in given forms. The mechanism channels the energy and chaos is reduced…so by employing this latter image, whenever the release of "inner pressure" takes place at an individual level, then it also required external mechanisms in order to configure the individual energy in forms that can be applied to the external individual domain, as facing one's irrational atavistic energies without any mechanism of external affirmation can cause huge levels of anxiety and existential trauma to the individual. The progressive propagation of new freedoms in the West inevitably caused deep conflict to those individuals who identified with Puritan lifestyles and ethics, and the repression of the instinctive individual energies by the moral superstructure caused many dysfunctional behaviors and psychological pathologies which became transposed or sublimated by the advertisement industry. It was by analyzing this conflict between the new sense of freedom as "release" and the potential loss of the sense of identity that made psychoanalysis a necessary field to be developed in the West.

The main assumptions of this branch of psychology (psychoanalysis) dogmatically assumed that all chaotic and irrational forces that disturb the human psyche and behavior emerge exclusively from the realm of the subconscious and unconscious. By reasonably arguing that the cultural constructs, religions, ideologies, and dialectical approaches to the universe existent during those times constituted altogether a rigid barrier that is in constant conflict with irrational forces, the only methods that appeared in psychoanalysis in order to reduce the psychological conflict of the individual were those of guiding the individual carefully in therapy through a series of *transfer* techniques, where the psychoanalyst is intended to become the spiritual "attractor" of all the repressed irrational tendencies of the patient, thus becoming the configuring axis of such tendencies. This technique is equivalent to the confessions of guilt that Catholic adepts practice with their priest. Even though Catholicism had also experienced a decline when it attempted integrating with the forces of the modern era, tradition was characterized during its splendid moments by a whole series of practices and rituals that aimed to form and spiritually select balanced and *homeostatic* priestly castes, as individuals who constituted the embodiment of metaphysical forces that could make effective the ritual techniques embedded in sacrament. In the case of psychoanalysts the formation of the psychoanalyst is highly discursive, abstract, and speculative, to the extent that some spiritually balanced individuals have been much more successful as therapists.[78] However, the powerlessness of psychoanalysis and its decadence as a branch of psychology became very clear in recent times when irrational and unconscious tendencies of the masses existent in urban/industrial societies were no longer treated and configured in order to build individuals free from the fear of themselves, and fears became transposed and sublimated by the advertisement techniques developed by motivational research,[79] where according to Jünger *"there is nothing but demonic evil."*[80] These techniques provided gratification of the self and reduction of the feeling of guilt when obeying unconscious drives triggered by publicity and propaganda techniques. On one hand these techniques contributed to social order and the rigid configuration of social classes, yet on the other hand, they also fostered a deep passivity of the individual to social and industrial determinisms where the

---

[78] See Hans Eysenck *Decline and Fall of the Freudian Empire* (1985; second edition 2004).

[79] Vance Packard, *The Hidden Persuaders*: "What the persuaders are trying to do in many cases was well summed up by one of their leaders, the president of the Public Relations Society of America, when he said in an address to members: "The stuff with which we work is the fabric of men's minds." (…) Motivation research is the type of research that seeks to lean what motivates people in making choices. It employs techniques designed to reach the unconscious or subconscious mind because preferences generally are determined by factors of which the individual is not conscious. Actually in the buying situation the consumer generally acts emotionally and compulsively, unconsciously reacting to the images and designs which in the subconscious are associated with the product."(…) "For each man, woman, and child in America in 1955 roughly $53 was spent to persuade him or her to buy products of industry. Some cosmetics firms began spending a fourth of all their income from sales on advertising and promotion. A cosmetics tycoon, probably mythical, was quoted as saying: "We don't sell lipstick, we buy customers." Vance Packard, *The Hidden Persuaders*. Mark Crispin Miller. Ig Publishing 2007.

[80] Ernst Jünger. *Der Waldgang*. Klett-Cotta. Stuttgart 1980, pg 48.

question in regard to the aims remained unanswered, thus fostering the attachment to economic determinism at all levels. It cannot be denied that the crucial influence exerted by marketing as a *technique* that, assisted by a wide variety of psychological experiments on behaviorism, induces stimulus-response actions in the individual that resort to the coarsest elements present in the human instinctive nature. The use of these techniques degrades the chance of any responsible reflection on the specific causes or roots of one's actions and also impedes the acknowledgment of their effects, triggering higher and more insatiable levels of passivity and mobilization. Arthur Koestler, when referring to the influence caused by a whole school of thought developed in the US that attributes animal characteristics to the human condition writes: *"by trading with an anthropomorphic evaluation of the rat, north American psychology has accomplished a ratimorphic evaluation of men."*[81]

In one of the very few documentaries on the life and work of Jacques Ellul titled "*The Betrayal of Technology*," the French philosopher exposes a simple yet shocking example of how the concept of freedom itself became "customized" during the last century due to the rationalization and mechanization of human factors by the emerging of new urban-industrial infrastructures: he explains how the car, having already corresponded to one of the first technical objects that clearly succeeded when inducing a dazzling sensation of freedom to modern men, ended up becoming an actual treason to the ideal. This is because, as Ellul points out, it became more and more common to observe in the summer periods that the average citizen of Paris, by commencing to enjoy holidays after months of intense work, felt released once getting into his car in order to drive towards the Mediterranean coast. From an individual viewpoint, the action in the case of the man of Paris can be understood as an individual action based on freedom. But if one, however, observes the same fact from a wider and more systemic viewpoint, one can verify that not only the individual we are mentioning but also millions of people from Paris began to drive

---

[81] Koestler, A., *The Act of Creation*, New York: Macmillan, 1964, pg 560.

in unison to the Mediterranean coast. It is precisely when achieving this wider and systemic perspective that the concept of freedom becomes more problematic. How is possible to talk about individual freedom provided by the car if all the drivers were all heading towards the same *destiny*?

In this specific sociological context of Ellul's example, "freedom" is conceived of as the "letting go" of any residual or atavistic moral constraint during action or when taking a decision, thus allowing external social conditionings and determinations to channel individual desire, by acting as "blow-off valves" that articulate in social terms the individual's drives. McLuhan, when showing the liberating aspects of technique—in this case, the telephone—also remarks: *"An extraordinary instance of the power of the telephone to involve the whole person is recorded by psychiatrists, who report that neurotic children lose all neurotic symptoms when telephoning."*[82] In this specific form of individual "freedom," the individual passively reacts to external stimuli, and furthermore, he is necessarily incapable of judging the psychological impact of surrendering to the stimuli. By showing that the development of the technosphere surpasses all moral delimitations, Ernst Jünger presents us with a very appealing image of the process when writing: *"The increase of traffic not only approximate more quickly the good Europeans, but also the bad Europeans."*[83] So if we apply the formerly explained concept of *homeostasis* at an individual psychosomatic level, it refers to the capacity of the individual to maintain a state of being that "filters" from external stimuli those that empower the homeostatic state of being that selects those stimuli that contribute to keeping up the balance. In order to accomplish this in practice, a given set of values and criteria are required on the part of the individual, and these values transcend morals and ethics; the values constitute the being and symbol that the individual represents. Morals and ethics merely correspond to "steps" of the inner disciplining of action required in order to attain such a state.

---

[82] Marshall McLuhan. *Understanding Media. The Extensions of Man.* The MIT Press Cambridge, Massachusetts. London, England 1994.

[83] "Die Steigerung des Verkehrs nicht nur die guten, sondern auch die bösen Europäer schneller aneinanderbringt" Ernst Jünger. *Der Arbeiter.* Ernst Klett, Stuttgart 1981. Klett-Cotta, pg 80.

We previously referred to homeostasis as a concept equivalent to health. If we particularize the former explanation to the domain of the individual's nutrition, a healthy individual who doesn't surrender constantly to the food that is constantly advertised—feeling "free"/"released" do to so—but he who has a very particular criteria on the kind of food that empower his state of being and proper body constitution, not out of a rigid "diet" program, but naturally, as a spontaneous habit. Before attaining this homeostatic state of being, many years are required in order to create a synthesis between the irrational and rational elements of the psychosomatic constitution, that is, the development of the capacity to channel all desires and drives into stable rational forms. When this occurs in any organism or system we can refer to this process as a *neguentropic* process. Ultimately all forms of qualitative growth are characterized by *negative entropy*, that is, by the capacity to constitute inner order out of a chaos of external perturbations. As Erwin Shrödinger wrote in 1944:

> *"What then is that precious something contained in our food which keeps us from death? That is easily answered. Every process, event, happening—call it what you will; in a word, everything that is going on in Nature means an increase of the entropy of the part of the world where it is going on (...) What an organism feeds upon is negative entropy."*[84]

In modern societies, the former case of local or individual *neguentropy* are extremely rare. Both at an individual, community, or even country level, there is less and less presence of selection criteria when facing change or external perturbations. To some extent it is because all changes in our environment are extremely novel and there is no "program" in our genes or historical record available for handling properly the change, as in the genetic evolution of humans or History. As McLuhan points out: *"We are no more prepared to encounter radio and TV in our literate milieu than*

---

[84] Erwin Schrodinger. 1944. *What is life? The Physical Aspect of the Living Cell.*

*the native of Ghana is able to cope with the literacy that takes him out of his collective tribal world and beaches him in individual isolation. We are as numb in our new electric world as the native involved in our literate and mechanical culture.*"[85] So this progressive incapacity necessarily entails a higher production of entropy at all levels, more chaos, dissipation, and mobilization of all factors.

If there is an adequate way of diagnosing a crisis in the West, and by extension, in the entire planet, such a diagnosis refers to the *progressive decline of neguentropic cultural processes*, at the level of individuals, communities, and State-nations. In all living forms, when this lack of neguentropy takes place, it inevitably entails that entropy production starts to gain predominance, that the coldness of death, passivity, momentum, and inertness starts to prevail. When this process takes place before our eyes there emerges a dynamics that Alexander Dugin refers to as *monotonic* processes. By resorting to the work of American scientist Gregory Bateson, the Russian author writes:

> *"The monotonic process is the idea of constant growth, constant accumulation, development, steady progress, all accompanied by the increase of only one specific indicator. In mathematics, this is associated with the ideas of the monotonic value; in other words, the ever-increasing value—hence, monotonic functions. Monotonic processes are the type that always proceed in only one direction: for example, all their indicators consistently increase without cyclical fluctuations and oscillations (…) at three levels—at the level of biology (life), at the level of mechanics (steam engines, internal combustion engines), and at the level of social phenomena."*[86]

Even though the decline of the West had been already diagnosed more than a century ago, it is 2016 and here we still are already. Some say alive and kicking—and to most eyes, there appears to be more signs of life and vitality than ever. Is, therefore, the decline

---

[85] Marshall McLuhan. *Understanding Media. The Extensions of Man.* The MIT Press Cambridge, Massachusetts. London, England 1994.
[86] Alexander Dugin. *The Fourth Political Theory.* Arktos 2012, pg 59.

of the West pure delusion? If by supposing that the West lost its Historical mission more than a century ago and we still enjoying the maximum comforts more than ever, isn't there a contradiction in all of this? It is very clear that during the last century, all areas of knowledge, history, science, philosophy, culture, art, and politics became progressively more and more disconnected from each other, more fragmented and isolated. The inevitable "coldness" that emerges in any cultural framework that is losing its capacity for synthesis impels more and more individuals to surrender to activities characterized by a narrower and narrower perspective of all things.[87] The more all these domains were approached scientifically by intending to establish laws, norms or work hypothesis upon the domains, the more the second law of thermodynamics expressed its inevitable presence, and the more uncertainties, chaos, and dissipation of efforts impeded to recover any synthesis. As Robert M. Pirsig tells: *"Through multiplication upon multiplication of facts, information, theories and hypotheses, it is science itself that is leading mankind from single absolute truths to multiple, indeterminate, relative ones."*[88] For instance, the more the domain of History was approached through rational, dialectical or conceptual frameworks or by verification of historical facts, the more History rebelled by exposing highly irrational elements through legend, myth, and saga. The more politics became a science, the more challenged politicians became by the irruption of the irrational behavior of the masses. The more art was understood in purely intellectual or sociological forms, the more the hand of the artist rebelled anarchically in the most irrational forms of canon subversion. The more scientific groups and international organizations warned about the devastating

---

[87] "Progress is possible only by passing from a state of undifferentiated wholeness to differentiation of parts. This implies, however, that the parts become fixed with respect to a certain action. Therefore progressive segregation also means progressive mechanization. Progressive mechanization, however, implies loss of regulability. (…) Progress is possible only by subdivision of an initially unitary action into actions of specialized parts. This, however, means at the same time impoverishment, loss of performances still possible in the undetermined state. The more parts are specialized in a certain way, the more they are irreplaceable, and loss of parts may lead to the breakdown of the total system" Ludwig Von Bertalanffy. *General Systems Theory*. George Braziller. New York. 1968, pg 85.

[88] Robert M. Pirsig. *Zen and the Art of Motorcycle Maintenance: An Inquiry into Values*. 1974 (William Morrow & Company).

consequences of resource depletion and climate change, the more accelerated became the destructive planetary processes. The more people talked about peace, the more the risk of an abrupt explosion of war emerged on the horizon.

Western culture was very much challenged and compromised by all these processes taking place. In the case of any culture that is *healthy,* culture is characterised by being highly homeostatic with the conditions of reality and power, by having an optimal capacity for synthesis, self-regulation, and high levels of entropy dissipation regulated by specific nonlinear attractors that have a symbolic character, thus constituting a culture that is highly symbolic at many levels. Yet when the generation of entropy flows outwards, it is as if the organism "freezes," like if it starts "dying." It is very important to recall at this point the words of the French physicist, Leon Brioullin, who solved Maxwell's paradox—a paradox that shall allow us to later penetrate into the key aspects of *poiesis*—who wrote that *"the diminishing of entropy [of a system] can be taken as a measurement of the amount of information."*[89] For instance, in the case of a laboratory, living beings can be studied and rendered into scientific information to the extent their levels of self-organization (autopoiesis) are highly minimized, and thus to the extent their behavior obeys cause-effect and mechanistic laws, as a response to the specific stimulus present in the laboratory. And in order for the latter to occur, the production of entropy (and thus the intrusion of chaotic behavior) in the living being has to diminish.

Hence, due to its progressive incapacity to "recycle" its own productions the decline of the culture of the West was demonstrated in its incapacity to generate new synthetic symbols adapted to the new conditions of existence, and an inevitable consequence of this is that, when approached by a new system or power that integrates the decline, former cultural symbols become transformed into *ciphers, signs or information*, which all correspond to the particular projection of the former symbols upon the new emerging symbols

---

[89] Brillouin, L., 1956, *Science and Information Theory*, Nueva York, Academic Press; trad. fr., 1959, La Science et la Théorie de l'information, París Masson.

of the new State, a State that "sees" the cultural forms based on the characteristics of its own eyes, its own *Zeitgeist*, a completely new principle of development. Jünger writes in this regard: *"A new principle is credited for creating new facts, to create unique and effective ways—and these forms are profound because they are existentially referred to such principle."*[90] So returning to our little example of a given living being, being used for laboratory experimentation, the scientific information obtained from the being is not that of the actual being (since the true behavior of the being takes place outside the laboratory and in its natural environment), but it is rather the behavior of the being projected upon the conditions of the experiment. In other words, the information obtained corresponds to a partial representation that depends on the conditions of the experiment and on the eyes of the observer, as corroborated by Werner Heisenberg's statement *"The method can't be separated from its object."*[91] Another example of this can be pointed out in the domain of language: the idea and concept that is pointed out by a given term in Chinese, when translated into English has to become necessarily projected into the particular syntactic structure embedded in the English language, and therefore filtering the Chinese idea or concept through the syntactical structure of a language very easily risks to impoverish the *meaning* of the Chinese idea based on the *means* of another language. Spengler also presents the issue of a feasible sharing of cultural symbols by writing:

> *"red and yellow do or do not mean the same for others as for themselves. It is particularly the common symbolic of language that nourishes the illusion of a homogeneous constitution of human inner life and an identical world-form; in this respect, the great thinkers of one and another Culture resemble the colour-blind in*

---

[90] "Ein neues Prinzip weist sich durch die Schaffung neuer Tatsachen, eigentümlicher und wirksamer Formen aus – und diese Formen sind tief, weil sie existentiell auf dieses Prinzip bezogen sind." Ernst Jünger. *Der Arbeiter*. Ernst Klett, Stuttgart 1981. Klett-Cotta, pg 80.

[91] *Physics and Philosophy: The Revolution in Modern Science* (Pelican). Werner Heisenberg Published by Penguin Books Ltd (1989).

*that each is unaware of his own condition and smiles at the errors of the rest."*[92]

In the case of the analysis of nature and its necessary surrendering to the conditions and structures of a given experiment in the laboratory, there is also an inevitable degradation of the higher symbols of all living phenomena towards the generation of information structured based on space, time and causality. Not even geometrically, but even at a cultural level the "squaring of the circle" is just impossible. Hence, whenever we approach a domain of nature that is highly mechanistic, like a billiard pool for instance, the functional language provided by mechanistic and reductionist science is adequate for the "translation" and comprehension of the phenomena through formal laws, but when approaching any organism or system that is characterized by dissipative structures and non-linear configurations, then the language ends up impoverishing the original phenomena. This "gap" between the qualitative aspects of phenomena and the quantitative comprehension and measurement of it through mechanistic or reductionist science can be very wide when intending to relate to living organisms,[93] and subatomic or galactic domains. The extreme case of such inadequacy of the categories of time, space, and causality to address given phenomena taking place at the atomic level was announced in Werner Heisenberg's *principle of uncertainty* in 1929, where it is stated that the position and momentum of a given atomic particle can not be known simultaneously with the same level of certainty. But it is not even necessary to delve into atomic domains to realize that, even in the case of the string of a guitar, there is no way of determining simultaneously in practice its frequency and position.[94]

---

[92] Oswald Spengler. *The Decline of the West*. Alfred A Knopf. 1926. New York, pg 179.

[93] "A botanist does not measure the difference between two species on the dial of an instrument; nor does an entomologist recognize butterflies by means of a machine; nor an anatomist bones; nor a histologist cells. All these forms are recognized directly." Rupert Sheldrake. *Morphic Resonance. The Nature of Formative Causation*. Park Street Press.

[94] In order to capture the "wave" of the idea I've just mentioned I'll resort to a quite "solid" example that represents an electron with a guitar string. Let's see: When the string of a guitar vibrates and produces a sound (wave) there exists an energetic dissipation in the form of an attenuated sound-wave.

This principle of uncertainty had also been experienced directly by Eugen Herrigel when intending to grab with his fingers the string of the arc at the point of maximum tension, and when after intending to release the bow in a way Master Kenzo Awa did, softly and with absolutely no apparent dissipation and friction during the release. The uncertainty Eugen Herrigel experienced was the following: the more tension he applied with his fingers, the more the tension impeded the release of the arrow, and whenever he consciously attempted to release the arrow, the friction caused by his fingers on the bow was followed by a chaotic reaction on the part of the entire weapon, which diminished to a large extent the potential speed of the arrow. So ultimately, the more Herrigel wanted to control conscious and simultaneously tension and release, the more he failed and the more the arc was destabilized and sort of "tailspin." If Herrigel focused exclusively on tension, then he could not control the release; and if he managed to control the release, then the tension had to be necessarily insufficient. This situation became very likely the most challenging part of Eugen Herrigel's learning. On one hand because his intellect was still "fixed" into a comprehension of the cosmos based on the laws derived from space, time and causality which were substantially powerless when handling the uncertainty principle applied to the arc and bow, and on the other hand because by observing Master Kenzo Awa, Herrigel realized that there was actually a solution to this problem, which he unsuccessfully tried to intellectually understand, since as we formerly explained, the attempt of structuring metaphysical domains into conceptual and discursive language frameworks is always irreversibly prone to "freeze" the genuine character of the phenomena in regard to its metaphysical significance.

---

This dissipation generates entropy. While this wave is perceived with a specific frequency, let's say 2000Hz, this frequency is a function of the vibratory attribute of the string, and not a function of its specific position. Therefore it is the variables of frequency and amplitude that define the specific sound, independent of the exact position of the string. And if I stop the string abruptly and I determine its exact position, this action will abruptly end the emission of sound waves. In the case of determining the problematic wave-particle character of an electron, the physics of the last century encountered equivalent experimental difficulties.

So at this point in Herrigel's accounts, Master Awa Kenzo only advised his extremely willful German student to keep on practicing, and to breathe adequately during the entire cycle of tension-release. Yet at some point the Great Master intended to provide Herrigel with a symbolic comprehension of the whole arc-bow configuration; Awa Kenzo pointed out to his disciple that the arc was equivalent to a "bridge" between heaven and earth, and that the friction type of sound of the string during the bow release corresponded to a perfect symbol of an "interruption" present between the metaphysical domain and the physical domain, mostly due to the limitations of the conceptual paradigms that Eugen Herrigel had inherited from the culture of the West.

Eugen Herrigel finally—as we shall show in this book—found a temporary way out of this dilemma. But what happened to the Western view of the cosmos when faced with the principle of entropy and uncertainty? We shall deal with this temporary solution in the pages ahead.

\* \* \*

During those years with the principle of uncertainty that was becoming expressed effectively in countless cultural domains, there was a point during Eugen Herrigel's training in the "Great Doctrine" when the German teacher intended to solve cleverly and intellectually the conflict arisen by the principle of uncertainty. Herrigel experimented moments of deep lack of faith in the "Great Doctrine," and eventually surrendered to the need of developing a "smart" and clever solution... this "solution" consisted of developing a clever technique that permitted his finger to progressively "slip" through the bow, so the arrow could be released in a smoother way. Surprised by the sudden accomplishment of Herrigel, Master Awa Kenzo, finally realized the "trick" developed by his student and decided immediately to expel Herrigel from any further teachings on Zen Buddhism.

The decline of the West, which was expressed to a great extent by the incapacity to develop and provide to the European peoples a synthetic and convergent view of the new emerging States, and that was also expressed in the powerlessness to provide any absolute certainty in terms of knowledge—due to the principle of uncertainty and entropy—was however "solved" due to a renewed over-emphasis on the attainment of clever and smart resolutions that, from the viewpoint of any Operative Tradition, were treasonous to pure wisdom.

It is a mysterious matter of historical uncertainty why the West, in parallel to Eugen Herrigel's own developments, adopted a standpoint of "tricky cleverness" in order to avoid revealing the intensity of its synthetic impotence. It can be hypothesized that during the 1900s the over-emphasis on action, and the incapacity for *otium* on the part of the Western man, impeded any attempt to reconnect with metaphysical sources of stability and balance. In effect, countless developments were taking place at huge paces, thus allowing no time for devotion to calm reflection, nor for contemplating the perennial symbols of Tradition, in order to relate such symbols to the actual conditions of the world. Some like René Guénon, who however, were capable of developing a bridge between the metaphysical symbols and the overall sociological, historical and psychological conditions of modern times, asserting constantly in their writings the existence of a cyclical determinism taking place based on the doctrine of the Yugas. The Kali-Yuga corresponds to an inevitable process of fall that burdens the human condition at irreversible and more intense rates as the Kali-Yuga reaches its conclusion.

If we consider the characteristics of the spirit of the times during the last century (its Zeitgeist) it's hard not to agree with Guénon that there was an overall tendency which empowered mobilization, at the expense of their symbolic synthesis. And in the case of the West, such mobilization of factors and progressive disempowerment of the old State-nation and dynastic configurations of power provide us with the strong impression of an ever-increasing fragmentation and mobilization of factors, which was becoming more and more separated from their primordial sources and origins. Alexander Dugin writes in regard to the political decadence: *"In the end, all forms of vertical symmetry (the orientation of a 'top to bottom' hierarchy) are subject to destruction, and everything becomes horizontal."*[95] In this context, it is time to refer to a new concept that became more and more prevailing in all domains: *complexity*.

---

[95] Alexander Dugin. *The Fourth Political Theory*. Arktos 2012, pg 179.

# THE GROWTH OF COMPLEXITY

> Any fool can make things complicated, but it requires a genius to make things simple.
> E. F. Schumacher[96]

When approaching the concept of *complexity*, American anthropologist and historian Joseph Tainter, author of *The Collapse of Complex Societies*, by drawing upon the works of Peter M. Blau[97] and Randall H. McGuire[98] states that *"a society with a great deal of heterogeneity, then, is one that is complex"* by considering heterogeneity as *"the number of distinctive parts or components to a society, and at the same time to the ways in which a population is distributed among these parts."* By referring to the work of Gregory J. Johnson,[99] Tainter refers to complexity as the *"growth in the amount of information that must be processed by a society, with greater quantity and variety of information requiring greater social complexity,"* and at last by referring to H. Simon[100] defines complexity at that characterized by *"nearly decomposable systems."*

---

[96] E. F. Schumacher. *This I Believe and Other Essays*. 1997 Dartington, UK: Green Books Ltd, pg 8.

[97] Texas Press, Austin. Blau, Peter M. (1977). *Inequality and Heterogeneity: a Primitive Theory of Social Structure*. Free Press, New York.

[98] McGuire, Randall H. Breaking Down Cultural Complexity: Inequality and Heterogeneity. In Advances in *Archaeological Method and Theory*. Academic Press, New York (1983).

[99] Johnson, Gregory J. *Information Sources and the Development of Decision-Making Organizations*, 1978.

[100] Simon, H. (1965). *The Architecture of Complexity*. General Systems. F. Schumacher. *This I Believe and Other Essays*. 1997. Dartington, UK: Green Books Ltd, pg 8.
 Texas Press, Austin. Blau, Peter M. (1977). *Inequality and Heterogeneity: a Primitive Theory of Social Structure*. Free Press, New York.
McGuire, Randall H. Breaking Down Cultural Complexity: Inequality and Heterogeneity. In Advances in *Archaeological Method and Theory*. Academic Press, New York. (1983).
 Johnson, Gregory J. *Information Sources and the Development of Decision-Making Organizations* 1978.

One conceptual barrier that appears whenever defining the concept of *complexity* is that it is often opposed to *simplicity*. However, such opposition is very reductionist and misleading whenever intending to come up with a wide and applicable notion of complexity. By referring to the concepts formerly defined (homeostasis, attractor, etc.), we shall attain a more integrating perspective on complexity and its typical character.

First of all, complexity is not an intrinsic characteristic of a system but rather a *perception* or *interpretation* of a given system based on an analytical and radically rational perspective. The more we perceive things separated from each other, the more a character of complexity is attached to the perception, even in domains such as philosophy. Evola writes: *"There isn't a philosophy because there is rationality in things, but philosophy provides things with a rationality they formerly did not have—this is the fundamental point."*[101] For instance, if I am capable of perceiving an arc in my surroundings and I can find its geometric center, I have the chance of defining the arc as a symbol (a circle) defined by position coordinates, angle and a radius. Hence, if I am capable of perceiving symbols or figures amidst reality, the definition of the objects of reality is simple, even though in the definition of the object there is always embedded an irrational "residue," since any symbol (like in this case a circle, is characterized both by rational and irrational elements). However, if my perception is exclusively analytical and I'm thus unable to integrate the irrational elements of reality into the analysis, then whenever I perceive an arc and I'm unable of proceeding by synthesis, that is, unable to induce or imagine the center of the object of reality, I'll have to be necessarily forced to consider the arc as an unclosed polygon... so the more I intend to capture the form of the polygon, the more amount of information (coordinates) I'll require in order to define the object, and as I cannot perceive symbolically the irrational elements, the amount of information required in order to "capture" the object tends to infinite.

---

[101] Non vi è filosofia perché vi è una razionalità nelle cose, ma la filosofia dà alle cose una razionalità che prima non avevano—questo è il punto fondamentale. Julius Evola. *Fenomenologia Dell'individuo Assoluto Iii* ed. corretta: Edizioni Mediterranee, Roma 2007.

By visualizing the former example, it is easy to conclude that the perception of complexity in a given system is directly related to the presence of analytical and highly rationalistic approaches to external objects that are inherent to such perception. If complexity has to be in opposition to any concept, the most adequate opposition is not simplicity but rather *νοῦς*, or *intellectual intuition*, where all centers and immobile movers of objects are perceived and where all representations of objects are not only analytical but synthetic and symbolic. When the focus of perception is in the means, the complexity of perception increases; when the focus of perception is the meanings the νοῦς of perception increases.

As formerly explained, the decline of the West can be equated to the decline of the cultural capacity to synthesize all the new developments and novel configurations taking place during the last two centuries in order to attain a symbolic, homeostatic and stable configuration of power. Without the synthesis, any moral or ethical discipline in the realm of action was becoming more and more disempowered. And the incapacity for cultural synthesis caused inevitable fragmentation and mobilization at all levels. The diminishing capacity to sustain homeostatically the old structures caused—as Leon Brioullin pointed out—the inevitable generation of information, which corresponds to the projection of the old symbols upon the novel and yet highly mechanistic structures. Hence, the decline of the West was accompanied by a massive production of information in the most varied fields, and as the worldwide cultures progressively lost the capacity for synthesis, the process required handling and organizing the production of information and mobilization of factors in the most efficient way.

In modern times the increase of complexity and the increase of information production due to the decline of the homeostatic character of the Western culture became handled by very specific new configurations of power. Even though it became progressively difficult to visualize the *meaning of the means*, Western societies to a large extent managed to transform the chaos into a new order and infrastructure. In effect, there was a very important development that took place during the 1920s-1950s, which has been mostly dismissed an imperceivable by most historians, due to the inevitable factor that whenever approaching any historical facts, the facts can only become integrated into our language, concepts and representations to the extent the facts can be recognized by the intrinsic structure of the concepts and language. In many cases, the conceptual constructs employed in order to analyze the events of the 20th century are based on political and sociological concepts such as nationalism, liberalism, capitalism, and a division of society into classes, etc. that rather emerged in previous centuries. And yet qualitative modifications of phenomena always require the emergence of a qualitatively new language. On these grounds, if we keep up with the initial approach of Tradition that we've presented in the former pages, we can more easily pursue the developments taken place in the "thread" of Tradition by attempting to locate, first of all, the emergence of novel modes of relating to means of power by men. This "learning to relate to the means" is also essentially one of the key aspects that Eugen Herrigel had to learn through his spiritual disciplining.

Based on this assumption, it can be considered that, as we shall see immediately, strictly revolutionary events took place during those decades, away from the domain of public opinion, official historiography, and that of everyday politics. Ernst Jünger, in the preface to *Der Arbeiter,* writes:

> *"In all times it are the episodic and accidental ingredients of a problem, its political and controversial facade, and not its substantial core, what captivates more powerfully the attention."*[102]

Far from the public eye, transcendent changes would completely alter worldwide the infrastructures and configurations of power in the world. We can recall at this point Nietzsche's words when saying:

---

[102] "Das Episodische und Akzidentielle, der politische und polemische Vordergrund eines Problems, zu allen Zeiten die Aufmerksamkeit stärker fesseln als sein substantieller Kern. Dieser wirkt jedoch auf die Dauer, wenn auch in stets wechselnden Verkleidungen" Ernst Jünger. *Der Arbeiter*. Ernst Klett, Stuttgart 1981. Klett-Cotta, pg 3.

*"It is the stillest words which bring the storm. Thoughts that come with doves' footsteps guide the world."*[103]

In effect, in the case of the almost invisible thread of an integral Tradition, we must first acknowledge that such thread is extremely delicate, and whenever intending to capture its changes we have to resort all the time to the *subtle* directions that take place in the material configurations and developments.

Based on this premise, it was during the year 1922[104] that a Russian-American mathematician, Nicolas Minorski (1885-1970) who worked in the US Navy in tasks related to the design of stabilizing systems for the routes of the ships, became inspired by the typical behavior of the helmsman, who basically has to compensate perturbations caused by the sea waves with opposite reactions. Minorski's aim was to achieve stabilization in the case of the guiding of ships, that is, to achieve *automatic piloting*. The whole idea was to outsource the pivotal responsibility of inducing homeostasis by the pilot to an external set of devices. The technical contribution of Minorski to the issue of self-governing by technical devices would be later developed and advanced by the engineer Harry Nyquist (1889-1976) and then by electric engineer Harold Locke Hazen (1901–1980) when implementing methodologies of research and the assessment of a considerable number of servomechanisms within diverse areas of engineering. The technical innovations implemented by these three men became highly decisive in the way control and power were exerted over given technological means. The industrial and military developments taking place during the two World Wars can not even be conceived in practice without the technical innovations.

The innovations responded technically as well as to the crisis that modern scientific reductionism was already going through when approaching reality, and the language that emerged in order to overcome the gap was the characteristic language of a novel

---

[103] Friedrich Nietzsche. *Thus Spoke Zarathustra*. The Stillest Hour.

[104] Minorsky, N. (1922). "Directional stability of automatically steered bodies." J. Amer. Soc of Naval Engineers 34: 280–309.

scientific paradigm: *cybernetics*. Hence, during the decade of the 30s all areas of science, technique, thought, art, economy and politics were all sieved by a new way of conceiving the deepest essence of reality, a new way of *seeing* reality, a new way of dealing with complex perceptions of reality. Thomas Kuhn, when intending to define the crucial concept of *paradigm*, directly relates it to a specific way of seeing in terms of problems, models and empiric *praxis*. He writes: *"paradigms: These I take to be universally recognized scientific achievements that for a time provide model problems and solutions to a community of practitioners."*[105] This corresponds to a clear mutation in the spirit of the times or *Zeitgeist*: it implies a new world-view, a new form of power, and the actual power of the mutation is honestly expressed in the fact that we are not referring here to an abstract, idealist, ideological or romantic world-view, which as Julius Evola would have said, *"leaves one's internal throne empty or occupied by ghosts,"*[106] but rather referring to a new set of concepts that allow active men to relate to means of power. Absolutely novel concepts such as *net, computation, memory, processing, feedback, information, control, system, code, program, translation, direction,* and *inhibition* started to be equally applied to all fields, that is to say, they were applied to biology, economy, sociology, philosophy, psychology, natural sciences, politics, education, etc. The new modes of objective, effective and material power were thus structured through a new language that began to invade everything, modifying inevitably the nature of power and the intrinsic structure of the States itself. As Oswald Spengler remarks, when referring to the specific artistic modes of a given culture and the form of its State:

> *"Who amongst them [present-day historians] realizes that between the Differential Calculus and the dynastic principle of politics in the age of Louis XIV, between the Classical city-state and the Euclidean geometry, between the space perspective of Western oil painting and the conquest of space by railroad, telephone and long-*

---

[105] Thomas Kuhn. *The Structure of Scientific Revolutions.*

[106] Julius Evola. An Intellectual Autobiography. *Path of Cinnabar* Arktos 2009, pg 54.

*range weapon, between contrapuntal music and credit economics, there are deep uniformities?"*[107]

Thus during those years a new form of power developed a new form of material configuration characterized by new symbols. Those individuals who embraced these new symbols gained an individual participation in such power, and those who did not embrace them had to necessarily remain in the "wagon" of History's rails, being pulled by "railway engines" that in some cases they were aware of, and in many cases they were not. It is a well-known fact that some trains are so comfortable that the passengers completely forget that they are being impelled at huge speeds... Jünger writes:

> *"from the moment that humans restrict their own decisions for the benefit of technological facilities, they attain numerous comforts. (...) What emerges is optimism, an awareness of power generated by speed."*[108]

Yet the "railway engines," by being characterised by operative procedures that were radically different to those that governed previously the issues of the State, implied necessarily a renewed process of circulation of elites (in Vilfred Pareto's[109] terms) where it is precisely the technician or technocrat (this is, a highly pragmatic figure who is not determined by political passion and who is familiar with the new emergence of technical symbols) who becomes the best candidate to handle a "train" of power. Already Henri Comte de Saint-Simon argued during the French Revolution that management of "the governmental machine" ought to be turned over to people who could properly administer it, namely industrialists, scientists, and technicians.[110] Also, Augusto Comte (1798-1857) was one of the

---

[107] Oswald Spengler. *The Decline of the West*. Alfred A Knopf. 1926. New York, pg 7.

[108] "und zwar insofern, als der Mensch zugunsten technischer Erleichterungen sich in der Entscheidung beschränkt. Das führt zu mannigfaltiger Bequemlichkeit. Notwendig muß aber auch der Verlust an Freiheit zunehmen. (...) Es tritt im Gegenteil ein Optimismus auf, ein Machtbewußtsein, das die Geschwindigkeit erzeugt." Ernst Jünger. *Der Waldgang*. Klett-Cotta. Stuttgart 1980, pg 30-31.

[109] *The Mind and Society*, Vol. IV, New York: Harcourt, Brace and Company, 1935.

[110] Winner, Autonomous Technology, p. 141; Scharff and Dusek, *The Technological Condition*, pg 6.

first to envision a technical and no longer political administration of society. Contrary to the political Statesman of former times, the technocrat is not asked to provide political objectives. Under the new configuration of powers, his function is that of allowing the "train" to work efficiently and the "passengers to be comfortable." As we shall see in Operative Traditions, within the new configuration of "rails" and networks emerged worldwide due to the irruption of new power configurations, the technician is in no case free to decide where he takes the "train"...

The fact that the development of cybernetics first emerged in the US and then spread so quickly around the world doesn't mean that a clear causal link can be established among such events. Some people might argue that it was actually the techno-scientific information on cybernetics developed by the US what spread very quickly to other countries, thus facilitating R+D in the area, but no causal link can be established here in the very similar way as in the processes of metastasis in cancer there has not yet been discovered any cell that actually "transports" the supposed deadly illness pattern code around the body, no matter how much the hypothesis has been but forth.

The idea of the *Zeitgeist* transcends our common conceptions on processes and phenomena that we characterized as conditioned by time or space causal relations. We can point to the idea by recovering as well some important postulates of English biologist, Rupert Sheldrake, who in his book *A New Science of Life: The Hypothesis of Morphic Resonance* focused much more on the concepts of habits and interactions than those of rigid and immutable laws. This focus fits very well with the approach we are following through Eugen Herrigel's learning, where the devotion on *habit* and *ritual* on the part of the German professor was much more important for his Traditional learning process than discovering the physical laws of his particular interaction with the arc and bow, in order to eventually *resonate* with Tradition, in order to *aim* to Tradition.[111]

---

[111] Di là dalla nozione di leggi del divenire si riaffermerebbe quella di un informulabile di-venire delle leggi. [Beyond the notion of laws of becoming, should be rather affirmed that of a becoming of the laws that can not be formulated.] Julius Evola. *Teoria Dell'individuo Assoluto*. Edizioni Mediterranee,

Whenever Herrigel attempted intellectual approaches to his activity and discipline he was reproached by the Master.

In *The Hypothesis of Morphic Resonance*, Sheldrake exposes examples of great relevance, like the case of the tit bird, that during the 1940s in England learnt to open milk bottles at a prodigious rate, where no cause-effect relation can be established. He also describes how the formation of original anamorphic crystals and proteins in the laboratory is achieved for the first time, the genesis ended up becoming easier for other laboratories around the world that hadn't previously exchanged any material, data or procedures with the original laboratories. As a source of relevant data, Sheldrake also makes use of the experiments of Ivan Pavlov (1849-1936) regarding the natural-born learning capacities of rats after the second or third generation.

The Zeitgeist acts in the microcosmos and in the macrocosmos. In terms of cybernetics, it is easy to perceive anagogic similarities between, for instance, an integrated circuit and a vast megalopolis of our times. Especially at night, the highly regulated traffic of automobiles in a city resembles that of the packs of bits we can imagine moving and being processed along an integrated circuit, while the chips resemble the buildings themselves. In the case of the development of cybernetics, we can resort to a similar approach. It is the correspondence or *resonance* with a set of technical habits what allowed other countries of the world apart from the US (such as the UK, USSR, or Japan) to "magnetize" and "attract" the Zeitgeist of those times into their domains.[112] It was not only about diffusing science in the sense of acknowledgment of abstract laws and conceptual discourse but rather it was the continuous repetition of very specific interactions with the material domain by men that allowed this "morphic resonance"—as defined by Sheldrake—to take place, furthering its development. State-nations such as the US, the

---

1988, pg 111.

[112] Jacques Ellul. *The Technological Society*. Vintage books, 1958: "I feel that this transformation of civilization can be explained by the conjunction in time of five phenomena: the fruition of a long technical experience; population expansion; the suitability of the economic environment; the plasticity of the social milieu; and the appearance of a clear technical intention."

UK, USSR, and Germany gained the overall international effective power to the extent they participated in the development; a highly revolutionary development at an operative level. As Ernst Jünger points out: *"what is at the center of the confrontation is certainly not the way of being of two nations, but rather a different way of being of two ages, one of which, the nascent, devours the other that sinks. This is what determines the true depth of the landscape, it is what determines its revolutionary character."*[113] Hence, the fact that the configuration of space, the new cybernetic "landscape," became so vastly reproduced around the planet, devouring all rural domains, natural landscapes and traditional spaces can already provide us with a clue of the powerful macrocosmic and microcosmic character of the Zeitgeist of a given age. Cybernetics corresponds to that which Spengler referred as *"inherent in the form of the State"* where:

> *"the religious myths and cults, the ethical ideals, the forms of painting and music and poetry, the fundamental notions of each science but it is not presented by these. Consequently, it is not presentable by words, for language and words are themselves derived symbols. Every individual symbol tells of it, but only to the inner feelings, not to the understanding."*[114]

In the Middle-Ages similar symbolic correspondences appeared between cultures that were separated in time and space, analogies where the Gothic symbols coined the cultures maximum expressions. Yet what is necessary to point out at this point is the absolute opposition existent between the Gothic sacred symbols, and the functional symbols characteristic of cybernetics. As we shall see, the opposition is such as that of the Day and Night.

∗ ∗ ∗

---

[113] "Im Mittelpunkte der Auseinandersetzung steht nicht etwa die Verschiedenartigkeit der Nationen, sondern die Verschiedenartigkeit zweier Zeitalter, von denen ein werdendes ein untergehendes verschlingt. Hierdurch wird die eigentliche Tiefe, der revolutionäre Charakter dieser Landschaft bestimmt" Ernst Jünger. *Der Arbeiter.* Ernst Klett, Stuttgart 1981. Klett-Cotta, pg 77.

[114] Oswald Spengler. *The Decline of the West.* Alfred A Knopf. 1926. New York, pg 175.

In terms of the very delicate "thread" that exists in a living and integral Tradition, it is doctrinally assumed that it is the selfsame artist who has to constitute the bridge between the means available and the realm of the supernatural. In this context, the artist is conceived as a "captain" or "pilot" who is deeply aware of who he is in convergence with the development of all things. This human-cosmos convergence corresponds to the ancient Olympic idea of *fatum* or the Indo-European of *rta*. In Operative Traditions the artist is therefore deeply aware of his destiny and the destiny of all things he relates to. His mission is to serve as the "adjuster" or "corrector" of the development of objects whenever the latter are affected by chaos or irrationality. The artist is considered therefore as the "bridge" between "heaven" and "earth," rendering all material chaos into heavenly order. This idea can also be applied to the realm of politics, economics, society, and culture: he who is one with *fatum* knows which are the optimal corrections to apply upon specific economic, technical, political, social decisions. This is the core idea of the art of *governance*. If from the standpoint of an Operative Tradition there is any regal dignity granted to any form of mastery or governance, such dignity is due to the fact that in order to master the governing functions, the governor has to partially incarnate metaphysical dictums, and this aspect alone is what impregnates functions with a regal or sacred character. As formerly exposed, one of the main signs of decadence in the case of the decline in the West was that the sacred conception of the State became secularized, and thus eventually many attempts were put forth by many schools of thought in order to dialectically approach the ultimate power that was driving all economic and social forces. Due to the nature of the dialectical approaches, the idea of *fatum* or destiny was eroded from its sacred or metaphysical character and began to be accessible to the discursive intellect. Hence, the State then became a *thing*; a *thing* that no longer required a regal dignity in order to relate to it, but just required conceptual instruction through books and documents. Consequently, a "map" of the configuration of the State in all its diverse aspects (legal, judicial, administrative, executive, economical, etc.) was accessible to anybody capable of reading the political "maps" of the State. And yet *this dialectical approach of the State is to the true State what a map of reality is to the forces that drive the reality.* Spengler wrote in this regard:

> *"There is no best, or true, or right State that could possibly be actualized according to plan. Every State that emerges in history exists as it is but once and for a moment; the next moment it has, unperceived, become different, whatever the rigidity of its legal-constitutional crust. Therefore, words like "republic," "absolutism," "democracy," mean something different in every instance, and what turns them into catchwords is their use as definite concepts by philosophers and ideologues. A history of States is physiognomic and not systematic."*[115]

No matter how discursively or conceptually logic can become the approaches to power or to the State, the domain of chaos and the irrational, or that which Ernst Jünger referred to as "elementary powers" are always present, veiled by the mists which emerged during modern times between our eyes and naked reality. The German WW I hero writes:

> *"The elemental sources are of two species. On the one hand they are in the world, which is always dangerous, like the sea, which always contains within it the danger, even in the moments when the wind is not blowing. And on the other hand they are in the human heart, which always longs for games and adventures, loves and hates, triumphs and defeat."*[116]

Hence, the sinking of the Titanic ship in the Atlantic Ocean during 1912 can be pointed out as one of the deepest and most shocking symbols of this entire decline experienced in the West. The

---

[115] Oswald Spengler. *The Decline of the West*. Alfred A Knopf. 1926. New York, pg 370.

[116] "Die Quellen des Elementaren sind zwiefacher Art. Sie liegen einmal in der Welt, die immer gefährlich ist, so wie das Meer auch während der tiefsten Windstille die Gefahr in sich verbirgt. Sie liegen zum zweiten im menschlichen Herzen, das sich nach Spielen und Abenteuern, nach Haß und Liebe, nach Triumphen und Abstürzen sehnt" Ernst Jünger. *Der Arbeiter*. Ernst Klett, Stuttgart 1981. Klett-Cotta, pg 25.

destination of the transatlantic was very clear for all the passengers embarking in Southampton, as by merely having a look at the geographic maps they all knew that their destiny was New York. Many of the passengers had never been to New York, yet they had heard about this destiny countless times and had been emotionally affected by the image of the statue of liberty. The emerging "American way of life" corresponded to the seductive image they were waiting for, thus dreaming to arrive to a continent where individual freedom—released from the rigid constrictions of the States of the old European continent—was in the process of becoming a thriving lifestyle. Yet since the times of the Industrial Revolution the massive development of machinery and techniques had corresponded clearly to a process embedded with Titanic undertones, as the power progressively gained by the techno-scientific ensemble of means took place in parallel to the progressive disempowerment of the State-nations in order to master the international development in the sense even the mythical Zeus would have conceived mastery, that is, by all titanic developments "orbiting" around the ultimate and highest power. Yet titanic or Promethean developments are always rebellious and are intimately incapable of surrendering to the sacred. Even Lenin himself, by having incarnated the main Statesman, architect and "captain" of the Soviet Union, became shocked and overwhelmed by the powerful momentum and unstoppable growth of the economic, industrial, administrative, and social processes that took over the State of power and dominion he had originally conceived.

During 1912, the Titanic, by initially intending to follow a clear and direct route from Southampton to New York—the shortcut of less resistance and maximum performance—had also powerfully integrated within its metallic infrastructure a microcosmic representation of the European people of those times. Most passengers, still conceived the idea of power as strongly dependent on economic and socio-political classes, as a speculative and momentous superstructure. However, Titanic power was actually the effective source of all decisive movements. As Ernst Jünger points out, the new modes of power were overthrowing all the *"provinces of the 19th century"*[117] created by liberalism and, as he writes: *"none of the places where one tries to sort and group individuals in classes based on ethical, social and political principles, in none of those places one is in the decisive places of the Front."*[118] Apart from the engineers and technicians of the crew, very few of the passengers were aware that, in contrast to the speculative interpretations of human power mostly derived from modern ideologies, in the Titanic power was *operatively* and *technically* dependent on the momentum caused by the 50,000 tons of the ship, and more importantly, power was incarnated and represented by the *governing* or steering capacity on the part of the captain of the ship, who was responsible for directing the powerful momentum, whenever chaos appeared along the misty horizon.

The tragic sinking of the Titanic when crashing against the unexpected iceberg and finally sinking clearly expressed the risky aspect present when intending to direct or "pilot" any development of mechanistic and industrial factors through mere human means, especially when the momentum of the entire process is so extreme and overwhelming. Even though the catastrophe symbolized the powerlessness of human dialectics in order to "pilot" the titanic process of industrial and technological development—which was not only expressed in the case of the Titanic, but in that of all State-nation. The aforementioned technical developments implemented by Minorski, Nyquist and Hazen, where *the "piloting" of the means is externalized to the means themselves*, and the "ability to respond" began to be outsourced from men to a *system of control*. Let's recall that the word cybernetics has its origin in the Greek term χυβερνητηζ, which means "helmsman." The term had previously been employed by Plato in 5 BC when referring to the act of directing (ruling or governing) a ship, and later it was recovered by André-Marie Ampère (1775-1836) when referring to government ruling. The term

---

[117] "Überall wo man sich ethisch, gesellschaftlich oder politisch klassenmäßig zu ordnen und einzuordnen sucht, steht man nicht an den entscheidenden Stellen der Front – man bewegt sich in einer Provinz des 19. Jahrhunderts, die der Liberalismus in jahrzehntelanger Tätigkeit" Ernst Jünger. *Der Arbeiter*. Ernst Klett, Stuttgart 1981. Klett-Cotta, pg 50.

[118] Idem.

*cybernetics* was re-established in the 1940s mainly in the U.S. and it is not outrageous to think that the specific concept was chosen due to the extraordinary results obtained by the aforementioned engineer Nicolas Minorsky in regard to his accomplishments in developing self-governing automatic pilots when inspired by the behaviour of a helmsman. The "ship" of the West, navigating in decline for a long time, had finally been governed once again. But in this case, running amok, no longer governed by *anybody*…

As we previously pointed out, any living organism from the microscopic of bacteria to the macrocosmic level of cultures/civilizations can be modeled as *dissipative systems*; systems that can be considered to be characterized by higher or lesser levels of vitality and homeostasis to the extent that on one hand they generate entropy, and on the other hand they are capable of creating an "island of negative entropy" that propitiates self-regulation and autonomy, in harmonic correspondence with the characteristics of the cosmos, or that experiential domain that Evola refers to as *"a diverse system of determinations."*[119] It must be clear as well that dissipation of energy itself or entropy production in a given system doesn't necessarily mean that the system can be characterized as autonomous. For instance, the dissipative system constituted by a given helix amidst a gas, once impelled to rotate, at last stops rotating because all its initial energy has been dissipated into entropy, and no "recycling" of entropic currents in the air has been possible. But let's suppose that we could design a servo-mechanism that allowed the helix to position itself exactly in the direction of the external wind currents. In this case a "recycling" of external chaos into energy would be possible, that is, *neguentropy* could be achieved. In order for the servo-mechanism to work properly, it has to be capable of establishing a communication with the characteristics of the atmosphere that surrounds the helix. It is only when this "double-circuit" is achieved that the system can be considered to be as autonomous.

---

[119] "un sistema vario di determinazioni" Julius Evola. *Teoria Dell'individuo Assoluto*. Edizioni Mediterranee, 1988, pg 35.

The capacity of a given system to endure a "double-circuit" among a wide range of perturbations is what defines the homeostasis or State of the system. Any being that attains power and growth does so to the extent that the assimilation or "recycling" of the external factors is effective, and it experiences decline to the extent that the "recycling" becomes weak. So in short, any autonomous or self-organizing being requires on one hand entropy production as a requirement for maintaining its stability and integral functions, and on the other hand it is also required for the system to be embedded somehow with a knowledge in regard to the very nature of the chaotic perturbations or noise, that is, the patterns that exist within noise. As we formerly said, cybernetic infrastructures are rather *alien* to the always mysterious, yet integral and symbolic character present in the subtle forms of chaos that entropy generation causes in a given organization. Cybernetics can only deal with visions of *complexity*, but is totally alien and in raw opposition to νοῦς; it marginalizes νοῦς. As Ludwig Von Bertalanffy writes: *"the cybernetic model—i.e., a system open to information but closed with respect to entropy transfer."* One of the main characteristics of any cybernetic configuration is that entropy—conceived here in the sense of the randomness and chaos that appears in any measurement framed on time, space, and causality—can only be treated statistically under the cybernetic framework. So by facing this required approach to noise in cybernetics, American mathematician and philosopher, Norbert Wiener, developed a filter in the 1940s that permitted noise reduction within signal processing,[120] thus discarding the intrusion of any entropic perturbations from outside the cybernetic configuration. And after having worked during the Second World War in the detection of mobile targets with firearms, Norbert Wiener published in 1948: *"Cybernetics: Control and Communication in the Animal and the Machine."* In this work, Wiener puts forth the foundations of the collaboration between mathematics and the diverse disciplines in the most varied fields, and by means of

---

[120] Wiener, Norbert (1949). *Extrapolation, Interpolation, and Smoothing of Stationary Time Series*. New York: Wiley. ISBN 0-262-73005-7..

encountering similarities with neuroscience, he believes feasible the *"ideal or actual mechanization of processes of thought (...) bound to devalue the human brain, at least in its simpler and more routine decisions."*[121] Within all this new world-view, metaphysics is then necessarily considered as an imperfection or subjective residue, also implying eventually that the validity or power assigned to any ethic or moral valuation to start definitely trembling.[122]

All these important operative, paradigmatic and technical developments, in addition to the first construction of computers like the ENIAC or EDVAC in the 1940s, had its global epicenter in the Massachusetts Institute of Technology (MIT) and the University of Pennsylvania. Hence, new power configurations had rapidly emerged in that part of the world, that is, qualitatively new configurations that necessarily attracted vast amounts of capital, human intellectual dedication and workforce. The economic and cultural power gained by countries such as the USA and the USSR during those times can not be separated from their participation in the development of these novel techniques, which could be vastly applicable to all servomechanisms and systems of control existent in all sectors of the worldwide industry. We are generally all much more familiarised with the famous Manhattan Project (the development of the atomic bomb during those years) and for instance scientists such as Albert Einstein have gained much greater popularity than Norbert Wiener... And yet *cybernetics* corresponds to a paradigm or way of seeing the world that all scientist, technician or engineer who deals in our times with systems has to resort to inevitably, no matter

---

[121] *Cybernetics: Control and Communication in the Animal and the Machine*, Norbert Wiener. John Wiley & Sons Inc (1961).

[122] However, Wiener was already aware of this human moral dilemma and writes: "this new development has unbounded possibilities for good and for evil. For one thing, it makes the metaphorical dominance of the machines (...) I have heard the statement that learning machines cannot subject us to any new dangers, because we can turn them off when we feel like it. But can we?... To turn a machine off effectively, we must be in possession of information as to whether the danger point has come. The mere fact that we have made the machine does not guarantee that we shall have the proper information to do this (...) the checker-playing machine can defeat the man who has programmed it, and this after a very limited time of working in. Moreover, the very speed of operation of modern digital machines stands in the way of our ability to perceive and think through the indications of danger." *Cybernetics: Control and Communication in the Animal and the Machine*, Norbert Wiener. John Wiley & Sons Inc (1961).

if the professional is aware or not of the fact that the set of procedures he relies on corresponds to a very specific paradigm, or no matter if he rather associates the term cybernetics to the more widespread conception of "cyberspace" and "science-fiction cyborgs."

The fact that cybernetics has been much more revolutionary in social terms than the paradigms of general relativity or quantum mechanics is because, contrary to the latter fields of knowledge, the paradigm of cybernetics is used whenever we operate with today's technological systems, from the lowest to the highest echelons of operation. And this factor already took place many decades ago, in order to manage problems that involved progressively more complex arrays of operative factors. For instance, already in 1966, Luther J. Carter wrote:

> *"The political authorities are asking for a "systemic approach" to problems such as environmental pollution, the traffic jams, the repression of crime, public health ant urban calamities (...) The systemic approach is a new revolutionary concept, the only true solution to the problems that actually count among the most important and disturbing of the nation."* [123]

*"Progressive segregation is often connected with progressive centralization,"* writes Ludwig Von Bertalanffy.[124] The technical character of such international developments on the military, economic realm and industry all implied the emergence of vast amounts of documents and reports that, based on a new objective language, express a new form of dominion over territorial constraints of a political, economic and social nature. This unstoppable process worldwide implied the inevitable constitution of "central nervous systems" that, as powerful encroaching networks, overcame the former juridical boundaries of the previous State-nation configurations. Hence, in 1947 was founded in the US the first central agency of intelligence, the CIA, and in 1954 the KGB was founded in

---

[123] Luther J. Carter. "Systems Approach: Political Interest Rises," *Science*, 153 (1966), 1222-1224. http://www.sciencemag.org/content/153/3741/1222.citation.

[124] Ludwig Von Bertalanffy. *General Systems Theory*. George Braziller. New York. 1968, pg 86.

the Soviet Union, an agency that was precisely called "the center."[125] The competition that emerged between both centers for worldwide dominion cannot be equated merely with a political struggle as long as we conceive here the term "politics" in a partisan way, or by intending to have the monopoly of public opinion. The struggle that emerged between the agencies was operative, technical, and technological, constituting a new mode of global power structure having completely different characteristics than all former States in History. Some authors have defined these power configurations as *States within the States* or *Deep States*.[126]

The intelligence agencies and networks constituted the inevitable centralization of all the operative developments taken place at a vast worldwide scale, due to the new means of technical power that all countries of the world were relating to. The monopoly of power acquired by the agencies is demonstrated by the way that they managed to organize and establish different operative echelons of functionality all around the world in the form of a new spatial configuration of power: the *network*. The rise of the power of technocracy and the technician profile was channeled to a large extent by these vast organizations. And because of the abrupt change of all former landscapes and communities into the urban/industrial/technological/corporate, all problems or conflicts began to progressively be more and more of a technical character, and in this domain, those groups or sectors that gain the capacity to centralize all data are necessarily those in charge of taking decisions, that is, they constitute the new elite in power; a power that is not political in the classical sense nor in the modern sense, but a power that is strictly *operative*.

It can be hypothesized that the intelligence agencies gained more and more power and predominance as they were the ones that firstly followed in practical terms the Zeitgeist of the 20th century, an interpretation and manipulation of the material conditions based

---

[125] Frattini, Eric. *KGB Historia del Centro*. Editorial: EDAF, ISBN 84-414-1708-3.

[126] A very insightful film that approaches the idea of a developing transnational power structure is *Three Days of the Condor*, a 1975 American political thriller film directed by Sydney Pollack and starring Robert Redford, Faye Dunaway, Cliff Robertson, and Max von Sydow.

on the paradigm of cybernetics. And any professional around the world in whatever field (civil or military) who had important and outstanding skills in handling power based on the new concepts of *net, computation, memory, processing, feedback, information, control, system, code, program, translation, direction,* and *inhibition* would eventually work, directly or indirectly, for the agencies and networks. And yet the controversial question remains here: *do these agencies constitute a political power, in the classical sense?*

In classical terms, politics refers to the domain of aims established by the political elite of a given society or people; it is about providing finality to the most diverse groups based on the particular internal and external conditions of society or people. Even in its most eminent sense, politics aims to establish a hierarchy of functions in society in order to empower a homeostatic political organism, where every social sector and individual can develop an activity and life that is embedded with a meaning and service. Yet as we've seen, in the case of power infrastructures based on the cybernetic paradigm, any question referred to the aim constitutes "residual metaphysics." In terms of cybernetics, the technician job is to diminish the production of entropy in all and every single process. As Claudio Finzi points out[127] the technocratic mindset and activity, in order to expel entropy, is determined by radical rationalism:

> *"In the technocratic mentality rationality and "truth" are inextricably linked, according to a almost universally recognized scheme in contemporary thought, where besides rationality is based on purely quantitative elements, deferring the world of the irrational, and therefore by definition, discarding anything that is not quantifiable. It is obvious that there will be no room for value judgments, that is, for judgments which by their very substance can not be based on quantitative elements."*

This is the *modus operandi* of anyone who works for these agencies or networks. Yet as we've formerly explained, any living being, community, or State that follows autonomous aims that

---
[127] Claudio Finzi, *Il potere tecnocratico*, Bulzoni, Roma 1977.

are *autopoietic* (it produces its ultimate figure or destiny), and requires entropy production and dissipation as a condition of existence. By considering that the technician or technocrat's job is to diminish entropy amidst highly networked processes, this entails the diminishing of any informational chaos that might emerge in the process. Hence, the technician or technocrat has no need to be responsible and has no need to have the *ability to respond* to unexpected perturbations since this task is outsourced to a complex network of control systems. The only responsibility thus demanded of the technician or technocrat is a very partial one, that is, to comply adequately with the functional training and skills formerly received, in order to apply purely engineering and scientific approaches to all matters and issues. And whenever any technician or technocrat takes any decision based on an individual or liberal sense of interest, it is in these situations where he risks not complying with the policies of technocratic modes of power and thus can be legally persecuted by the very specific juridical establishments set up by these networks.

The incapacity on the part of cybernetic infrastructures and paradigms to address goals and aims, causes the operative determinisms to induce that which Langdon Winner referred to as the strange phenomenon of reverse adaptation which he explains as follows:

> *"the adjustment of human ends to match the character of available means... People come to accept the norms and standards of technical processes as central to their lives as a whole... Efficiency, speed, precise measurement, rationality, productivity, and technical improvements become ends in themselves applied obsessively to areas of life in which they would previously have been rejected as inappropriate... Reverse adaptation is an interesting situation which is the exact opposite of the idealized relationship of means and ends... Ends are adapted to suit the means available (instead of the reverse)... Reverse-adapted systems represent the most flagrant violation of rationality."*[128]

---

[128] Winner, L. 1977. *Autonomous Technology—Technics-out-of-Control as a Theme in Political Thought.*

So whenever any nation, corporation, bank, administration, community or individual establishes their power and economic interests by relating to and by developing infrastructures and techno-industrial means of power based on the cybernetic paradigms, the more they necessarily outsource their capacity for decision to such purely *horizontal* power structure. This is a necessary consequence of the intrinsic character of any infrastructure of power based on cybernetics. The more such infrastructures and networks encroach upon a given organization, the more such organization loses strict political power for autonomous operative decision, at a societal, financial or economic level. What had formerly constituted a power that might have been configured symbolically as a brand, herald, flag, *idea*, etc., eventually lost its capacity for being an attractor of human developments and decisions, and eventually became a mere *image*; the image of the surrendering to the cybernetic network. It is reminiscent of the manner in which the transformation of things into images is considered a bad omen, which is expressed in the way many tribes in their local environment still fear being photographed. Ernst Jünger writes in this regard that: *"The adoption of a foreign technique constitutes an act of submission whose consequences are more dangerous as it is firstly carried out with the spirit."*[129] This process can easily be observed in the case of how many State-nations progressively lost their political powers during the last century, when allowing in their countries the development of networks, corporations, and infrastructures of an industrial-technological kind, which all required the assistance of technocrats, technicians, and engineers who can't allow their work to be interfered with by national preferences. This doesn't mean of course that in the State-nation there are still "democratic" elections and constitutions, but whenever a State reaches its pinnacle of power, develops its Historical and cultural mission, and finally experiences the inevitable decline in the elites for serving as a bridge between the metaphysical and the

---

Cambridge, MA: MIT Press, pg 229, 242.

[129] "Die Annahme einer fremden Technik ist ein Unterwerfungsakt, dessen Folgen um so gefährlicher sind, als er sich zunächst im Geiste vollzieht." Ernst Jünger. *Der Arbeiter*. Ernst Klett, Stuttgart 1981. Klett-Cotta, pg 37.

material, then the elites can only maintain their political privileges through the conservation of the temporary means of power, which refers to the national resources available in the economic, military, technical, and religious domains. In the particular case of the State-nations in the West, political representatives are forced to resort to the propaganda of their political image in order to have higher chances of becoming elected. Yet the propaganda required in order to gain privileges in the State, a propaganda that Ernst Jünger referred to as *"the panacea offered by the merchants of political opium"*[130] corresponds to a technique which itself requires techniques and technologies highly based on cybernetic functionalism, so ultimately those politicians who are best assisted by technicians and experts when intending to efficiently transmit their image to the elector through the "bridge" of the mass-media technological infrastructure, become eventually the politicians who have greater chances to become elected during the State-nation elections. As Spengler remarked:

> *"the ambitions of all revolutionaries expend themselves in playing the game of rights, principles, and franchises on the surface of history. But the statesman knows that the extension of a franchise is quite unimportant in comparison with the technique—Athenian or Roman, Jacobin or American or present-day German—of operating the votes."*[131]

It should be clear at this point that here we are presented not with a political problem, but rather a technical one which can be defined as follows: *at a given budget what is the most optimal set and arrangement of cost-efficient means of communication required in a political campaign to provide to a given segment of potential or actual voters a slogan or image that can promote their vote to the party?* This technical problem requires the teamwork of experts in many fields (advertisement experts, marketing professionals, sociologists,

---

[130] "das Allheilmittel der politischen Opiumkrämer ." Ernst Jünger. *Der Arbeiter*. Ernst Klett, Stuttgart 1981. Klett-Cotta, pg 35.

[131] Oswald Spengler. *The Decline of the West*. Alfred A Knopf. 1926. New York, pg 429.

financial advisors, campaign designers, TV producers, graphic designers, surveying technicians, etc.) and the more optimally this problem is solved by the actual technical means and resources, the higher the chances of the political party to win the election and become an "arm" of the State-nation.

The power of ideological *mobilization* is what ultimately counts, as Jünger points out when writing: *"the accuracy and speed by which any party newspaper reaches its readers are much more significant than the differences between parties we can consider."*[132] Yet the "arm," once attained a privileged position in the government of the State-nation, had only feasibly attained the position by formerly resorting to technical, technological, and financial means that transcend national borders, which entails that—in terms of handling and managing the conditions of existence of a given society—actual power has been *outsourced* to an external network and array. As Marshall McLuhan writes: *"Once we have surrendered our senses and nervous systems to the private manipulation of those who would try to benefit from taking a lease on our eyes and ears and nerves, we don't really have any rights left."*[133] Hence, the more politicians owe their privileges and presence in the public opinion within the State-nations to the new configuration of transnational cybernetic infrastructures, the more their actions in the government cannot be exclusively "channeled" to their own nation, due to the international character of the networks. And whenever the political groups intend to force the development of national barriers in order to reduce the exchange of information, goods, and workforce, the higher are the economic costs that such going against the Zeitgeist implies.

An issue arises here which refers to the existence of any group at any scale to be capable of dominating the cybernetic means. In order to tackle this it is important to recall that History shows us on many occasions how cultures became capable of providing higher

---

[132] "die Präzision und die Geschwindigkeit, mit der jedes beliebige Parteiblatt an seine Leser gelangt, als alle Parteiunterschiede, die man sich ausdenken mag." Ernst Jünger. *Der Arbeiter.* Ernst Klett, Stuttgart 1981. Klett-Cotta, pg 136.

[133] Marshall McLuhan. *Understanding Media. The Extensions of Man.* The MIT Press Cambridge, Massachusetts. London, England 1994.

expressions of dominion upon the most diverse technical means, and the dominion required the establishment of well-defined operative hierarchies which were capable of projecting their technical and artistic activity amidst their territorial surroundings. The case of Ancient Rome is very meaningful in this regard, where political sovereignty maintained for centuries its capacity to establish control over the technical means and integrate them harmonically within the Empire. Yet technical configurations based on cybernetics correspond *to a very specific form of technique* that has developed at a scale never seen before in History, a development that might resemble accounts of the legendary Atlantis. Due to the nature itself of cybernetic infrastructures, the concept of dominion by an individual, group or nation is simply unrealistic, due to reasons that shall be explained more thoroughly in the both volumes of *Operative Traditions*.

As we shall present in this book, there are many different forms of technique and many conceptions of technology. In the case of the cybernetic paradigm, it provides the technician with a framework for efficiently relating to those technologies that are mostly of a cybernetic kind. The main characteristic of cybernetics is that the "piloting" of the functions of a given device or means is no longer assigned to the person who uses the device but rather to a system or network. This characteristic of cybernetics is very important to point out here, since in the case of an integral and living Tradition, all the symbols and disciplines aim to concentrate the responsibility of the user *to one single point*. The most magnificent symbol of this process is the Sun, which Buddhism and the Tradition of the Samurai referred to as "*the glorious Sun of Buda's nature that sparkles in the Zenith of the illuminated consciousness, even though men still dream a dream of illusion (…) Make your heart be as pure as possible so you never become unworthy of the Sun which reflects upon you the Universal Spirit.*"[134] To some extent, we can point to this idea through the symbol of the arrow which was present in the developments that Eugen Herrigel had to experience during his initiatory process. And in the case of learning to play a musical instrument, the whole

---

[134] K. Nukariya. *The Religión of the Samurai*. London, ed 1973, pg 93.

concept is not much different. However, in the case of the networked nature of cybernetics, control and power are no longer assigned to humans, *but to the system of control itself*. This assignation or outsourcing of human responsibility was initially the whole idea that drove the development of the techniques in the first place by Minorski, Nyquist, and Hazen.

So how does this specific Zeitgeist affect Tradition? It has to be clear at this point that Tradition in its integral and living sense has to be highly compromised. When Julius Evola referred to the main attributes of Tradition, he links it to *centrality, polarity, stability,* and *peace,*[135] which impregnates all elements of a civilization, embodied in the virtuous Master or *pontifex*, who is also characterized by regal attributes. This figure of stability and regality is what Eugen Herrigel perceived in Master Kenzo Awa, a Master who embodied power in the centre of his own Being, in resonance with the centre of the Tradition he served. *"Dominion and Service are the same things"*[136]— wrote Ernst Jünger. And the German war hero goes on by saying: *"if today there is still among us true greatness, if somewhere is hidden a poet, an artist, a believer, it will be recognized that he feels responsible and strives to serve."*[137]

In the case of the development of cybernetics, the infrastructure of power is not capable of granting any centrality, polarity or stability to the user, since *it is actually the system itself what responds to external perturbation*, and the user or technician remains not only in a subservient position, but becomes determined and *reified* by the system in his activity. Besides, in a cybernetic power framework, it is impossible to determine any absolute and stable center of power. This is so because cybernetic techniques correspond to a *transitional* form of technique; in no case do they correspond to a finished and

---

[135] Julius Evola. *Revolt Against the Modern World*. Inner Traditions. Polar Symbolism; the Lord of Peace and Justice, pg 16.

[136] "daß Herrschaft und Dienst ein und dasselbe sind." Ernst Jünger. *Der Arbeiter*. Ernst Klett, Stuttgart 1981. Klett-Cotta, pg 6.

[137] "Seien wir aber davon überzeugt: wenn heute unter uns noch wahre Größe besteht, wenn irgendwo ein Dichter, ein Künstler, ein Gläubiger verborgen ist, so wird man ihn daran erkennen, daß er sich hier verantwortlich fühlt und zu dienen bemüht." Ernst Jünger. *Der Arbeiter*. Ernst Klett, Stuttgart 1981. Klett-Cotta, pg 103.

stable creation, such as other techniques of previous cultures that were embedded with deep artistic and monumental connotations of *techné* and *poiesis,* which we shall explain more in depth in the third part of this book.

We are in the presence here of a realization of an event which took place during the 1920s-1950s in regard to the specific "bridge" that connects symbols of power to given means and resources. It is useful to recall at this point that metahistorical Tradition has always been identified symbolically with the carrying of a *flame* by groups of initiates who embody the power of such a flame in their lives. Even by resorting to this symbol in the physical domain, it is only when heat fluxes are concentrated at a specific point and a thermodynamic energy activation limit is surpassed, that a flame is finally born and maintained. It is then an effort of centripetal convergence of all heat fluxes—of "inner work"—allowing a new relationship to be established between the material domain and the ethereal; a gradient of different substances emerges both in the burning material as in the flame, which is affected by a stable gradient of temperatures. In this context, the noblest and the most slag elements are all alchemically synthesized into more subtle and gaseous substances, flowing in a direction that goes vertically, upwards. In the case of a flame, we can also perceive the *centrality, polarity,* and *stability* which assist those who aim to see in the darkness. Hence in the flame there is stillness, but also there is a powerful intrinsic dynamics, and groups of initiates, consciously or unconsciously, are prone to gather around the fire, as the planets revolving around the Sun. When contemplating a flame, it is like contemplating a process that is truly alive, that appears to have an autonomous life of its own.

It is also very important that the intensity of the flame shall eventually be exhausted. The laws of thermodynamics and entropy shall inevitably win once again; dispersion of energy shall eventually become the norm, and darkness shall prevail. Inevitably the material domain becomes more "solid," more quantitative, as the qualitative aspect has vanished and can no longer be seen. The dispersion of heat also makes the material domain be felt as rather "frozen" and rigid. Chaos, by no longer being perceptible, becomes discarded in all calculations, actions, and commitments.

So in this dark stage the material domain can no longer be illuminated by turning on the switch of an electrical bulb or lamp. The age of technological comfort emerges…this action of turning on the switch is already of a highly Promethean character, as it relies on the power provided by a technological infrastructure in order to perceive and manipulate the material domain, yet rebels against the central firepower of Zeus. However, perception is now different too. It is no longer perception caused by a qualitative view on the material domain with all its diverse gradients and hierarchies—which is the kind of illumination that is irradiated by the spiritual flame—but when assisted by the bulb, the way we perceive material objects is directly related to the same technical characteristics of the bulb, which produces a very specific kind of light irradiation and set of electromagnetic frequencies. At this stage, our view of reality inevitably loses depth…we can only see what the technical infrastructure allows us to see…and what the technical infrastructure is capable of integrating as information are those elements in reality that are strictly enclosed in categories of time, space, and causality. Anything that transcends these categories *can simply not be seen* by any technical infrastructure based on cybernetic paradigms. However, when Tradition is integral and alive, its highest representatives are capable of perceiving and relating to different levels of *physis* where elements that are dependent on the categories of space, time, and causality appear as the lowest, most quantitative, functional, and uniform foundation in order to further grasp the higher metaphysical levels through "the third eye," νοῦς, that destroys all veils; the perception of the cosmos that in Tradition is referred to as *intuitio intellectualis*.

Yet through the filter of cybernetic paradigms, perception loses the depth of νοῦς and hypertrophies a purely functional and progressively more complex view of reality. In what appears to be an ever-developing and complex network of cybernetic infrastructures that developed since the last century, it seems practically impossible to perceive any hierarchical "center." Ernst Jünger also once wrote: *"The men who explored the moon could not tell that there was life on it, for they brought wasteland along,"*[138] which is a very appealing way to present the issue addressed here. It can be argued that all infrastructure (distribution of energy, resources, people, information, etc.) has central offices of management, and as in this particular layer of reality, processes of centralization are what favor the efficiency and functionality of the entire systems. However, these kinds of cybernetic centers, which started to proliferate in all productive sectors since the times of the Industrial Revolution do not correspond in any way to the metaphysical centers that can be perceived by an initiate of a given Tradition. The lack of correspondence between both kinds of centers relies on the fact that the symbols embraced in the headquarters of the networks are the same as in the periphery; what constitutes the difference in these cases is not the nature of the operations, but rather the higher level of systematic organization of the operations and functions. This can be clearly observed in today's centralized infrastructures aiming to monopolize data (data centers such as those that support the functioning of Google, Facebook, etc.) where basic devices of information storage (hard drives, for instance) are configured in a more complex and concentrated way, yet are the same as any other device in the world connected to the network.[139] An analogy of this can be established in anatomy: the central nervous system of any animal being has a much higher density of neural activity than in the case of the peripheral nervous system, and yet it is naïve to believe that the spiritual center of an individual relies on the organ of the brain, as it has been shown by many cases of individuals who

---

[138] Ernst Jünger. *Eumeswil*. The Eridanos Library. 1980.

[139] "A cybernetic system cannot be "self-organizing," that is, they can't evolve from one state to another that is more differentiated. Its true that when are installed memory devices in the cybernetic systems they can learn, this is, they can change and increase their organization based on the information they receive. They can't however develop differentiation processes that require energy and material input. In other words, cybernetic systems can only increase in terms of their entropic content and diminish in terms of their informational." Ludwig Von Bertalanffy. *Towards a New "Natural Philosophy."* Clark University (Worcester. Massachusetts) January 1966.

suffered severe brain damage without alteration of their personality.[140] However, the traits of a given personality are more prone to change when individuals go through heart transplants.[141] As Nietzsche said: *"There is more sagacity in thy body than in thy best wisdom."*[142]

The traditional symbols of the flame, the sun and the heart, both relate to the idea of *spiritual centers*. The Sun, in almost all ancient traditions, has been conceived as the symbol of the divine and based on what we were formerly describing, it is a perfect expression of energy production, entropy production, self-regulation, and neguentropy, feeding itself constantly by a huge percentage of its own waste. Such a thermodynamic miracle can be traced by realizing that even though our sun rigorously follows a pattern of eleven year solar cycles, the actual behavior of solar spots is still a mystery to scientists. It is an autonomous form of life that also provides the potential for the development of all life forms in the solar system.

However, the developments triggered by the irruption of the cybernetic and quantum paradigms at the beginning of the 20th century, impelled the men of the West to act on reality strictly like the titan Prometheus, truly believing to be stealing the flame and power of the divine by exclusively relating to a very functional and material infrastructure, highly based on very novel technological forms. As described earlier, these material developments have increased at a rapid pace since the 1920s, and if we intend at this point to establish a specific date where finally we can encounter the definitive symbol that confirms when the cosmic powers of darkness defeated finally the cosmic powers of light, and the bridge between immanence and transcendence was broken as the key sign of arrival of the darkest of all darkest times, the symbol can be expressed in the *atomic bomb*.

* * *

---

[140] http://www.dailymail.co.uk/news/article-2102461/Man-half-head-Carlos-Halfy-Rodriguez-explains-got-bizarre-injury.html

[141] http://www.ncbi.nlm.nih.gov/pubmed/10882878

[142] Friedrich Nietzsche. *Thus Spoke Zarathustra*. The Despisers of the Body.

René Guénon once wrote:

> *"The Traditional spirit can not die, because, in essence, it is above death and change; but it can be removed entirely from the outside world, and then it will be really the "end of the world" (...) We can see the "beginning of the end," the harbinger of time, according to Hindu tradition, when the sacred doctrine must be enclosed all whole in a shell, to leave it intact at the dawn of the new world"*[143]...

In 1945 the last remnants of integral Tradition that existed on the entire planet had tragically faced their temporary death. Having once constituted one of the most magnificent and regal countries where the traditional idea of a Sun Empire persisted to its last breath by fighting courageously to its death like a kamikaze pilot, Japan waked up one morning with the sudden and shocking disappearance of its core identity, its tradition, its integral tradition and its devotion for attaining *immanent transcendence*. Suddenly what became immanent for the Japanese people was the disintegration of any transcendence and its expression in an imminent destruction of unimaginable consequences. The bombings of Hiroshima and Nagasaki express the temporary victory of the centrifugal powers of material destruction and nuclear disintegration against the highly constructive and integrating idea embedded in the sacred symbols of Japan's traditions and its devotion to the metaphysical idea of the Sun Empire. The Sun, always symbolizing in all integral Tradition the nuclear reactions of fusion and synthesis that provide a higher form of life in the cosmos, eventually was defeated by the industrial machinery of nuclear technology which detonates the power of the atom in an uncontrolled way, despite any short-term or long-term consideration of the human consequences. And yet, despite the former, the atomic bomb actually expresses better than any other symbolic figure the defeat on the part of the Promethean man of the West when aiming to handle, master and dominate a release of power in the cosmos that dis-*integrates* the core, that splits the

---

[143] René Guénon. *The Crisis of the Modern World.*

mysterious "center" and the *spiritual axis*, thus releasing all its inner energy *in all outer directions*, finally taking in a pure centrifugal dynamic where all particles centrifugally escape from the center they own their existence, their *being*. By constituting an imperial State of stability and dominion that contains and affirms a superior form of stable balance all configurations and all surrounding orbiting of particles, the *atomic bomb* ultimately detonates the explosion and the dispersion of its energy around the cosmos. No other cores can remain unaltered as they all also disintegrate the closer they are to other disintegrations. Hence, the chain reaction is unstoppable. Or shall it eventually stop? The only chance available to revert the process is by establishing once again an integrating process in opposition to a disintegrating process. The laws of thermodynamics shall win in this domain too, and the energies that were released in the cosmos by defeating all remnants of any integral Tradition in 1945, shall eventually exhaust their capacity for expansion, and at that point it shall be the moment to centripetally absorb all the fluxes of subtle entropy characterized by the character of renewal, like the reversal of an hourglass. The old speculative world-views are powerless to embrace and concentrate the forces since speculative world-views owe their own existence to such forces. The process of centripetal concentration can only be accomplished by the integration of all dispersed chaotic elements, and only the operative character of Tradition can assist us in this endeavor. Julius Evola once wrote in *Ride the Tiger*: *"The energies that have been liberated, or which are in the course of liberation, are not such as can be reconfined within the structures of yesterday's world."* Yesterday's world is the world of the Promethean spirit that controls but doesn't master, that promotes but doesn't *attract*…the future belongs to the Masters; the operative Masters of Tradition.

Ernst Jünger: *"One shall only speak of Mastery then when art no longer consists of learning more and more things, but rather to learn something deeply."*[144]

---

[144] "Ferner leuchtet ein, daß erst dann von Meisterschaft die Rede sein kann – dann nämlich, wenn die Kunst nicht mehr im Umlernen, sondern im Auslernen besteht." Ernst Jünger. *Der Arbeiter*. Ernst Klett, Stuttgart 1981. Klett-Cotta, pg 89.

# PART II

# INTRODUCTION TO JULIUS EVOLA'S "THEORY & PHENOMENOLOGY OF THE ABSOLUTE INDIVIDUAL" (TAI-PAI)

> *"Existence, truth and certainty are not to be found in the past but in the future: they are tasks."*
> - Julius Evola[145]
>
> *"Precisely by attaining an ideal, we surpass it."*
> - Friedrich Nietzsche[146]
>
> *"Perhaps nobody has ever been truthful enough about what "truthfulness" is."*
> - Friedrich Nietzsche[147]
>
> *"What matters is not to despise the intellect, what matters is to submit it."*
> - Ernst Jünger[148]
>
> *"The greater the man, the truer the philosophy."*
> - Oswald Spengler[149]
>
> *"The world we live is a product of perception, not its cause."*
> - Hadley Cantril[150]

The *Theory and Phenomenology of the Absolute Individual* masterly developed by Julius Evola aims to forge an *Individual* who from now on we shall write with a capital "I." This "I" is extremely in opposition to the individual with a small "i," that is, in extreme opposition to an individual [*individuum*] who is a product of the civilizational modes that predominated after 1717, during the times of the enlightenment [*Aufklärung*]. At this

---

[145] Julius Evola. *An Intellectual Autobiography. Path of Cinnabar*. Arktos 2009, pg 51-52.

[146] Friedrich Nietzsche. *Beyond Good and Evil. Prelude to a Philosophy of the Future*. Cambridge texts in the history of Philosophy. Epigrams and Entreacts.

[147] Friedrich Nietzsche. *Beyond Good and Evil. Prelude to a Philosophy of the Future*. Cambridge texts in the history of Philosophy. Epigrams and Entreacts.

[148] "Nicht auf die Verachtung, sondern auf die Unterstellung des Verstandes kommt es an." Ernst Jünger. *Der Arbeiter*. Ernst Klett, Stuttgart 1981. Klett-Cotta, pg 100.

[149] Oswald Spengler. *The Decline of the West*. Alfred A Knopf. 1926. New York, pg 41.

[150] Hadley Cantril. *Perception: a Transactional Approach*, 1954 (co-authored with William H. Ittelson).

moment in the History of the West, the individual started to gain more and more predominance as the main sociological side-effect of the establishment of political constitutions during the French Revolution, where the ideology of equality and the surrendering to economic thought became an inevitable propagandistic factor required in order to justify the new individual traits emerging along the Western horizon, in times when the twilight of the Operative Tradition's spirit had to a large extent obscured all disciplines that permitted the restoration of the Individual. The forging of an "I" based on progressive modes of dominion, mastery, and cosmic integration. Whereas the "i" or ego is a product of civilizations in decline characterized by highly mechanistic traits, the "I" Evola presents us with is not the product of a civilization, but rather constitutes a creative *source*[151] of mastery and dominion aiming not only to be free from the forces and determinisms of civilizations in decline, but also aiming to set the progressive levels of self-development required for the constitution of a new State, a mode of regal power capable of integrating all former functions and developments, justifying all social and economic phenomena, no matter how tragic or destructive these might be, and as Jünger's figure—the Anarch—points out: *"Equalization and the cult of collective ideas do not exclude the power of the individual. Quite the opposite: he concentrates the wishful thinking of millions like the focal point of a concave mirror."*[152] The emphasis Evola puts in establishing a set of categories in terms of experience—instead of establishing like most philosophers, spiritualists, and thinkers a system of categories of speculative thought—provides a chance for any individual to become an Individual who, instead of defining his sense of being in terms of an identification with an ideology, spiritual belief, political

---

[151] Individualità esprime lo stato della persona in quanto non più è la relazione che la definisce, ma invece la domina e pone secondo libertà assoluta, onde ha la prima esperienza del punto di una non più semplicemente esistente, ma mediata, positiva causa sui [Individuality expresses the status of the person when the relation that defines it no longer exists, but instead dominates it and places it according to absolute freedom, because it has the first experience from the starting point of no longer simply existing, but being mediated, positive causa sui] Julius Evola. *Fenomenologia Dell'individuo Assoluto* III ed. corretta: Edizioni Mediterranee, Roma 2007.

[152] Ernst Jünger. *Eumeswil*. The Eridanos Library. 1980.

standpoint, religion or image, can rather find a whole new realm of the most extraordinary elements included in his experience *here and now*. In the case of the individual, the leap towards the Individual is always uncertain. In this quest, as many others: *"many are called, but few are chosen"*[153] yet those who finally decide to go for this quest shall firstly learn to properly perceive what specific experiential elements support their individuality and sense of self. This phase is the most difficult and perilous for the individual since the realization of the contingency of the elements can trigger the existential anguish caused by awareness of the instability, ephemeral character and weakness of the individual. At this critical point of practice, the death of the individual is to be overcome, and the recovery of all classical warrior/kshatriya attitudes constitutes the most precious virtue.

If during the death of the individual there is what Ernst Jünger referred to as "heroic realism"—which in combination with Evola's magical idealism we can refer to it as *magical realism*—then the individual shall find himself in a completely new domain where, progressively having cleansed the spirit from all the speculative, ideological, spiritualistic and abstract constructs that are promoted by the civilizations in decline, he shall be then capable of perceiving the experiential ground where he, now as an Individual, stands on. Still deprived of all individual identifications, the whole landscape surrounding the Individual during the first phases of the process is necessarily dark. The great question at this point is "Where to go? Is there actually anywhere to go?" But now as an Individual, a new process emerges before his eyes, a *Magical* process, a process where an interplay of cosmic forces develop and integrate the configurations promoted by the conscious and unconscious actions of individuals. Individuals themselves then appear before the Individual as impelled by operative, technical, and objective forces; like colorful and kaleidoscopic traits attached to the rigorous mechanisms of phenomenological cogs. The Individual can perceive this underlying process to the extent that he himself has operatively

---

[153] Matthew 22:14.

mastered in his own experience such forces. As Ernst Jünger marvelously points out: *"Thus the singular person carries himself the norm; and the supreme art of life emerges to the extent the singular person lives as such, perceiving himself as the norm."*[154] Hence for the Individual a new world of phenomena emerges, a new experience of the world that is as expansive as when a child stands up for the first time and encounters a feeling of self-mastery and balance; the self-mastery and balance is then projected in the conditions of experience of the outer world. So the mutation that takes place from the individual to the Individual corresponds therefore to a re-birth, to a new way of relating to the world as it appears in a raw manner before our senses and without surrendering to any comforting interpretation of phenomena that impedes us *touch* the domains of chaos and danger where the individual identifications are always threatened. Following an alchemical principle where the fire *"increases the virtue of the wise and the corruption of the perverse"* the Individual, by going through a progressive sense of experiential mastery and dominion like that of an alchemical and synthesizing fire, begins to perceive, in parallel to this process, a new path emerge before him. This path, only perceptible to the Individual, is forged by the spirit of the times from the modes of chaos and entropy, aiming towards the synthesis of symbolic configurations that not only determine the direction of the objective processes of reality (economy, science, cultural frameworks, etc.) but also determine the actions, decisions, and commitments established by all individuals. This process can't be perceived by the individuals who are determined by it, whose existential role is that of constituting mobilization factors of a process that becomes individualistically self-justified in ideological, speculative, religious, economic, and spectacular forms. The Individual, once placed in the magical realm, not only has developed the faculty of perceiving the process but is also potentially capable of perceiving the direction of development of all means, techniques, and objects, that to relate to

---

[154] "So trägt er den Maßstab in sich, und die höchste Lebenskunst, insofern er als Einzelner lebt, besteht darin, daß er sich selbst zum Maßstab nimmt." Ernst Jünger. *Der Arbeiter.* Ernst Klett, Stuttgart 1981. Klett-Cotta, pg 17.

their *meaning*. This is possible to the extent that the Individual finds meaning to his own life, a meaning that has to be absolutely original and not "copy-pasted" from the representation of lives attached to the character of the individual. By experiencing the different "steps" towards the height of freedom and mastery at an existential nonespeculative way, the Individual leaves as a "cultural trace" for other Individuals the different levels and operative faculties required to attain the State of highest balance and power, the Absolute State, a form of State that disappeared progressively in the world after the 18th century. The Individual who by carrying the transmuting heroic fire all the way up to the metaphysical domain—thus bridging Olympically the domain to the actual worldly conditions— is he who establishes the human bridge between an Imperial form of State and the Individual; he is the *Absolute Individual, "where all forms find their essential life,"*[155] and no individual ever shall have the chance of relating to, unless the individual also risks leaping towards the experiential, magical and operative "I" that aims towards the Absolute. The path towards greatness corresponds to a path that goes along a very dangerous razor's edge where the risk of realizing our petty individual weaknesses is always very present, so in this regard it makes sense that Julius Evola's *Theory and Phenomenology of the Absolute Individual* (TAI-PAI) has not been approached by the representatives of modern culture for about a century. Verifying the petty morality, self-interest, mediocrity, and lack of regal valor in the mirror of the Absolute would be too shocking for the interests of all those popular individualities and dilettantes who have spread worldwide a distorted, corrupt and degraded ideological world-view of the relations of men to the cosmos.

---

[155] Individuo assoluto, nel quale tutte le forme trovano la loro vita essenziale. Julius Evola. *Fenomenologia Dell'individuo Assoluto* III ed. corretta: Edizioni Mediterranee, Roma 2007.

# MAIN STRUCTURE OF TAI-PAI

> We are interested in something hidden under the theory and practice, somewhere in the common root they both grow from.
> Alexander Dugin[156]

> The estrangement of body and spirit in modern society is an almost universal phenomenon
> Yukio Mishima[157]

One of the great contributions of Julius Evola's "*Theory of the Absolute Individual and Phenomenology of the Absolute Individual*" (TAI-PAI) is that it defines the presuppositions and stages required for the transition of the individual towards the Individual, and afterwards from the Individual towards the Absolute Individual. Even though Evola didn't resort to the *(I-i)* notation, in this text we shall make use of the capital I letter for "Individual" whenever referring to the rise of an operative and magic "I" that a given human embraces, in order to distinguish the magical "I" from any sense of attachment of a given human being to a sense of individuality based on a subjective psychological conception (the self) on the physical "I" (feelings, emotions, properties, sentiments, even intuitions) or a thinking "I." In other words, the sense of individuality and separation of a given human being can be due to the attachment to the physical "I" (our bodily traits, sentiments, emotions, our material possessions, mental image or

---

[156] Alexander Dugin. *The Fourth Political Theory*. Arktos 2012, pg 190.

[157] Yukio Mishima, John Bester. *Sun and Steel*. Kodansha USA (2003), pp 15.

concept of "my body") to the *self* of modern psychology (character traits, personality tendencies, psychological predispositions, etc) or to a thinking "I" (cultural attachments, religious beliefs, scientific or other world-views, ideologies, formal education, artistic/aesthetic taste) and finally to an "ego," which is an artificial barrier of separation not necessarily connected to the former and in many cases corresponding to a compensation of the lack of intensity of the "I"s in a given human being.

So ultimately there is the Magical and Operative "I" of the Individual, and the arbitrary and contingent "i" of the individual. Due to the specific cosmic traits of our time, the pure transitional aspects that emerge from the individual to the Individual cannot be established by any norm, discipline, rule, method or technique, and however, diverse properties can be assigned to each stage or state in the same way there are properties that define water in a solid or vaporous state, though the ways water can become ice are countless and cannot be hence established by a unique norm. So as a brief introduction, we present here the three basic stages in the development towards the magical and operative "I."

The first stage corresponds to that of spontaneity. Most individuals fall into this category. As Evola points out by referring to it,

> *"it is not so much he who thinks, speaks, and asserts himself, as much as that various forces and impulses think, speak and assert themselves in him. He, therefore, is only a type of medium, a passive instrument that has its own life outside of himself."*[158]

This stage might seem difficult to conceive by the individual himself, but if we look closer to our life in general, it is not difficult to realize that it is not so easy to deduce in terms of our actions, thoughts and impulses which are exactly those that I can say are mine, that are my possession, or what Evola refers to as those free actions

---

[158] Julius Evola. *L'individuo e il divenire del mondo*, Libreria di Scienze e Lettere, 1926, pg 21.

where "the power to act always precedes and rules the act itself."[159] A huge percentage of the opinions, ideas and concepts we express in our daily lives have been passively absorbed in order to justify irrational impulses and predispositions. Jünger wrote on opinions that they are "harmless, and in times when everyone likes to be described as "revolutionary," freedom to produce real and effective change is narrower than ever."[160] We might even believe that we think independently, but even the thinking processes are still not enough to define a magical and operative "I." In terms of actions and decisions, it is as difficult to choose one's actions as to choose the characteristics of one's children. In most cases, actions, decisions and choices are also decided by external configurations or by irrational impulses, "something mine without being me."[161] Winston Churchill came up with a very appealing statement to express the state of life affairs when saying that life was but: "one damn thing after another." Hence, in this stage where thoughts that represent things are provided with an autonomous and isolated reality,[162] the individual mostly owes a sense of identity, ego, or self to all external stimuli, and there emerges a desire for the external provision of stimuli, which causes economic competition with other individuals in the same stage. Albert Einstein once referred to egotistical traits as "an optical illusion of consciousness" and Pietr Ouspensky, when recalling Gurdjieff's words: "I am sucked in by my thoughts, my memories, my desires, my sensations, by the steak I eat, the cigarette I smoke, the love I make, by the sunshine, the rain, by this tree, by that passing car, by this book."[163] This stage can be referred to as a sort of infantile stage where the individual deep down is floating around in a world of chaotic circumstance which can only

---

[159] Julius Evola. *An Intellectual Autobiography. Path of Cinnabar*. Arktos 2009, pg 43.

[160] "Alles, was Meinung ist, ist unbedenklich; und in einer Zeit, in der jedermann sich als revolutionär zu bezeichnen liebt, ist die Freiheit zu wirklichen Veränderungen begrenzter als je." Ernst Jünger. *Der Arbeiter*. Ernst Klett, Stuttgart 1981. Klett-Cotta, pg 129.

[161] Idem.

[162] "Of all the rest—of the boundless ocean of names, forms and beings—there is no real certainty" Julius Evola. *L'individuo e il divenire del mondo*, Libreria di Scienze e Lettere, 1926, pg 15.

[163] *Our Life with Mr. Gurdjieff*. Thomas de Hartmann, Olga de Hartmann.

be channeled through specific control from the outside, both in terms of activity, work as in that of thought contents (propaganda, advertisement techniques, etc). It can be affirmed that in this stage the individual is determined by the past. The rather extroverted state that characterizes this state is well expressed by a character in Gustav Meyrink's novel *Walpurgis Nights*, when saying:

> *"You truly believe that all those who often mill around in the streets possess an I. They do not even possess anything. They are rather possessed at every moment by a phantasm that plays the part of an I in them."*

What causes the individual to emerge from this dream state and realize its true nature might be the need to establish some stability and sense of possession beyond the chaotic nature of all life phenomena. It is at this point that the individual resorts to gain a position of his own, intending to address elements of the world he can say *they are mine, I've placed them*, and to distinguish that which Evola writes as: *"as 'placing in something else one's principle' respect to 'placing in oneself the principle'."*[164] This phase corresponds to the *second stage*, where the individual "filters" the chaos of events through a specific framework and the more the individual affirms the rationalized world-views the more powerful becomes what Ernst Jünger referred to as the "elementary powers" of the irrational which constantly disturb constructions. As Evola wrote: *"irrationality is grasped in its existential raw character only at the point the sense of the subjective principle is totally exposed."*[165] It can be affirmed that when Eugen Herrigel met Master Awa Kenzo his individuality was immersed in this second stage, as he mostly perceived

---

[164] Nel riferimento agli elementi della coscienza empirica queste opzioni possiamo definirle come «porre-in-altro-il-proprio-principio» e «porre-in-sé-il-proprio-principio» [When referring to the elements of empirical conscience such options can be defined as "placing in something else one's principle" and "placing in oneself the principle"] Julius Evola. *Teoria Dell'individuo Assoluto*. Edizioni Mediterranee, 1988, pg 60.

[165] "l'irrazionalità è avvertita nella sua crudezza esistenziale solo nel punto in cui il senso del principio soggettivo è messo a nudo" Julius Evola. *Teoria Dell'individuo Assoluto*. Edizioni Mediterranee, 1988, pg 37.

and represented the world through the "lens" of Neo-Kantian philosophical assumptions, which corresponded to those were the "I" is conceived to be a "transcendental idealist I," practically an abstract "thinking machine" that is supposed to develop knowledge on the world based on strict intellectual categories. This stage corresponds mostly to that of thinkers, scientists and philosophers who identify their individuality with an attachment to a set of axioms, principles and laws. It wouldn't be completely misleading to equate this stage with that of the *bourgeoisie* historical character and with that of individuals who identify with moral laws, scientific world-views, aesthetic conceptions, and ideologies. Hence the individual here is still attached to a speculative plane of consciousness. The word "identification" derives from the Latin word *idem*, which means "the same," as well as *facere* which means "to make." So in the case of all identification, there is an attachment to an ego or thinking "I" which is embedded with the character of individual *identity*.

Eugen Herrigel had the chance, however, to introduce himself in a realm that could allow him to transcend this speculative plane and enter into a new operative kingdom which is much more primordial. Yet one of the most important issues that arise whenever approaching a traditional doctrine that aims to provide a new significance and meaning to the life of a given individual is that of the level of credibility provided by a given Master, in addition to the credibility granted by the techniques, methods, and perspectives he provides to the student. The Master, as an embodiment of the Doctrine and its ultimate truth, is therefore required to *be* the human expression of absolute certainty, or otherwise, the student who approaches the Master would not be willing to spend the many years required to develop the entire series of methods, sacrifices and disciplines.

So this brings us to the key issue of *certainty*, which has to be addressed before presenting the main ideas and methods presented by Julius Evola in his philosophical works. Oswald Spengler presents this key problem as following:

> *"Every professed philosopher is forced to believe, without serious examination, in the existence of a something that in his opinion is capable of being handled by the reason, for his whole spiritual existence depends on the possibility of such a something. For every logician and psychologist, therefore, however skeptical he may be, there is a point at which criticism falls silent and faith begins, a point at which even the strictest analytical thinker must cease to employ his method the point, namely, at which analysis is confronted with itself and with the question of whether its problem is soluble or even exists at all."*[166]

As was shown in the first part of this book, the cultural decline of the West was to a large expressed in the incapacity to reach an absolute certitude in terms of knowledge. As formerly described, despite all hopes projected in modern science, all the certitudes of science can be only applied satisfactorily to a rather limited domain of reality, and all elements of nature (atoms, systems, galaxies, etc) from being conceived initially by modern science as "solid" realities all became "fluffy" and diffuse entities after the set of developments taken place, namely: the principle of entropy, Heisenberg's principle of indetermination, Einstein's relativity and quantum mechanics, where ultimately a highly perspectivistic standpoint is inevitable to resort to, as the technical conditions of the observer determine the character of what is observed.

In the hypothetical case that a student approaches the *nucleus*, a Master who is the embodiment of the ultimate truth, how could the student gain certainty about the qualitative attributes on the part of the Master? In terms of Evola's viewpoint, the Italian author would state at first that nothing exists outside the "I" that is not present potentially by the "I" itself, and that such potentials and objects of experience are *placed* by the "I." This act of *placing* corresponds to the main contribution of Idealist philosophy, in terms of realizing the basic conditions for knowledge to even be possible.

---

[166] Oswald Spengler. *The Decline of the West*. Alfred A Knopf. 1926. New York, pg 299.

This position inevitably takes us to an absolutely crucial element in the philosophy: that of pointing to the "I" that Evola is referring to. In effect, when Eugen Herrigel traveled to Japan and reached Master Awa Kenzo to learn the Doctrine of Zen, he trusted that he had encountered an actual Master. But where did this trust arise from in the case of the German student? To a large extent Herrigel believed Awa Kenzo to be a Master because he was highly regarded as a prestigious Master by all those who surrounded him.[167] But in this context the issue of certainty arises. How could Herrigel really *know* if Awa Kenzo was a Master? Why not consider the hypothesis that all the beautiful and artistic developments, rituals, and social behaviors that surrounded the Master weren't but a perfect staging and seduction, a mirage of beauty, a world of illusion or something *virtual*? To a large extent this is the issue present in many of the remaining traditions of the past, and one of the most important traditionalist authors, René Guénon, after being initiated in 1910 into Islamic esotericism (obtaining the name Abd al-Wāhid Yahyá) eventually considered, especially when dealing with Freemasonry, that not all the rituals and ceremonies had an initiatory effect on the members, which impelled him to consider the organizations as rather virtual, only providing an exoteric framework. Though Guénon considered Catholicism to have an effective initiatory character in some cases, Evola considered that was not actually a correct approach. Despite who was closer to the truth, it is obvious that even in the case of these authors there was a divergent view in terms of the issue of certainty and the absolute truth we are referring to.

All that Eugen Herrigel could perceive surrounding the Master was a highly ritualistic configuration of elements that he could not relate to but by just experiencing feelings of awe and amazement before such an extent of traditional beauty impregnating practically all elements and objects of life on a daily basis. From the perspective

---

[167] Eugen Herrigel: "I begged one of my colleagues, Sozo Komachiya, a professor of jurisprudence who had been taking lessons in archery for twenty years and who was rightly regarded as the best exponent of this art at the university, to enter my name as a pupil with his former teacher, the celebrated Master Kenzo Awa" Eugen Herrigel. *Zen and the Art of Archery*.

of TAI-PAI it is the "I"—the entity that *places* all those elements in the world—but when intending to start to properly introduce the idea of the "I" we are not equating the "I" with Eugen Herrigel *as an individual*, or as a self. Hence, it is not appropriate to state that "Eugen Herrigel" *placed* the elements in the cosmos, since the individual "Eugen Herrigel" is itself a construct based on the idea of separation between the aggregate of psychic/physic elements that compose the individual "Eugen Herrigel" and the elements present in the external world. This sense of individuality is described by Robert M. Pirsig as the,

> *"thinking of a personality as some sort of possession, like a suit of clothes, which a person wears. But apart from a personality what is there? Some bones and flesh. A collection of legal statistics, perhaps, but surely no person. The bones and flesh and legal statistics are the garments worn by the personality, not the other way around."*[168]

The "I" that we are introducing here transcends all limitations that define an "individual"; it corresponds to an impersonal and immortal *principle of development* of the cosmos. From this particular standpoint, it is as illusory to refer to an "individual" as to refer, for instance, to a smartphone as an "individual" device; if we refer to a smartphone device as a "unit," it is merely due to functional purposes as smartphones are manufactured in the mode of batch production, but if we look a little closer to the configuration of the technical cosmos, it appears that many electronic components of a given model of smartphone are the same as other similar models, so then we have to ask ourselves: What allows us to actually define the separation between one model and another close model? In most cases elements such as different software versions, cases, screen resolution, etc. which are rather minimal in terms of effective difference. So, in this case, we are considering that what creates the separation among models is the existence of minimal different attributes. But, if we look closer, we see that the

---

[168] Robert M. Pirsig. *Zen and the Art of Motorcycle Maintenance: An Inquiry into Values.* 1974 (William Morrow & Company).

functional resemblances between models are extremely higher than their differences, and yet we still consider them in our minds as "different models." From an industrial viewpoint of the organization of production, the model denominations are absolutely necessary for batch control, etc. in the same way that it was necessary for Eugen Herrigel—as any other individual in this world—to have a name and be called "Eugen Herrigel." If we look at the human condition—especially in modern times—there are also many more resemblances among humans in general than substantial differential attributes among them, which we in our times is highly controversial to agree with. For instance, we can affirm that a given "individual" corresponds to a given "race," "class," "gender," "culture," "ethnic background," etc. yet all these issues have clearly shown during the last decades highly controversial outcomes, since as we pointed out in the beginning, the decline of the West was clearly linked to the incapacity to define absolute referential points for knowledge, to the extent that it was spread to many schools of thought where no absolute referential point was established, but instead promoted an *absolutism*, that of *absolute relativism* in all fields. In the case of the smartphone industry, or any industry where things *function* and *develop* beyond all relativisms, it is undeniable that when a given device is assigned a model identification, it is because there has been an absolute agreement on the part of all agents and parties of the industry involved that in the device there is *a correspondence between form and function*. This is one of the core purposes of language: to name things where there exists a correspondence between form and function.

As most of the psycho-somatic predispositions have a highly hereditary character, also dependent on diet, custom, habits, and the determining elements of the cosmos that transcend the human domain, this already implies the challenge of defining an "individual," as an element supposedly separated from many overall elements. The dualism implicit in Judeo-Christianity consoled the disconnection of the singular being with his body (mortification, etc) and with his tradition by putting extreme emphasis in the power of the subjective contents of the mind, as determining factors of reality. To assume the latter as true is like believing that we can freely choose the drivers and controllers of a PC independent of the hardware configuration. Even though human nature is much more mysterious than that of a computer, the "hardware" element (bodily constitution, age, health, diet, ancestors, psycho-somatic capacities, etc.) have a very decisive effect on consciousness, in the way we perceive and relate to the world in operative terms. With human "hardware" the same issue of individuality takes place, it is extremely abstract to isolate the object from the techno-industrial structure that produced it, as the device corresponds to a material symbol and the production of its configuration. In our minds, we can establish a separation, but when it comes to mastery and development, we can verify that the characteristics of a given technological device are much more dependent on the overall techno-industrial infrastructure than dependent on our individual will to modify its configuration, which is extremely marginal. Can we "upgrade" our primitive instincts the same way we upgrade our PC? Yes, and in fact this is the aim of all spiritual traditions, but it can never be accomplished by exclusively focusing in the psyche, in a similar way as how in a given hardware configuration of a PC it is much more crucial to install adequate drivers and controllers in order to maximize the operative capacity of the computer, rather than in stuffing the hard drive with specific data packs (mp3 audio files for instance) that can be rendered useless if the specific audio driver/controller is lacking.

In the case of our psyche it is much more important in operative terms to implement an adequate "operating system" and "drivers," to develop those "codes" and "languages" that allow our naturalistic conditions to bridge and connect to the "upgrading" conditions, which in relation to human spiritual development implies connecting and participating in the cultural symbols of a given territory or civilization. The codes and languages that bridge the natural predispositions of a given individual with the overall development of the cosmos correspond to the cultural symbols.

And the symbols are necessarily operative. Today our "drivers" and "controllers" correspond to the techno-scientific paradigms.

Aside from the case of computers, other domains of reality can be captured in an extremely superficial way whenever approached from a specific subjective viewpoint that lacks firm reference points; hence our subjective mindsets can only see the "screen" of *physis*, like when interacting with a computer we can only relate to the information presented before our eyes on the screen, and are completely unaware of the complex electronic phenomena going on "under the surface" of the superficial representation. As Nietzsche points out:

> *"We set up a word at the point at which our ignorance begins, at which we can see no further, e.g., the word "I," the word "do," the word "suffer": these are perhaps the horizon of our knowledge, but not "truths.""*[169]

The concept of the individual or that of the ego refers to any of the arbitrary identifications that a given being can establish with the "surface" of *physis*, as it is presented to the contents of our subjective minds. Yet the "I" that Evola refers to in TAI-PAI transcends the "i" that defines the individual through subjective representation. In the case of the "I" of the Absolute Individual, the subjective mind becomes just one aspect among many others that are capable of relating to the world and comprehending the specific characteristics of *physis*. Evola's "I" is multidimensional;[170] it is also physical, yet it is not enclosed in the limitations established by the physical body; it is driven by physical and subtle determinations that the subjective mind can't even grasp, in a similar way to how a CPU can increase its temperature without the user having any idea of this phenomena by looking at the screen.

---

[169] Friedrich Nietzsche, Walter Arnold Kaufmann, R. J. Hollingdale. *The Will to Power* 1968, pg 267.

[170] As Nietzsche also points out: "The assumption of one single subject is perhaps unnecessary; perhaps it is just as permissible to assume a multiplicity of subjects, whose interaction and struggle is the basis of our thought and our consciousness in general? A kind of aristocracy of 'cells' in which dominion resides? To be sure, an aristocracy of equals, used to ruling jointly and understanding how to command?" My hypotheses: The subject as multiplicity." Friedrich Nietzsche, Walter Arnold Kaufmann, R. J. Hollingdale, *The Will to Power* 1968, pg 270.

However, in many cases with the installing of particular drivers, the user can willfully be capable of visualizing in the screen the CPU temperature or be aware of the underlying processes taken place at the level of hardware. As Julius Evola also remarks:

> *"The "spirituality" of "psychological man" is nonessential, contingent, and there are only too many circumstances that speak so us of this contingency, of the dependency of the "superior faculties" and individual consciousness itself on the body. The body is truly the root and origin of the soul and its faculties, but without producing them directly; the situation is almost analogous to a drum that without producing the sound itself, is the necessary condition for the sound to be manifested. And so also life, consciousness and self-consciousness cannot be manifested in man except through corporeal reality."*[171]

Similarly, our conscious mind may or may not be enabled to detect specific physical and physiological phenomena taken place in our bodies, and yet these phenomena clearly define the mode we relate to the world, in the same way that a computer running background processes might have less available processing power to deal with all the flow of network data. Our psyche can become easily distracted with all sorts of contents and information, diversions or entertainments, yet in no case do all these elements relate to the modes we link operatively to physis. So, in terms of applying to this the ideas of Evola's *Theory and Phenomenology of the Absolute Individual*, we can assume that none of the contents of our minds, such as the contents that can appear on the PC monitor constitute knowledge from TAI-PAI's gnoseological point of view, since the "I" of the Individual that aims to the Absolute not only perceives the objects of the world depending on its inner configuration and paradigms, it is also the source of the creation of new developments and determinations in the exterior objects of physis based on inner creative processes and determinations.

---

[171] Julius Evola. *The Hermetic Tradition*. Inner Traditions, pg 77.

In the case of the human condition, whenever there emerges a correspondence between a mode of being and an actual function, then we can refer to the activation of an "I," which has nothing to do with the self of modern psychology, nor the "ego," but overall it is an *operative magnitude*. Hence, from the viewpoint of an Operative Tradition and Doctrine, it doesn't matter if one is a man or a woman, a white or a black. What matters is *how one does what one does*, the form of a given activity.[172] It is precisely the lack of discipline in perfecting a given activity while aiming to provide it with a form that hypertrophies the illusory presence of the "individual" and its separation from the cosmic development.

Like any other student during the initial phases of spiritual development, when Eugen Herrigel approached the Master for the first time, he had no access to the "tools" that allowed him to verify the particular qualitative attributes of the Zen Master... His consciousness and state of mind were still in a formless condition, and all concepts he had previously learnt during his philosophical studies in Germany were constantly "floating" in his thoughts, so he could not help but allow his conscious mind to be passively driven by these psychological currents. This state of inner spiritual agitation is what triggers the need to identify all forms of knowledge through concepts, discursive constructs, and ideologies. During initial states of mind such as the latter, it wouldn't be too difficult for a smart cultural critic to convince Eugen Herrigel that everything the Master *represented* was just a very well staged lie, a "superstructure"; in a similar way as those who are akin to Marxism can't help thinking that a Gothic Cathedral was a sort of "representation intended to brainwash the working class about the economic superiority of the Papacy as owners of the means of production." The initial stages of approach to the Doctrine, or in the case of this book, to Evola's TAI-PAI, are generally quite challenging to withstand, as there is still no certainty in regard to anything at all, and thus Eugen Herrigel

---

[172] In generale, dato che la forma quale valore sia, in un qualsiasi modo e misura, presente in una esperienza, essa di diritto condiziona l'intera esperienza (In general, a form with a given value, in whatever mode and intensity, present in an experience, rightly conditions the entire experience) Julius Evola. *Teoria Dell'individuo Assoluto*. Edizioni Mediterranee, 1988, pg 28.

was unable to "measure" the diverse levels of development of truthfulness, since he still hadn't disciplined the crucial "tools" of his senses. Yet not having access to the realm wasn't an impediment for Herrigel, because though he didn't know the truth, *he had the heart*: he trusted the Master, he had faith in the Master and respected him. *"The intelligence has its seat in the heart because that is what precedes all the other organs,"*[173] says Geber.

Obviously, the faith, as with all faiths, grew from irrational grounds… But Herrigel knew one thing for sure: that he aimed to reach *the absolute, the centre, the heart*. Herrigel could perceive the sanctity emanating from the Master, which is a very specific allure that itself requires a fine perception. Jünger writes:

> *"The characteristic feature of the great saints—of whom there are very few—is that they get at the very heart of the matter. The most obvious things are invisible because they are concealed in human beings; nothing is harder to evince than what is self-evident. Once it is uncovered or rediscovered, it develops explosive strength."*[174]

This transparent perception of the divine in what is human was all Herrigel needed during the initial stages.

So this absolute bond had been firstly established between Eugen Herrigel and Master Awa Kenzo, the latter then provided the former with the tools: the *arc and bow*.

\* \* \*

The arc and bow provided to Eugen Herrigel by the Master correspond to an artifact or instrument where many dispersed elements present in nature become artistically fixed into a form. This process of fixating all chaotic and dispersed elements into a form is conceived in physics as a *neguentropic* process (negative entropy) since it goes "countercurrent" to the spontaneous and irreversible process defined by the first and second laws of thermodynamics,

---

[173] Geber, *Libro delle Billancie*, CMA, III, pg 140.

[174] Ernst Jünger. *Eumeswil*. The Eridanos Library. 1980.

wherein any closed system there is always a tendency to positive entropy production, to increase disorder and chaos. All the scientific laws that intend to enclose all phenomena of nature into closed and analytical systems correspond to an approach to phenomena *in the case of the individual as a passive observer*, that is, in the case of a particular subjective understanding of human nature that is also separated from the cosmos.[175] However, while the scientific approach is adequate when addressing mechanistic phenomena typical of inert bodies passively surrendering to the determinations expressed by scientific laws, the laws are inadequate to express and therefore predict the developments of those systems that are impelled by forces that eventually fixate a given form in a creative and synthetic way.[176] Life itself is one of those very unpredictable systems that can not be entirely coped with by resorting to scientific laws based on mechanistic premises. From a statistical viewpoint that only addresses or registers the phenomena explained by the laws, life is a highly improbable phenomenon. In other words, if life cannot be "enclosed" by the laws, it means that from the perspective of the laws, life can not exist, or rather that life itself is an unexplainable miracle.

All modes of life are essentially neguentropic, from the domain of microorganisms to that of high cultures. Therefore, based on the

---

[175] "According to widespread opinion, there is a fundamental distinction between "observed facts" on the one hand—which are the unquestionable rock bottom of science and should be collected in the greatest possible number and printed in scientific journals—and "mere theory" on the other hand, which is the product of speculation and more or less suspect. I think the first point I should emphasize is that such antithesis does not exist. As a matter of fact, when you take supposedly simple data in our field—say, determination of $CO_2$, basal metabolic rates or temperature coefficients—it would take hours to unravel the enormous amount of theoretical presuppositions which are necessary to form these concepts, to arrange suitable experimental designs, to create machines doing the job—and this all is implied in your supposedly raw data of observation. If you have obtained a series of such values, the most "empirical" thing you can do is to present them in a table of mean values and standard deviations. This presupposes the model of a binomial distribution—and with this, the whole theory of probability, a profound and to a large extent unsolved problem of mathematics, philosophy and even metaphysics. If you are lucky, your data can be plotted in a simple fashion, obtaining the graph of a straight line" Ludwig Von Bertalanffy. *General Systems Theory*. George Braziller. New York, 1968, pg 162.

[176] la persona per ora può prendere coscienza di sé come potenza oggettiva solamente a patto di alienare la propria persuasione [then the person can become aware of itself as an objective power only to the extent it alienates its persuasion] Julius Evola. *Fenomenologia Dell'individuo Assoluto* III ed. corretta: Edizioni Mediterranee, Roma 2007.

former explanation, if the individual aims to gain knowledge on the dynamics of life, scientific laws act as "boats" floating along a chaotic surface of non-graspable phenomena, and hence no dominion on the part of the individual over the forces is feasible. *"Physical laws aren't but the channel where the torrent of facts are drained, a channel that is formed by them, though they afterwards release from them"*—writes E. Boutroux.[177] It is rather obvious that a scientist might have control over the natural phenomena in the domain of the laboratory, hence being capable of framing the most diverse phenomena into universally valid scientific laws, but from the viewpoint of TAI-PAI, the laboratory itself constitutes a rigid fixation where the phenomena of nature are not grasped in regards to their *essence* but only graspable to the extent that the phenomena can be projected on the intrinsic technical characteristics of the laboratory. Inside the "boat" determined by scientific laws, all phenomena can be "frozen" (controlled) and explained, but outside the "boat," just like an open sea, *samsaric* chaos and unpredictability reigns. In other words, as soon as the scientist leaves the laboratory, nothing can assure us that he has better dominion over the events of his life (the relations with his friends, family, his health, etc) than someone who has no scientific preparation at all. The scientist can get easily drowned by irrational urges and behaviors, just like anybody else. If he has some control over the *world of becoming*[178] it is to the extent that he integrates his individuality into the technical procedures of the laboratory. As a scientist, he has to leave outside of the laboratory all his private and personal affairs and to act according to the determinations, policies, and rules of the laboratory.

---

[177] E. Boutroux, *De la contingence*, cit., p. 39: E. Mverson. De l'explication cians les sciences, Paris, 1921, voi. II, pg 285.

[178] Julius Evola: "The world of "becoming" is thus, in a manner of speaking, the truth Buddhism uses from the start. In the becoming nothing remains identical, there is nothing substantial, and nothing permanent. It is the becoming of experience itself, consuming itself in its own momentary content. Ceaseless and limitless, it is also conceived as nothing more than a succession of states that give place one to another according to an impersonal law, as in an eternal circle. We can here see an exact parallel of the Hellenic concept of the "cycle of generation" κύκλος της γενέσεως, and the "wheel of necessity." Εἱψαρψένης" Julius Evola. *The Doctrine of Awakening*. Inner Traditions, pg 44.

This context can allow us to penetrate a little further into the idea of the "I" presented by Julius Evola. In the former example our scientist becomes integrated into a closed domain, the laboratory (this is only an initial approach, since even a laboratory itself can hardly be conceived as a closed system), and by relating operatively to the techniques of the laboratory, the scientist progressively becomes akin to a given *paradigm* or set of *paradigms*; that is, to a way of relating to nature mediated by the technical means he has at his reach. Paradigms are always operative factors; they underlie the conscious mind and determine the developments of the conscious mind itself in terms of how the mind relates to effective and actual problems. Paradigms generally become ingrained in the activity of the scientist as the operative outcome of repeatedly the exposing the scientist to given situations and problems. Thus paradigms are a matter of *habit* rather than a matter of following laws, or as Rupert Sheldrake wrote by resorting to the formative causation hypothesis: *"learning is facilitated as the individual "tunes in" to specific morphic fields."*[179] The capacity to grasp the character of phenomena within a laboratory on the part of the scientist (namely, the capacity to elaborate work hypothesis) depends on the capacity of the scientist to intuitively perceive the configuration of a given object of study, and afterwards to translate the configuration into law-based data and information. As experience shows, whenever it comes to solving``` actual problems, the best engineers are those who have been trained according to a wide set of paradigms, and not those who are merely aware of the laws of physics. So to summarize all the former: the rather "fixed" environment of a laboratory or workplace corresponds to a "tool" that deep down "forges" and "fixates" operative predispositions on the part of the scientist or worker, and these operative predispositions are referred to in modern science as *paradigms*. Yet before modern epistemology discovered all these aspects, Nietzsche had already wrote that *"all activity enters our consciousness as consciousness of a "work."*[180]

---

[179] Rupert Sheldrake. *Morphic Resonance. The Nature of Formative Causation.* Park Street Press.

[180] Friedrich Nietzsche, Walter Arnold Kaufmann, R. J. Hollingdale *The Will to Power* 1968, pg 349.

Paradigms allow us to relate to the world, to provide structure to the vast array of stimulus that arrives at our senses. The structure of the particular vision upon the experiential domain is dependent on the structure of our paradigms, in the same way that a computer registers a sound from the environment is not only dependent on the hardware (its "senses," so to speak) but also on the software processing of the signal by the driver/controller (its "paradigm"). Our subjective and discursive mind can be stuffed with lots of different *images* and *concepts* of the world,[181] in the same way, our hard drive can be stuffed with hundred of different data files, but ultimately it is the paradigm (driver/controller) what can "translate" or "process" operatively the contents.

In terms of an Operative Tradition the principle is the same: in practical traditions the development of the psyche is focused almost exclusively in the disciplining of paradigms, that is, the construction of "bridges" ("operating systems" or "drivers") between the naturalistic conditions that determine the individual condition and the cultural overall symbols (the "goal"). Paradigms are always operative; they relate to a very specific way of doing things, of transforming things; hence they relate to *process* and *technique*. Paradigms are always revealed in gesture, in dynamic performance, and can be imagined as an operative *prosthesis*, a word that derives from the Greek *prosthenos*, for "extension." The first modern writer to explore systematically the connection between the operative faculties of men and the external material conditions was probably Ernst Kapp, in 1877. Kapp speculated that railroad systems unconsciously mimicked the circulatory system, while telegraph lines extended the nervous system. Along these lines, McLuhan writes:

---

[181] Ogni concetto, logica e sistema, materialiter ha sempre un valore ipotetico, poiché il principium individuationis, la potenza profonda che determina e afferma la varia realtà metafisica, è sempre il principio della persona [Every concept, logic and system, materialiter always has an hypothetical value, since the principium individuationis, the profound power that determines and affirms the diverse metaphysical reality is always the principle of the person] Julius Evola. *Fenomenologia Dell'individuo Assoluto* III ed. corretta: Edizioni Mediterranee, Roma 2007.

*"Everybody experiences far more than he understands. Yet it is experience, rather than understanding, that influences behavior, especially in collective matters of media and technology, where the individual is almost inevitably unaware of their effect upon him."*[182]

Ernst Jünger, relates these "extensions of men" to pain; he writes:

*"Not only are we working with artificial limbs as no other form of life did before us, but we are currently developing strange domains where the use of artificial sensory organs will create a high-level compliance with given types. This fact is in close relationship with the objectification of our image of the world, and our relationship to pain."*[183]

Paradigms have very little to do with personal opinions or religious beliefs; they are rather active and operative magnitudes. The implicit paradigms of a given individual are exposed *at work,* and mostly during conditions of stress and uncertainty in the work environments. The structure of paradigms, and even the structure of the selfsame scientific revolutions—as Thomas Kuhn demonstrated[184]—are not of a conceptual, subjective, or discursive nature, but instead have a symbolic and physiognomic character, referring to specific perceptions of the experiential domain which afterwards allow manipulation of the domain, as well as the particular development of

---

[182] Marshall McLuhan. *Understanding Media. The Extensions of Man.* The MIT Press Cambridge, Massachusetts. London, England, 1994.

[183] Über den Schmerz: David C. Durst, *On Pain*. New York: Telos Press Publishing (2008).

[184] Kuhn resorts to very insightful expositions that shed light on the idea of paradigms as operative faculties. He writes: "Turn now to another, more difficult, and more revealing aspect of the parallelism between puzzles and the problems of normal science. If it is to classify as a puzzle, a problem must be characterized by more than an assured solution. There must also be rules that limit both the nature of acceptable solutions and the steps by which they are to be obtained. To solve a jigsaw puzzle is not, for example, merely "to make a picture." Either a child or a contemporary artist could do that by scattering selected pieces, as abstract shapes, upon some neutral ground. The picture thus produced might be far better, and would certainly be more original, than the one from which the puzzle had been made. Nevertheless, such a picture would not be a solution. To achieve that all the pieces must be used, their plain sides must be turned down, and they must be interlocked without forcing until no holes remain. Those are among the rules that govern jigsaw-puzzle solutions. Similar restrictions upon the admissible solutions of crossword puzzles, riddles, chess problems, and so on, are readily discovered" Thomas S. Kuhn *The Structure of Scientific Revolutions* Third Edition The University of Chicago Press, pg 52.

means towards a known or unknown goal. Robert M. Pirsig points to this idea of the working hypothesis implicit in a given paradigm when writing: *"Skill at this point consists of using experiments that test only the hypothesis in question, nothing less, nothing more."*[185] Paradigms constitute unconscious determinations of which we can only be aware of them due to their effects in terms of application. In an Operative Tradition, the cultivation of paradigms is of crucial importance, as paradigms allows us to structure knowledge itself according to very specific values. The medieval idea of *Post Laborem Scientia* (first comes work, later comes science) refers directly to this order of ideas, which is expressed accurately in the words of Max Planck, who sees how the modes by which we become familiar with the material world condition the scientific truthfulness of given paradigms in historical terms. The German theoretical physicist who won the Nobel Prize in Physics in 1918 states that: *"a new scientific truth does not triumph by convincing its opponents and making them see the light, but rather because its opponents eventually die, and a new generation grows up that is familiar with it."*[186]

The paradigms that a given individual embraces can be visualized by experienced eyes whenever the individual has to resolve a practical problem within the material world, without any human interference. The specific patterns of behavior that an individual exposes with other individuals can also be meaningful in order to determine the unconscious predispositions of the individual, yet strictly speaking, paradigms can only be adequately traced in terms of the handling of specific means, once action is liberated from any interference arising from human emotion or sentimentalism. This doesn't imply resorting arid and emotionless character traits on the part of the individual who relates to the raw material domain, but rather a dominion of all emotions and sentiments that ultimately cause an arousal of psychological interferences during the operations.

---

[185] Robert M. Pirsig. *Zen and the Art of Motorcycle Maintenance: An Inquiry into Values.* 1974 (William Morrow & Company).

[186] Kuhn, T. S. 1996. *The Structure of Scientific Revolutions.* 3rd ed. Chicago, IL: The University of Chicago Press, pg 151.

Even though we are resorting to cybernetic or computer analogies in order to expose the crucial operative aspects that transcend the limits defined by prevailing psychological frameworks since the times of Freud, not all cases of human nature are passively subjected to the conditionings established by territorial configurations, or by what Marshall McLuhan referred to as "extensions of man." For instance, children in our times are subjected at many levels to a *training* (eventually in the third part of this book we shall resort to Heidegger's idea of "enframing" to define the training) of their unconscious paradigms by the intense integration of their actions and decisions between the limits defined by the operative capacities present in the gadgets and techniques they are linked to on a daily basis, yet this adaptation—which has very considerable effects on their character traits— does not imply that they don't have any freedom of choice, when it comes to being trained in their deep traits by other techniques.[187] However, the practical problem is the following: why would a child spend today more time learning to play a musical instrument, let's say, a violin, instead of playing with a gadget his parents bought him? We arrive here at the issue of *value*, which shall be addressed in the section '*The concept of value and the empirical state of being*'.

Ultimately the important aspect of paradigms is that they have to be experienced. If a teacher aims to teach a student a given paradigm, he'll have to take the student to an environment where problems constantly emerge, that is: a workplace or laboratory where things are *done in a very specific way*, and where nature is processed *in a very specific way*. Hence paradigms can only be *trained*. They act on the somatic aspect of an individual in the same way an individual drives a car without thinking or even dissociating completely his conscious thoughts from the driving activity. By considering at first

---

[187] Come non è che per uno scambio di padroni un servitore cessi di esser tale, così la facoltà di scegliere un motivo anziché un altro non dice nulla riguardo alla libertà, giacché di là dai particolari motivi resta sempre, nell'ipotesi di cui sopra, la forma della conformità in generale ad un motivo [As the change of a given chief doesn't make one less a servant, in the same way the faculty of choosing one motive or another doesn't say anything about freedom, since beyond particular motives always remains, according to the former hypothesis, the form when conforming in general to a motive] Julius Evola. *Teoria Dell'individuo Assoluto*. Edizioni Mediterranee, 1988, pg 104.

a workplace or laboratory as a closed system, it appears very intuitive at first that the knowledge that a closed system can gain over the outer universe can only be a *function* of the same paradigms implicit in the system. This is a key aspect of TAI-PAI that must be realized, and that shall be explained in the following pages.

*  *  *

Let's say someone tells me that they've read on Wikipedia that the distance from the earth to the moon is about 400,000 km. From the viewpoint embraced by Magical Idealism, the information can strictly represent for me a form of privation and self-alienation since the "I" cannot rigorously know anything that transcends its perceptive means of representation. In the tradition of Zen there is the saying of "not confusing the moon with the finger that points to it," and this saying can be interpreted in diverse ways by the Western mindsets,. When applied to the Theory and Phenomenology of the Absolute Individual it can be understood *in the sense* that one can constantly refer to the domain of the "ideal" and to "transcendental concepts" which point to the "spiritual domain" beyond the material domain, but that if "I" can not reach to "heavens" with my own particular means of representation, the concepts shall only justify self-defeat and a weak persuasion/dominion of my effective reality. Then we are on the verge of nihilism, for as Nietzsche shows, since *"an escape remains: to pass sentence on this whole world of becoming as a deception and to invent a world beyond it, a true world."*[188] Hence, if by disciplining my paradigms in the rigorous domains of mathematics and physics I can come up with an improvised instrument of measurement of my own making (a technique) and determine approximately the earth-moon distance, the dominion on the part of my "I," my new means of representation, my new senses, shall immediately transmute my "I" to a higher level of awareness, and *physis* shall be approached intuitively in a new qualitative manner. This procedure is pure operative magic. *Magical*

---

[188] Friedrich Nietzsche, Walter Arnold Kaufmann, R. J. Hollingdale *The Will to Power* 1968, pg 13.

*Idealism* refers to the core understanding that the procedures are an *ideal* approach in the sense that I have to *a priori* conceive the existence of ideal "supra-terrestrial bodies" beyond the reach of my physical senses, but it is also a *Magical* approach in the sense that an activity of dominion and persuasion of the means and objects provided by my immediate reality shall allow me to magically build the progressive "stairway" that extends my "I" ("my core eye," so to speak) to higher and even heavenly cosmic standpoints, not in a speculative way, but in operative and experiential ones.

This little example, extracted from one of the gnoseological peaks of Western philosophy, allows us to realize the importance of the techniques we are attached to as extensions of our perceptive capacities, and not only of our perceptive faculties or deep senses but also of our creative faculties. McLuhan also wrote on this issue:

> *"The effects of technology do not occur at the level of opinions or concepts, but alter sense ratios or patterns of perception steadily and without any resistance. The serious artist is the only person able to encounter technology with impunity, just because he is an expert aware of the changes in sense perception."*[189]

Since more than a century ago, science has absolutely dismissed all these phenomenological aspects, expelling the "I" of the scientist from any methodological training of the senses, at the expense of a hypertrophy of speculation and data production which constitute all sterile "photographs" of a complex technical reality over which there is also an absolute lack of dominion and persuasion.

To keep on smoothly introducing the ideas present in Magical Idealism and TAI-PAI, let's consider a simple example of a space observatory. In addition, let's consider as well that the artists, engineers, technicians, and scientists who work in the places have access to elaborate techniques and apparatuses by resorting to local means and resources. Let's suppose as well that in the organization there exists a hierarchy of integrated functions

---

[189] Marshall McLuhan. *Understanding Media. The Extensions of Man.* The MIT Press Cambridge, Massachusetts. London, England, 1994.

where the director of the space observatory embodies the spirit of the entire organization, that is, who justifies the entire activity of the organization. Let's suppose as well that the organization is specialized in producing telescopes with absolute mastery. The space observatory, as a hierarchical configuration of dominion, can be conceived as a particular "I," a fixed and homeostatic form of the cosmos which materializes an idea. Let's suppose that the main production of the organization is that of astronomical charts determining the characteristics of stars as far as 10.000 light-years away, due to the technical limitations of the optics they produce. Given this gnoseological subject-object context, the radical postulate of Evola's Magical Idealism and TAI-PAI would state that nothing can be known by the "I" that has not been placed firstly by the "I." This conclusion has also been realized by transcendental idealist philosophers such as Kant, Fichte, Shelling, and Hegel who conceived the world as "one's representation,"[190] or what Ernst Jünger pointed out when writing: *"For the sociologist the whole is sociological; for the biologist, biological; for the economist, economic; (...) Such absolutism is the undisputed privilege of the conceptual view of things."*[191] So the solution provided by the school of idealist thought was that of imagining the existence of an extremely abstract and clearly non-operative "I" (an Idealist "I"), an *abstraction*, determined by a discrete set of rigid categories (space, time, causality), hence concluding that the historical development and progress of knowledge would correspond to the further "filtering" of the entire universe through the categories. This process of accumulation of knowledge proposed by the transcendental idealists would (*ideally*, of course) correspond thus to an extremely impersonal and highly abstract development, aiming to reach the "Logos," the "Idea" or the "Pure Act." It was at this point during the historical development

---

[190] "in representing the real is *not* controlled by the possible, the I is passive in respect to its own act—not as it asserts things but rather it is as if things were asserting themselves in it" Julius Evola. L'individuo e il divenire del mondo, Libreria di Scienze e Lettere, 1926, pg 23.

[191] "Für den Soziologen ist das Ganze soziologisch, für den Biologen biologisch, für den Ökonomen ökonomisch in jeder Einzelheit, von den Systemen des Denkens bis zum Pfennigstück. Dieser Absolutismus ist das unbestreitbare Vorrecht der begrifflichen Anschauung" Ernst Jünger. Der Arbeiter. Ernst Klett, Stuttgart 1981. Klett-Cotta, pg 30.

of Western philosophy when the "I" had became uprooted from all immanent and operative factors, and started to fly totally astray into a cloud of wild speculation which eventually contained all domains of science, history, art, and politics.[192] It was at this point where discussion could arise practically on any topic, but all absolute certainties had vanished.

Based on the former example and according to the denominations expressed in Magical Idealism, we can affirm that the "I" *places* the material attributes of space up to the established 10, 000 light-year limitation. The charts of the observable universe, which correspond to the knowledge produced by the observatory, can be conceived very simply as a specific spectrum of colors that are refracted as the white light of the universe (or universal spirit) is projected in the specific optic "prisms" of the observatory. Therefore from an operative and empiric standpoint, the individuals who work in the observatory can only hypothesize the existence of an unknowable "white light" beyond the realm of the universe encompassed by their perceptive techniques. However, they could obviously listen to what idealist transcendental philosophers have to say about the universe, when the philosophers affirm that "white light" corresponds to the "thing-in-itself," the "Absolute Spirit" (Kant), or could even listen to what monotheistic religions have to say when inducing peoples to believe that beyond the perceptive realm is the domain of god or the devils; or they could listen to nihilist philosophers who would state that beyond the realm is the domain of pure nothing. Also, if someone provided the observatory workers with charts presenting the characteristics of stars beyond the range of 10, 000 light-years, workers would be powerless to actually realize if there is any truth in the new charts through their own operative paradigms and optics.

---

[192] Proprio a questa ulteriore posizione o assunzione è il sopprimere ogni antecedente eterologico o empirico dell'attività determinante del logos e l'intendere questa nei termini di un sistema chiuso e autonomo, che si sviluppa in un certo modo da sé. Ora, ciò significa solo introdurre il nemico ancor di più dentro la propria cittadella. [Typical of such standpoint or assumption is that of discarding all former eterology and empiricism within the activity that determines logos, and to conceive such in terms of a closed and autonomous system, that develops somehow on its own. Yet, this entails introducing the enemy even more inside the domains of the city] Julius Evola. *Teoria dell'individuo assoluto*. Edizioni Mediterranee, 1988, pg 107.

Workers could passively accept the information,[193] and make use of it, but this action would inevitably testify the decadence of the "I" they work for, and the forging of a debt (surrendering to the determinations) for another unknown "I." They could even think that because they were provided with charts of the universe that are rational, scientific, and objectively based on space, time, and causality, that the charts are necessarily "real," and that they have to be trusted due to the intrinsic character of objectiveness. However, even if the workers accepted to read the new and sophisticated charts provided to them, the charts would not be enriching the empiric, magical, and concrete "I" of the observatory, but merely hypertrophying the abstract, transcendental, and discursive "I." So the issue that would arise here is that of verifying the potency, autonomy, and absolute freedom on the part of the observatory.

If the workers of the observatory start feeling tired in their lives and decide to profit from adding to their own charts those they were provided from the outside, this decision would correspond to the revelation of an *impotence*, a *privation*, and a *lack of dominion and persuasion of reality*, or a sense of impurity, very much related to the term *steresis*.[194] Yet if the workers decide not to profit from the new information and to keep up focusing on what they can truly demonstrate through their own means,[195] then they would follow the

---

[193] la spiegazione del fatto che si è impotenti in determinate zone dell'attività rappresentativa mediante il riferimento ad un «altro» è una pseudo spiegazione, anzi un circolo vizioso [the explanation of the fact one is powerless in a given determining zone of representing activity by means to the referral to "other" – corresponds to a pseudo-explanation and a vicious cycle] Julius Evola. *Teoria Dell'individuo Assoluto*. Edizioni Mediterranee, 1988, pg 142.

[194] [Greek steresis, from steresthai, to lack or to be deprived of] Normally, something suffers privation when it lacks an attribute that, according to its nature, it should possess. According to Aristotle's analysis, privation, substratum, and form are the three basic elements in the process of change. Privation at the beginning of change is the absence of a character which the change will provide at its completion and which the substratum is capable of receiving. Aristotle, *Physics* The Blackwell Dictionary of Western Philosophy.

[195] Vi è tuttavia un'altra possibilità: spiegare mediante l'azione, cioè risolvendo, facendo passare all'atto ciò che è in potenza, creando sufficienza là dove vi è privazione, determinando una nuova situazione fra le potenze dell'Io, trasformando la stessa coscienza. [There is still another possibility: to explain by means of action, that is, by solving, by allowing to become act what is potential, by creating sufficiency there where there is privation, determining a new situation between the power of the "I," transforming the selfsame consciousness] Julius Evola. *Teoria Dell'individuo Assoluto*. Edizioni Mediterranee, 1988, pg 144.

path of the *Absolute Individual, the path of persuasion and dominion*, that is, the path that allows the stars they perceive to revolve around the "I." In the same way, all workers of the observatory are hierarchically configured around the Individual present in the top of the observatory, the Absolute Individual in the centre of the cosmos.

In the latter case, if anyone asked the artists of the observatory what they considered to be "divine" or "transcendent," they would state that they themselves are partial embodiments of immanent transcendent powers,[196] having no existential need to believe in "gods," religions, ideologies, utopias, etc, which are connected to desire drives that need and expect given actions to be "compensated" in the future.[197] They would consider their telescopes to be divine creations, the stars to be the ultimate cause of their production, and the charts they produce to be the subtle language of the stars. They would even very likely state that *"What we "see" are merely light-indexes; what we comprehend are symbols of ourselves,"*[198] as Oswald Spengler wrote, by considering the telescope itself a typical expression of the Faustian traits of the West. If they were asked if they believed in "metaphysics" they would say that the whole observatory is a school of initiation produced by the "I,"[199] with every single individual integrated to participate in the creation of the vision. Some artists would even consider that the stars act as the ultimate attractors of all material and creative developments of their telescope production, such as the development of a sunflower pointing towards the sun, and that the only way the metaphysical

---

[196] Il vero Dio non deve essere dimostrato, pensato, determinato: esso o è fede e pura evidenza interna, o è nulla [The true God is not to be proved, thought, determined: it is either pure faith and internal evidence, or otherwise nothing] Julius Evola. *Fenomenologia Dell'individuo Assoluto* III ed. corretta: Edizioni Mediterranee, Roma, 2007.

[197] L'oggettività assoluta—il valore del mondo del desiderio—si distanzia cosi nei termini di un futuro: essa, in ogni punto, non vive che come sospesa all'atto successivo in cui l'Io ne vede la realizzazione compiuta. [Absolute objectivity—the value of the domain of desire—sets apart thus in terms of a future: at all levels it doesn't live but as dependent to the successive acts in which the "I" perceives the task as accomplished] Julius Evola. *Teoria Dell'individuo Assoluto*. Edizioni Mediterranee, 1988, pg 68.

[198] Oswald Spengler. *The Decline of the West*. Alfred A Knopf. 1926. New York, pg 331.

[199] La metafisica, in questi termini, è trascendentalismo, anzi trascendentismo, solo in quanto empirismo assoluto [In these terms metaphysics is transcendentalism, considered as absolute empiricism] Julius Evola. *Teoria Dell'individuo Assoluto*. Edizioni Mediterranee, 1988, pg 145.

powers can be materialized is by the progressive development taking place in the material domain, in order to create the "eye," the "lenses," and the "I." Some artists would even be bold enough as to equate their telescopes to flowers.

They would state that all they are doing is producing an "eye" ("I") which they would consider as the ultimate purpose of every single culture. Hence for them, each culture ultimately produces a way of looking at the universe, of experiencing the universe, and for them, it would be unrealistic to consider that there is "only one way of looking at the world or experiencing it." They would affirm that modern science is an adequate "starting point" for approaching the universe, but that it is unable of integration and convergence since the convergence requires operative and magical disciplines that disappeared from most cultural frameworks after the 1800s. This necessarily requires magical artists to consider their telescopes as to be magical objects, to constitute material mediations between men and the cosmos, symbolizing and thus pointing to the idea in order to see the heavens one has to get very close to the product of one's actions. As Jünger pointed out in his Diaries when writing:

> *"The visible contains all the signs that lead to the invisible. And the existence of the latter must be demonstrable in the visible model."*[200]

When artists look inside their magical objects, the material paradigms implicit in the selfsame modes of production of the objects filter the view of the observable cosmos in a very specific way. The universe for them is no longer filtered by discursive categories and concepts, but rather filtered effectively and empirically, through what Novalis referred to as *active empiricism*,[201] impeding a speculative dispersion of efforts, and allowing the effective attainment of absolute certitude. They would state that our eyes see one thing, and that the magical development of our senses and gestures can "see" another. From

---

[200] *Paris Diaries: 1941-1942*, trans. M. Hulse (London: Farrar, Straus & Giroux, 1992).

[201] Il metodo è, per usare un termine del Novalis, un empirismo attivo [The method is, by using a term of Novalis, an active empiricism] Julius Evola. *Teoria Dell'individuo Assoluto*. Edizioni Mediterranee, 1988, pg 145.

a subjective viewpoint, if we ever asked each worker what they actually knew, they would show us the books they are studying, where they learn to understand the phenomena taking place in their activity. The library of the observatory would be very extensive, but the workers would only assure us that they have competence in the books of their specific domain and that all other books are not necessarily forbidden, but that they disperse the focus of the mind into a specific activity.[202]

In the observatory, all workers, scientists, and artists manage to release themselves from their individuality—their human names—and aim to acknowledge their divine name defined by the constellation of stars that produces their being; a being they are impelled to discover. According to diverse levels of integration, they would not only filter all phenomena in a specific way but what's more, *produce all phenomena in a specific way*, which would be reproduced in their charts. If some smartass told the director of the observatory that all astronomers in the 21st century agree that the earth revolves around the sun and that it is not in the center of the universe and has no power over the cosmos, the director would reply that *HIS* universe—the domain of reality which we can empirically and effectively grasp—is actually revolving perfectly all around him, and that smartasses and astronomers are precisely the individuals more affected by the invisible currents they've learnt to smartly ignore.

Hence, the common idea that still exists about the supposed "power of man over nature," based on the development of modern science and technique is misleading. Because of this way of thinking, was linked to romantic or Rousseaunian views that separate "nature" from "civilization" are completely inapplicable in our times, as seen in the presence of any of our surrounding objects or phenomena. It is very difficult to discern if they are strictly a product of the current civilization or rather a product of what is of conceived as the "natural world." Even the human species, made of flesh and blood, is

---

[202] "Everything I cannot act on, everything that resists my will, is only a privation of this very will, something negative, not a being, but a non-being" Julius Evola. *L'individuo e il divenire del mondo*, Libreria di Scienze e Lettere, 1926, pg 25

to a large extent a product of civilization; hence the "separation from nature" or "alienation from nature" corresponds to common themes present even in the recent developments of the eco-psychological movements that point out conceptions about the root causes of insanity amidst the urban-industrial configurations, but the aim of reestablishing relations with the natural world cannot deny the importance of cultural dimensions present in the human condition. Most humanist cultural developments and anthropomorphic historiography have spread a view of men's relation to the cosmos as if considering that the minds of men can determine reality, and hence dominate nature. In exceptional cases in History it is observable that some groups of men appear as "riding" or "piloting" the whole development and integration of the natural phenomena at the top, but in this case we are referring to the emergence of an enlightening spiritual factor—a cultural singularity—where men correspond to other elements of the cosmos that surrender to higher determinations. In these cases, the psychological contents of men's minds correspond to a by-product of the civilization process, yet they are not determining factors of it.[203] The fluidity of the psyche adapts its forms to the symbols embraced by civilization, which overcome and integrate all conceptual and discursive mind-sets.

So when it is stated that "man has power over nature" not only do we encounter problems in order to define exactly what "nature" actually is, but we have the same problems when defining what "man" is, in operative/territorial terms. Eugene Schwartz, commenting on the limitations of science and technology in this regard, writes *"the concept of harnessing nature through conquest was in error because it failed to recognize that man was a part of nature and that what happened to nature would, in turn, rebound upon man."*[204] Even in the case of the scientist or engineer, who both correspond to

---

[203] Un imperativo o una legge morale non è, e non è determinante, che in quanto sia per me, in quanto io l'assuma nell'adesione e nel riconoscimento, oppure nell'avversione [an imperative or moral law is not determining, both from my viewpoint, or to the extent I assume it as an attachment, an acknowledgement, or something to reject] Julius Evola. *Teoria Dell'individuo Assoluto*. Edizioni Mediterranee, 1988, pg 182.

[204] Schwartz, E. 1971. *Overskill—The Decline of Technology in Modern Civilization*. Chicago, IL: Quadrangle Books, pg 171.

well-defined operative types, "nature" represents very different things: for the scientist "nature" corresponds to the sets of empiric data provided by instrumentation, and for the engineer "nature" corresponds to all physical elements susceptible to being technically configured.[205] The conceptual distinction of nature/civilization or organic/inorganic is also inapplicable when embracing this issue. For instance, cattle is made of flesh and blood, which can strictly be considered as "organic" living tissue, but in terms of how cattle react to given stimulus the responses are mostly mechanistic, what ultimately allows cattle to be dominated by the technician or scientist. Despite its symbol being a black stallion, a Ferrari is not an "organic" being as it has no living tissue, but its "life" at hundreds of HP cannot be dominated by the scientist or technician, but by a *pilot*, whose responses are not of a mechanistic kind, but determined by instinct. *"There is neither man nor nature, but only a process that produces the one in the other and that connects machines"*[206] states Deleuze and Guattari.

Deep down, it is not completely misleading to say that we can only exert dominion upon what we are, but this assumption already carries us to the thorny issue of "who we are," and if we are actually embracing the synthetic spirit capable of giving sense to all life phenomena, and even to all philosophical forms, as Spengler points out when writing: *"Alles Verganzliche ist nur em Gleichms [all what is transitory is symbol] applies also to every genuine Philosophy as the intellectual expression of a being, as the actualization of spiritual possibilities in a form-world of concepts, judgments and thought-structures comprised in the living phenomenon of its author."*[207] In the former example, both the scientist and the technician necessarily adopt an impersonal, objective and mechanistic perspective on reality which allows their work and developments to be shared with

---

[205] Non vi è una scienza perché vi è realmente una natura ordinata e oggettiva, ma vi è una natura ordinata e oggettiva perché vi è una scienza [There is no science because there is in truth an orderly and objective nature, but there is an orderly and objective nature because there is a science] Julius Evola. *Fenomenologia Dell'individuo Assoluto Iii* ed. corretta: Edizioni Mediterranee, Roma 2007.

[206] G. Deleuze y F. Guattari: *L'Anti-OEdipe*, Minuit, 1975, pg 8.

[207] Oswald Spengler. *The Decline of the West*. Alfred A Knopf. 1926. New York, pg 367.

other professionals in their field. In the case of developing systems for controlling "cattle-type" elements of nature, such an impersonal, mechanistic approach is adequate and efficient. Hence, whenever any phenomena are characterized by objective and mechanistic responses to external stimuli, the phenomena can be easily dominated and integrated, like in the case of the Pavlovian dogs, thus implying that mechanistic activities developed by corporate scientists and technicians can be controlled too. But, if this is so, *who is actually controlling the activities and developments carried out by the operative type of the scientist and the engineer?* It is at this point that Evola presents us with his concept of the "I" which is also impersonal but not necessarily mechanistic. Therefore Evola's "I" corresponds to a configuring power of the operative domain that cannot be thought, but only experienced through the senses. One can try to define this "I" with words, but it would be as impoverishing as intending to describe a music piece with concepts.

* * *

The Masters of the 21st century shall no longer provide books to the students, but techniques, methods, discipline, and ritual. Such *praxis* shall allow the individual to configure the senses so all mechanistic reactions are transcended and the attainment of progressively integrated levels of freedom can be accomplished. The petty "i" and ego of the modern individual, stuffed with many artificial elements and imageries, shall no longer have to constitute an expression of defeat and privation, but shall be burnt into a realistic and solid "I" connected to the vertical and centripetal developments that take place once the global Techno-System exhausts its historical operative function. Once comprehended, all issue of politics and ideological "-isms" correspond to a mere provincialism of the 19th and 20th century, necessary only for those individuals who still need to justify their defeat. For other individuals who aim to participate with an Operative "I" instead of a "i," the door shall be open for them for Politics, not politics. As Spengler points out: *"he who is obsessed with the idealism of a provincial and would pursue the ways of life of past ages must forgo all desire to comprehend history, to live through history or to make history.*"[208] Because of the nature of the development we are referring to, the historical processes will not be uploaded to the Internet, and definitely won't appear on CNN.

From the perspective adopted by Evola in his philosophy it is much more convenient to start getting accustomed to the idea that an individual doesn't *think* or *act*, but that specific modes of power *make use* of the individual in order to activate the presence of a given thought pattern or pattern of action.[209] Such modes of power are characterized by a scale and hierarchy (*gunas*[210]) characterized by laws and determinisms which condition *how* they are developed, their *form*, and the forms are determined by metaphysical *figures* or *attractors*. This can also be applied to cultural patterns as a whole, as expressed by Rupert Sheldrake when writing: *"All the patterns of activity characteristic of a given culture can be regarded as morphic fields."*[211] In the case of ritual and games (*ludi, certamina*) these morphogenetic powers (*numina*) emerge by expressing their particular attributes. Ultimately the Individual and the individual constitute *mediators* for the development of the cosmos, where the specific modes of manipulation and processing of the elements of nature are determined by the specific *type* that characterizes the

---

[208] Oswald Spengler. *The Decline of the West*. Alfred A Knopf. 1926. New York, pg 38.

[209] Leggi, norme, convenzioni, imperativi, doveri, ecc. sono come dande per sostenere l'atto di un Io non ancora interamente mediato, di un Io che è ancora fuori di sé, sotto la legge dell'idealità [Laws, rules, conventions, imperatives, duties, etc. are questions for supporting the act of an I not yet fully mediated, of an ego that is still beside itself, under idealistic laws] Julius Evola. *Fenomenologia Dell'individuo Assoluto Iii* ed. corretta: Edizioni Mediterranee, Roma, 2007.

[210] In Samkhya philosophy, a guna is one of three "tendencies, qualities": sattva, rajas and tamas. This category of qualities have been widely adopted by various schools of Hinduism for categorizing behavior and natural phenomena. The three qualities are:
Sattva: the quality of balance, harmony, goodness, purity, universalizing, holistic, constructive, creative, building, positive, peaceful, virtuous.
Rajas is the quality of passion, activity, neither good nor bad and sometimes either, self-centeredness, egoistic, individualizing, driven, moving, dynamic.
Tamas is the quality of imbalance, disorder, chaos, anxiety, impure, destructive, delusion, negative, dull or inactive, apathy, inertia or lethargy, violent, vicious, ignorant. Alter, Joseph S., *Yoga in modern India*, 2004, Princeton University Press, pg 55.

[211] Rupert Sheldrake. *Morphic Resonance. The Nature of Formative Causation*. Park Street Press.

individual, just like the actions developed by enzymes, bacteria, or any other micro-organisms that function to serve the life of a given being. Marshall McLuhan viewed this process in the following way: *"Man becomes, as it were, the sex organs of the machine world, as the bee of the plant world, enabling it to fecundate and to evolve ever new forms."*[212] Hence, it is misleading to think that the individual can decide where to carry the development of a given activity. For instance, within all those organizations that integrate and develop activities of production, the hiring of an individual by recruitment agents is focused on determining the experiential background of the Individual fits the purpose of the organization. Everything strictly individual is considered less relevant, to the point that what really matters is a correct integration of the individual through the application of specific techniques and methods. The so-called individual "creativity" of our times is also integrated under very strict limits; in the film industry, scripts came from original and creative minds, but scripts have to fit into standards of production and ideological requirements. In modern sports, the capriciousness of the individual is also eliminated.

* * *

In the experiential context of the observatory there is an inevitable idealist approach very which cannot be discarded. We are referring to the idea that dominion of reality on the part of the "I" still corresponds to an ideal, an ideal that is necessary, but can only be into certainty to the extent that the individual can empirically verify actual dominion of reality at a given level of development.[213] The important difference between Magical Idealism and Transcendental Idealism, is that in the case of those who follow the postulates of the latter, if they ever found themselves in the former observatory, they would be more prone to move to the library and study

---

[212] Marshall McLuhan. *Understanding Media. The Extensions of Man.* The MIT Press Cambridge, Massachusetts. London, England, 1994.

[213] *"What is first is precisely the finite and the particular"* Julius Evola. *L'individuo e il divenire del mondo*, Libreria di Scienze e Lettere, 1926, pg 27.

all volumes present, and their "research" would carry them to scholastic knowledge according to a set of cultural assumptions which could be then taught to those who have based their minds on the framework of *the "I" they are working for*. The key intellectual conditioning caused by devoting one's activity and time to a specific organization is so intense that Upton Sinclair once stated that: *"You'll never succeed in making someone understand something if their salary depends on them not understanding it."* Yet the extremely individualistic "I" we are referring to here is no longer a Magical "I" but that which has been called in philosophy as a "transcendental I," an "I" which is accessible to everyone, very much related to Kant's "practical reason." While the Magical "I" allows all phenomena to converge effectively, the position of the "transcendental I"[214] favors the production of texts and the dispersion of all concrete creativity, to the point that Yukio Mishima wrote that the *"purely intellectual convolutions were as yet nothing but the entangling of themes within the prelude to a human life that so far had achieved nothing."*[215]

The radical standpoint of Julius Evola's philosophy, which converges with the core principles of the ancient mysteries, Zen, Buddhism, and many other spiritual traditions, is that those scholars who penetrate the observatory and do not aim to gain any operative or experiential knowledge on the production of the telescopes, are following the *path of privation, the path of alterity, the path "of the object" or that of the other,* the path of *spiritual and cultural impotence.*

---

[214] Pragmatisticamente, il criterio della determinazione del logos sarebbe la comodità; vero e razionale risulterebbe essere quel sistema che si presenta più comodo o utile ai fini di una sistemazione complessiva dell'esperienza e di un coordinamento delle possibilità dell'azione. Ma cosi il problema è soltanto spostato. In primo luogo, bisogna domandare perché si debba preferire il comodo (a che basso livello si finisca per tale via, ognuno lo vede: si potrebbe parlare di una filosofia della pigrizia o del comfort come principio trascendentale) (...)«comodo» e «utile," non possono aver nessun significato autonomo (utile, in relazione a che e a chi?) [From a pragmatic viewpoint, the criteria of determination on the part of the logos would be that of comfort; true and rational would end up being whatever system that is presented as comfortable or useful aiming to a systematic complexity in terms of experience and a coordination of the action alternatives. Yet this way the problem is merely relegated. In the first place, it is require to ask oneself why the comfortable positions ought to be preferred (to the debased level one ends up attaining through such path is easy to see: one could thus refer to a philosophy of laziness or comfort as a transcendental principle (...) "comfortable" and "useful" are not characterized by any autonomous signification (useful, in relation to what and who?) Julius Evola. *Teoria Dell'individuo Assoluto*. Edizioni Mediterranee, 1988, pg 110.

[215] Yukio Mishima, John Bester. *Sun and Steel*. Kodansha USA (2003), pg 25.

This is the case, because according to Magical Idealism, *one has to be what one knows*, one has to transform the configuration of the senses in order to follow the perceptive determinations of a new "I." The degree and levels of experience are therefore determined by the specific character of the "I." Meister Eckhart describes this with an appealing example:

> *"suppose a burning coal is placed in my hand. If I say the coal burns me I do it a great injustice. To say precisely what does the burning, it is the "Not." The coal has something in it that my hand does not. Observe! It is just this "Not" that is burning me—for if my hand had in it what the coal has, and can do what the coal can do, it, too, would blaze with fire, in which case all the fire that ever burned might be spilled on this hand and I should not feel hurt."*[216]

The issue of absolute certainty, which was the main focus of the idealist philosophical currents of the West, could only resort to an abstract set of categories as the criteria of objective truth. To begin with, the main problem that appeared with transcendental idealism is that the relation between subject/object became more divergent and mechanistic. In other words: the individual life of the scholar and his private affairs were considered to be totally separate from his object of study (nature, history, art, science, etc).[217] From an integral viewpoint on culture both domains are not separate. Even though our observatory director knows he is at the centre of the cosmos, he could never demonstrate such thing to a scholar, for a very simple reason: because they are working for the development of different "I"'s that do not necessarily converge in function and form.

---

[216] Meister Eckhart, *The Essential Sermons, Commentaries, Treatises and Defense*, trans. and ed. by Bernard McGinn and Edmund College, New York: Paulist Press, 1981. Speech 5b.

[217] l'idealista come uomo di solito resta un piccolo borghese il quale prova l'orrore del vuoto, che non sa dare un valore indipendente al proprio essere, che può ammettere bensì che «il mondo è la mia rappresentazione," ma non fino al punto di sentirsi solo, di non aver bisogno che altri soggetti, intorno a lui, siano. [The idealist, considered as a man remains a petty-bourgeoisie, which proves the horror of the void, not knowing how to provide a value independent of one's being, capable of admitting that "the world is my representation," but not to the point of feeling alone, of not having the need of other surrounding subjects] Julius Evola. *Teoria Dell'individuo Assoluto*. Edizioni Mediterranee, 1988, pg 180.

Let's say we have a picture of a new species of flower that is scientifically described and all its phenomena are expressed through mathematical laws. How can we know if the flower corresponds to something real? Evola, when developing his philosophy, readdresses Hegel's claim that "all that is real is rational and all that is rational is real," and shows that an idealist Hegelian claim is extremely limited. This is because from the viewpoint of the transcendental "I" of the idealists, a flower is considered real because it is rationalized, therefore it not only exists, but also the "I" exists, verifying the classical statement of Descartes: *"I think, therefore I am."*[218] Yet this existing "I" corresponds to a very arbitrary way to grasp the reality of the flower.[219] The limitation of this "transcendental I" is that *it can not place the flower, it can not produce it.*[220] Mathematical laws are still powerless to explain the essence of the flower, that is, its *being*. Hence the "I" doesn't correspond to a Magical "I." In the case of the observatory, however, the "I" we were referring to is strictly Magical, since the workers actually *produce the stars, and place them.* The specific medium of a Magical "I" produces what is real, allowing both the subject and object to converge. By being capable of perceiving a representation of a new species of flower in a book, and to understand this flower through its scientific representations does not correspond to the activation of a Magical "I," but just to the appropriation of features based on the "I" of the

---

[218] Si deve dire non «Io penso,» bensì «L'Io pensa un pensato, pensa qualcosa, quindi io che penso e la cosa pensata sono» [One must not say "I think," but rather "The I thinks a thought, thinks something] Julius Evola. *Teoria Dell'individuo Assoluto*. Edizioni Mediterranee, 1988, pg 183.

[219] Il «cogito» interpretato in termini razionalistici esprime dunque, come senso, la formula della non-centralità. [the "cogito," interpreted in rationalistic terms expresses however, as a meaning, the formula of non-centrality] Julius Evola. *Teoria Dell'individuo Assoluto*. Edizioni Mediterranee, 1988, pg 45.

[220] Quale è la differenza—abbiamo detto—fra una cosa reale ed una imaginata? Rappresentate, lo sono tutte e due egualmente, e dal punto di vista logico od essenziale, non vi sono note nell'una, che l'altra non abbia. Ma di là da ciò, l'attività rappresentativa a cui corrisponde la cosa reale è una attività rispetto a cui sono impotente. Questo è tutto [What's the difference—we said—between a real thing and an imagined? Represented, they are both the same, and from the logic point of view or that of what is essential, there are no aspects in one, that the other hasn't got. But beyond that, the activity itself of representation that corresponds to the real thing is an activity with respect to which there is impotence. This is all] *Fenomenologia Dell'individuo Assoluto Iii ed. corretta*: Edizioni Mediterranee, Roma, 2007.

speculative idealists.²²¹ The only way a Magical "I" could arise out of a situation like this, would be if the individual who contemplates the representation of the flower can also identify completely with the *essence* of the flower, that is, to be capable of *loving* the flower due to what the flower *is*, and not due to how the flower is represented (the medium). The identification with the essence, whenever it miraculously takes place, has an initiatory capacity. The potential initiate inevitably realizes the illusory and deceptive character of all religious, ideological, and scientific constructs based on the rigid categories, and which all mediate the experiential cosmos for the average individual. Evola expresses this state of mind when writing:

> *"there is nowhere to go, nothing to expect, nothing to fear, nothing to ask. You yourself, such as you are, you are the eternal, are the Lord of the gods, the Aeon of aeons—all in all, composed of all powers."*²²²

Potential psychological and psychosomatic traumas can be just around the corner. As an antidote to this psychological state, noble traits, a profound sense of courage, and a natural predilection towards heroic realism can, however, come here to the rescue... As Ernst Jünger writes: *"There are no flags except those you wear on the body. Is it possible today a faith without dogmas, a world without gods, a knowledge without maxims and a homeland that can not be occupied by any worldly power?"*²²³

---

[221] Dunque, fin qui il concetto cercato e sfuggito: né il non-impedito; né la causa sui come spontaneità, né la natura cosciente, né il finalismo, né la determinazione morale, né le varietà della libertà razionale in sede di apriori od anche in sede semplicemente psicologica o fenomenologica sanno fondare una relazione, in virtù della quale l'idea di libertà acquisti un senso davvero definito di contro a quella di necessità. Occorre dunque tentare ancora un passo. [Hence, once the concept is captured and released: it is neither the non-prevented; neither the cause of itself as spontaneity, nor the conscious nature, nor finalism, nor moral determination, nor the varieties of rational freedom as to a priori or even when simply psychological or phenomenological in order to establish a relationship, whereby the idea of gained freedom acquires a defined when faced with necessity. therefore it is required to move forward] Julius Evola. *Teoria Dell'individuo Assoluto*. Edizioni Mediterranee, 1988, pg 117.

[222] Niente dove andare, niente da aspettare, niente da temere, niente da chiedere. Tu stesso, tale quale sei, sei l'eternità, sei il Signore degli dèi, l'Eone degli eoni—tutto in tutto, composto di tutti i poteri. Julius Evola. *Fenomenologia Dell'individuo Assoluto Iii* ed. Corretta: Edizioni Mediterranee, Roma, 2007.

[223] "Es gibt keine Fahnen außer denen, die man auf dem Leibe trägt. Ist es möglich, einen Glauben ohne Dogma zu besitzen, eine Welt ohne Götter, ein Wissen ohne Maximen und ein Vaterland, das

In the case of a non-traumatic activation of a Magical "I" caused by the identification with an essence present in the cosmos, the "I" is in a cosmic and homeostatic configuration, a divine "I" which is immanent and not speculative, or strictly speaking: a *personality*.[224] If the Magical development progresses adequately through practices and disciplines, then the senses of the individual shall substantially mutate according to a transformative process conceived as *metanoia*,[225] and the "I" shall not only capture new phenomena but also *produce new phenomena*, which from a human viewpoint can only be considered as miraculous. The Magical "I" develops new faculties of perception of the immanent domain: new relations, new connections, and new homeostatic organizations that couldn't be perceived formerly. The Magical "I" thus develops a *potency of the senses*, which causes the mind to surrender to the power. The discursive mind itself corresponds to the psychological attribute that develops all subjective representations, yet the representations correspond only to the "shadows" of the objects of the cosmos, and these objects can be characterized by countless forms, depending on the nature of the light that illuminates them, thas is, the nature and values of the "I." Evola writes:

> *"What is the difference between a real and an imaginary thing? Represented, they are both the same; but beyond that, the representing activity to which the real thing corresponds is an*

---

durch keine Macht der Welt besetzt werden kann?" Ernst Jünger. *Der Arbeiter*. Ernst Klett, Stuttgart 1981. Klett-Cotta, pg 46.

[224] La personalità implichi un organismo, centro delle relazioni dinamiche e spazio-temporali – e ciò non nel senso che prima o separatamente sia data una organizzazione, bensì nel senso che se l'oggetto dell'affermazione trascendentale è il valore relativo alla personalità, essa porrà e comprenderà altresì una organizzazione, sia pure come momento ed astratta materia [personality implies an organism, a centre of the dynamic and space-time relations, and not in the sense that in the first place or separately is provided an organisation, but rather in the sense that the object of transcendental affirmation is the value relative to such personality, which can place and integrate as well an organisation, in a pure sense as in that of abstract issues] Julius Evola. *Fenomenologia Dell'individuo Assoluto Iii* ed. corretta: Edizioni Mediterranee, Roma 2007.

[225] From the Greek term μετανοῖεν. For Peter Senge (Peter Senge. *La quinta disciplina*. Barcelona: Granica. 1995) to grasp the meaning of metanoia is to understand what it means to learn in relation to metacognition. It is a change of focus a change of perspective from one to another, which in turn is related to perception.

*activity in respect to which they are impotent. There are elements on which I cannot act. This is all."*[226]

This causes us to conclude that Magical Idealism is only slightly concerned with *understanding*, and more concerned with *doing*, by action, and what's more the *underlying forms of action*, that is, in developing operative *techniques* which express correlations between a given "I" and the reality the "I" dominates through the centripetal power of its own essence. Wishful thinking or idealisms are tossed to the fire here, for *"Life is too short and too beautiful to sacrifice it for ideas,"*[227] as wrote Jünger. This standpoint appears radical to a huge percentage of scientists, scholars, and other groups. But the magical "I" has no concern with opposition, since it is beyond time and space. An oak shall always be an oak, a rose a rose, a lion a lion, and an eagle an eagle.

\* \* \*

All the previous explanations allow us to introduce ourselves to how traditions that are based on operative procedures approach the issue of certainty. The issue does not primarily resort to philosophical, religious, or scientific discourse since all these subjective domains are considered secondary and can only be developed coherently later.[228] Based on the medieval operative premise *Post Laborem Scientia*, which can also be expressed by resorting to Nietzsche's idea:

---

[226] "The Individual and the Becoming of the World" was originally presented as two lectures to the "Lega teosofica indipendente di Roma" ("The independent theosophical league of Rome") in 1925.

[227] Ernst Jünger. *Eumeswil*. The Eridanos Library, 1980.

[228] Quanto alle varie «ragioni,» esse non vengono che dopo, sono elementi determinati e non determinanti, ovvero determinanti solo accessoriamente. «Spiegare» quella posizione originaria equivale dunque a negarla, a non ammetterla, cioè a pretendere che tutta la serie poggi sul nulla. In tali casi si può ben dire che il filosofo non pensa il proprio sistema, ma viene pensato [in terms of all the multiple "reasons," they don't appear but afterwards, they are determined and not determining elements, or determining only in an extremely peripheral way. To "explain" such primordial position is thus equivalent to deny it, to not admit it, that is, to pretend that all the entire series is based on nothing. In such case, one could affirm that the philosopher hasn't thought the system, but has rather been thought by it] Julius Evola. *Teoria Dell'individuo Assoluto*. Edizioni Mediterranee, 1988, pg 134.

*"We can comprehend only a world that we ourselves have made."*[229] The focus is therefore on developing techniques, methods, and disciplines that allow the individual to operate under the magical character of the "I." When referring to Aristotle's idea on *presencing (energeia)*, Heidegger refers to the operative (*operatio*) aspect of work [*wirken*] which unveils the essence of reality [*Wirklichkeit*]. The German philosopher writes:

> *"that which is brought hither and brought forth now appears as that which results from an operatio. A result is that which follows out of and follows upon an actio: the consequence, the outcome [Erfolg]. The real is now that which has followed as consequence."*[230]

When introducing the Magical "I," we demonstrated that the "I" is an *operative factor*. Its existence can only be proven in terms of the form of actions and developments[231] exerted by the individual in the domain of the immanent. In this sense all egotistic modes of individuality and separation have to be progressively "burnt"—a sacrificial dynamic Evola referred to as the *"tragic principle"*[232]—allowing consciousness to expand towards the objects of direct experience. To think that there is a separation between the individual and the cosmic determinations of surrounding objects still constitutes the burden of an "I," which is still conceived of as a "thinking I" and not a Magical "I." As Evola shows, even *"the domain of the spiritual may well look like the kingdom of order,*

---

[229] Friedrich Nietzsche, Walter Arnold Kaufmann, R. J. Hollingdale *The Will to Power* 1968, pg 272.

[230] Martin Heidegger. *The Question Concerning Technology.* Garland Publishing, Inc. New York & London 1977, pg 158.

[231] Si hanno due concetti di libertà e necessità interconnessi, ma anche, nello stesso punto, contrapposti. Essi risultano da due opposti rapporti funzionali tra la forma e la materia dell'atto, i quali, trascendentalmente, riportano al valore come affermazione e al valore come negazione [there are two concepts of freedom and necessity interconnected, yet also, in the same point, opposed. Such results from the two opposed functional relations between the form and matter of the act, which transcendentally refer to value as affirmation and value as denial] Julius Evola. *Teoria Dell'individuo Assoluto.* Edizioni Mediterranee, 1988 pg 128

[232] "tragic principle (...) to drag out from his own interior a principle that can secure a new reality beyond the order of appearance and mere representation, in which every thing up to now had to be submerged" Julius Evola. *L'individuo e il divenire del mondo,* Libreria di Scienze e Lettere, 1926, pg 19.

*peace, eternal oneness."* Evola goes on to remark that, *"to instead he who conceives what is universal as a test, as a point of no difference beyond which, wider and renovated, made free and divine, must be found the mode of the "I," such domain appears rather as a set of wild, raw, abyssal, blissful and terrible powers not governed by any law, reason or providential plan, but taken as an interplay of tensions, with respect to which every struggle in the material plane is not but a reflection."*[233] Hence the importance granted to the "I" as an operative factor might cause some readers to think that in development of the Magical "I" the personality of the individual doesn't count at all. However, the thought itself reveals a separation on the part of the individual who presupposes the hypothesis, since the traits that determine our individual actions have practically no relation to an egotistic construct of the individual or the self, but to a large extent are inherited by lineage/family traits and confirmed by social conditioning. As we know, one is a "mask," and the "face" is another. The individual who aims to relate to the Magical "I" is invited to discard all individualistic masks, and act on his or her true nature. By approaching an operative/magical discipline, the pupil who creates harmonies with the true origins of his life will facilitate the relation to the operative aims, which becomes present operatively in the modes of gesture, action, speed of movement, strength, and accuracy of the senses, confirming what Ernst Jünger pointed out when writing: *"the secret meaning of an animal is revealed most clearly in its movement."*[234]

Modern psychology, focused on the self and by not on established criteria of mental health beyond the character of the individual as a social construct, has completely discarded the transmuting

---

[233] Il mondo dello spirito può bene apparire come il regno dell'ordine, della pace, dell'eterna unità – a colui che invece l'universale ha come una prova, come un punto di non-differenza di là dal quale, più vasto ed intenso, rinnovato, fatto libero e divino, deve ritrovarsi il modo dell'Io, esso appare invece come un insieme di potenze allo stato libero, nude, voraginose, potenze beate e terribili ed un tempo non rette da alcuna legge, ragione o piano provvidenziale, ma riprese in un giuoco di tensioni, rispetto a cui tutto ciò che è lotta del piano materiale non è che riflesso. Julius Evola. *Fenomenologia Dell'individuo Assoluto Iii* ed. corretta: Edizioni Mediterranee, Roma 2007.

[234] "Die geheime Bedeutung eines Tieres am klarsten in seiner Bewegung offenbart" Ernst Jünger. *Der Arbeiter*. Ernst Klett, Stuttgart 1981. Klett-Cotta, pg 22.

power of developing the Magical "I." In literary accounts, this is expressed when European Knights and Eastern Samurai's confessed that their best friends were their weapons; objects constituted an immanent and magical expression of their "face" and not of their "mask." Even today, in the case of some bikers or pilots we can detect that same trait. Based on today's aesthetic criteria some warriors could be "good-looking," others could be "ugly"; at a deeper level, they also reveal somatic and physiological differences. In terms of the development of an operative/magical discipline, none of these elements are ultimately determining.

Whenever an individual, like in the case of Eugen Herrigel, aimed to have a *first-hand* experience of the operative power of Zen, the first issue is that of a "grounding" experience, impeding all thoughts that chaotically interrupt procedures and perceptions of reality. In order to make the grounding feasible, Master Awa Kenzo provided Herrigel with the arc and bow, as the first connection to a reality that from an operative viewpoint is pluridimensional, not in terms of categories of thought, but in terms of *categories of experience*.[235] When referring here to a "pluridimensional" character, the features can be equated to a "pluri-paradigmatic" character, by recalling the importance of *paradigms* as operative factors. Each way of disciplining a given habit that relates to the cosmos corresponds to a way of perceiving reality *through the configuration of the senses and the "extensions of man"* in McLuhan's terms, not through the vision provided by the eyes, nor the conceptual distinctions determined by the subjective mind, but rather through a set of relations where the qualitative attributes of subject and object have to be the same in order to create a sensible identification. Hence, in the same way, thirst or hunger are not caused by water nor food and are not a product of the condition of the mind. Modes of experience from the senses at a larger range are determined, *placed*, by the "I." For instance, if an individual enters a domain where due to specific weather

---

[235] Abbiamo detto che la fenomenologia tratta essenzialmente di significati, di modi secondo cui l'atto è vissuto [we had said that phenomenology is essentially about meanings, of the ways by which an act is lived] Julius Evola. *Fenomenologia Dell'individuo Assoluto Iii* ed. corretta: Edizioni Mediterranee, Roma 2007.

conditions water becomes scarce, such an environment will modify the "awakening" of given receptivities of the senses on the part of the individual, and the feeling of "thirst" shall be triggered, and other impositions of the mind become secondary to this experience. In the case of paradigms something very similar takes place: the individual is placed in a set of work conditions where the environment triggers, stimulates, and awakens specific configurations of the senses. If new trans-physiological currents and connections are established between the individual and the environment; this "intelligence of the senses" arises in operative ways whenever the conscious mind relaxes. These modes of intelligence constituted embedded reflexes or what Arnold Gehlen referred to as *automatisms*,[236] and have very specific configurations. In the case of these configurations we can state that "by their fruits ye shall know them," that is, they reveal their nature when body language relates to specific stimuli and environmental conditions. Yukio Mishima wrote in this regard: *"Both body and mind, through an inevitable tendency that one might almost call a natural law, are inclined to lapse into automatism, but I have found by experience that a large stream may be deflected by digging a small channel."*[237] As we shall explain later, these embedded reflexes are related to the concept of *technique* and *work*, where a given mode of processing and transforming reality is effectuated by the given character of an "I."

---

[236] Arnold Gehlen (1904-1976): "If we analyze what is truly fascinating in the two phenomena (magic and technology), such should be the automatism. It can be shown how little we understand technique when we apply on it the epithets of utility or power, in addition to the fact that the fascination of the automatism is independent of its achievement. (...) However, every machine has however has the appearance of an entirely understandable and rational automatism, and this poses a major problem. In the imagination of primitive peoples magical forces are not arbitrary or spontaneous, but constitute an animated automation inserted in all things which can be triggered with the right formula. Certainly there is still a remnant of this concept in astrology, which also involves the immense automatism of the stars that in their rotation determines destinies. (...) However, it seems that this fascination with automatism is not attributable to purely intellectual satisfaction or a somehow susceptible instinct defined. From all that happens in the unfathomable human soul, we can only scientifically rationalize some partial areas which we can not integrate theoretically. But from what we know of the mind, the intellect, of instinctual residues, etc., we can not explain the fascination caused by automatism, so that here we introduce a new psychological category: such fascination is a resonance phenomenon" Arnold Gehlen (1904-1976), *Antropologia Filosófica*, Editorial Paidós Argentina (1993) pg 117. Translated from Spanish edition by Miguel A. Fernandez.

[237] Yukio Mishima, John Bester, *Sun and Steel*, Kodansha USA (2003), pp 21-22.

When a scientist develops his activity in a laboratory, the interaction with the environment is "forging" or "molding" predispositions and automatisms in a very specific way. The scientist can obviously come up with many justifications about his activity, but even though there can be lots of different interpretations in this regard,[238] what is undeniable is that if he feels attachment to the activity itself, and this attachment reveals the establishment of fixed automatisms and operative correlations between him as an individual and the scientific institution as a whole. Ultimately, the institutional value and economic rewards granted to the individual in the scientific environment depends on the amount of social power and appropriation of resources the institution gains thanks to the work developed by the scientist; in this context, a given organization or institution corresponds to an overall mode of disciplining the paradigmatic traits of the individual.

When approaching operative training modes, it appears here quite clear that a given discipline is always integrated based on the determinisms of a given "I" embodied by tradition and in some cases even by a Individual or Master. We used the example of the scientist in order to show the importance of paradigm training under the configuring power of an organization, or in other words, the importance of *praxis*, especially in the case of those sectors of workers who are primarily considered to be producing knowledge. There is a difference between the diverse *praxis* or *techniques* and the way an Operative Tradition like Zen configures traits. In the case of an Operative Tradition such as Zen, the individual enters the discipline and in order to further progress, he cannot leave aside his being, his fears, and existential traumas. As we shall point out when dealing with the issue of *technique*, most of the operative factors that are highly active in modern corporations foster the predominance

---

[238] Nella corrente speculativa, cui ci riferiamo, si ammette che la serie sviluppata dalla riflessione, dal logos, ha valore di un ritrovamento e non di un vero porre , di semplice elevazione delle necessità di fatto a necessità di ragione , di una spiegazione e non dell'articolazione di un processo reale [In the speculative currents we referred to, it can be accepted that the development of series of reflections, or logos, has the value of a discovery and yet not of an authentic placing, from a simple elevation of the factual needs to reasonable needs, of an explanation and not of any articulation of an actual process] Julius Evola. *Teoria Dell'individuo Assoluto*. Edizioni Mediterranee, 1988, pg 106.

of work profiles where all private matters become sublimated so that private affairs don't interfere with the functional elements in the organization. When it comes to the individual's development in an Operative Tradition, the latter modes of psychological compensation do not favor, but rather fortify long-term stagnation. So the individual who has the chance to be admitted into a school based on traditional operative principles, must realize that the whole aim is to ultimately attain the center of his *being*, and to acknowledge that the specific disciplines and methods present in the tradition are intended to "distill" the essence, to empower it and to release it. Every single authentic life experience can be put into the operative "distiller," no matter how dramatic, tragic or painful the experiences are.[239] They are all affirmed, in symbolic movement, gesture, and art. Another difference is that while in most corporations it becomes more difficult for the worker to realize the core institutional value, and to thus relate to it in practice—causing the worker to surrender to economic determinism—in the case of an Operative Tradition, the center *is within the reach of the student's hand*. It's quite understandable that the loss of individuality imposed by highly systematized organizations causes most workers to identify with the "I" of the organization ("WE produce this," "WE develop this," etc.). Their world is ultimately perceived through such an "I." But do workers have the chance to realize first-hand if the work methods and policies "distill" their nature as human beings?... Only indirectly, through the internal campaigns promoted by staff and departments, which shows how the issue of value is mediated, and that there is uncertainty in terms of this value.

In the case of an Operative Tradition the issue of value is no longer speculative, but operative: it can be actually touched and sensed, the distance from the core center of the tradition can be perceived by the practitioner by learning to breathe, and just by practicing the disciplines. It can be affirmed that Operative Traditions do not aim for dispersion of the "I," impeding the natural traits of the individual

---

[239] Nulla saprebbe esser sperimentato che non incorpori già, in un certo grado, il momento del valore (nothing can be experienced which doesn't already incorporate, to some extent, the moment of value) Julius Evola. *Teoria Dell'individuo Assoluto*. Edizioni Mediterranee, 1988, pg 28.

to become passively mobilized by countless and dispersed centers of power that can be even contradictory, though it is biblically stated that *"No one can serve two masters"*[240] Rather an Operative Tradition defines a central value of the "I" around which the individual can progressively concentrate all efforts in one single direction, aiming towards his being.

In the case of these Operative Traditions, it always provides the student with a material object or set of objects as a support for developing the whole process, corresponding to elements that bridge consciousness to the world of becoming. It can be an arc and bow (like in the case of Master Awa Kenzo), a tool, weapon, or even the body like in martial arts and dance. Such development could be accomplished perfectly well without any external object, yet the manipulation of a given object allows one to relate to the concept of *technique* which is the key factor in all Operative Traditions.

Symbolically, there are considerable differences between the way a scientist relates to nature in a laboratory and the way an Operative Tradition provides mediations with nature to the student. Scientists are encouraged to work in highly controlled environments, where it is technically necessary to produce and share scientific information that can be reproduced in other research centers or laboratories. In the case of scientists or technicians, the ensemble of elements that mediates and determines their activity, hence their activity becomes enclosed into the specifics of such environments. Everything these professionals know about the cycles of the cosmos is distorted by the ensemble which reinforces mechanistic time-schedules and work patterns that crash against any potential resonance emerging in the individual that transcends the ensemble.[241] A typical example of this is the artificial lamps that are installed in buildings, providing light during a time that *in terms of non-mediated direct*

---

[240] Matthew 6:24 "No one can serve two masters. For you will hate one and love the other; you will be devoted to one and despise the other. You cannot serve both God and money."

[241] Per la scienza l'esperienza è ciò a cui il nostro spirito non comanda, ciò su cui i nostri desiderî e la nostra volontà personale non possono avere presa, ciò che è dato e che noi non facciamo [For science experience is that which our spirit does not command, that upon which our desires and our personal will can not grasp, what is given and that which we do not do] Julius Evola. *Fenomenologia Dell'individuo Assoluto Iii* ed. corretta: Edizioni Mediterranee, Roma 2007.

*experience* is characterized by darkness. This mediating ensemble of configurations between the qualitative attributes of the subject and the qualitative attributes of the cosmic objects is easy to discern, yet as we formerly explained, such environments tend to "mold" the paradigms of an individual to the extent that, even if this ensemble disappeared, the intrinsic character of these forged paradigms would still be operatively present in the individual, in his natural reflexes and nervous predispositions. Therefore, the mediation can have a visible or invisible character. The latter idea, which is very challenging to explain conceptually, is more real for those individuals who have been shaped by military institutions, and who have managed to feel what Jünger refers to as *"the unity of men with their means as the expression of a higher unified nature."*[242] In civil society, however, where economic and abstract thinking is the rule, this is extremely difficult to realize, even for the technicians, engineers, and scientists. However, there exists a quite good depiction of this deep phenomenology of the reflexes, which is that of the "cyborg." We can conceive of the "cyborg" as an anthropomorphic figure in which specific human functions are prolonged and made more efficient by means of biomechanical prosthesis. The first modern writer to explore this systematically was Ernst Kapp, in 1877. Kapp speculated that railroad systems unconsciously mimicked the circulatory system, while telegraph lines extended the nervous system. Marshall McLuhan expounded the same idea in his first major work on technology, *Understanding Media*, which bore the subtitle *The Extensions of Man*. *"It is a persistent theme of this book (…) that all technologies are extensions of our physical and nervous systems to increase power and speed."*[243] If we were willing to take a photograph of modern scientists, by using of a camera which reveals the physical "I" (the ego and the self of modern psychology), and also the operative and Magical "I," it would be figure of the cyborg what would emerge. This depiction clearly raises queries on the

---

[242] "Die Einheit des Menschen mit seinen Mitteln ist daher der Ausdruck einer Einheit von übergeordneter Art." Ernst Jünger. *Der Arbeiter*. Ernst Klett, Stuttgart 1981. Klett-Cotta, pg 119.

[243] Marshall McLuhan, *Understanding Media*, p. 90.

aforementioned ephemeral character of the "individual." *Are you what you think?* Well, that can be changed by the subtle introduction of cultural standards through the media. *Are you what you feel?* That can be changed by providing pharmaceuticals. *Are you what you like?* That can be changed by clicking "likes." *Are you your material possessions?* That can be changed by a global economic crisis. *Are you your status?* That can be changed by providing a disinformation campaign. Hence, in very few human beings, the realization of the ephemeral character of the individual and the of all mediations has the potential of establishing a completely new relation with cosmic determinations. The latter realization for almost 100% of individuals can be as traumatic as taking a strong entheogen, but for those who aim to develop a Magical "I" this realization is always the point of departure, the beginning of a new journey. One might feel completely "naked" in this regard, but the "nakedness" can be considered as that of a newborn. So, by releasing people from these individualistic "clothes," at this point, it is necessary to become grounded in what Evola refers to as the *"naked principle of the Individual, which transcends the character of thought."*[244]

So in terms of an Operative Tradition like Zen Buddhism, the awakening of the Magical "I" requires a progressive realization of the implicit character of all mediations, and the acknowledgment of how the mediations determine our experience of the world and our consciousness. This is why in the case of Operative Traditions, the mediations have to be highly minimized in order to favor direct experience with nature. This is what allows knowledge, as Julius Evola conceives the concept *knowledge,* when considering it as *"the awareness of a given static or dynamic determination.*[245] The mediation is provided by an instrument, a tool, or a weapon, which corresponds to a microcosmic material expression of the spirit of the tradition, and is therefore hence of *resonating* with the tradition.

---

[244] "Il nudo principio dell'individuale, il quale trascende l'ordine del pensiero" Julius Evola. *Teoria Dell'individuo Assoluto.* Edizioni Mediterranee, 1988, pg 41.

[245] Conoscere, nel suo senso più generico di consapevolezza di una certa determinazione statica o dinamica (to acknowledge, in the most generic sense of awareness of a given static or dynamic determination) Julius Evola. *Teoria Dell'individuo Assoluto.* Edizioni Mediterranee, 1988, pg 31.

Contrary to the usual scientist, technician or engineer of our times, the student of an Operative Tradition is situated *in an open system* where all mediations that enclose the individual are minimized, and exposure to new phenomena is favored. Symbolically, the change of context is significant, since the mediation is so noticeably reduced, the student can actually feel in his hand the power of the Tradition, and, assisted by this tool, he can learn to dominate and persuade all elements of the cosmos, as an external expression of domination over himself.

The basic assumption of these preliminary requirements is that by regaining contact with the material domain and the body by minimizing the interferences caused by mediations, it favors direct experience. Even once the latter is accomplished, the most challenging mediations will always be the "invisible" ones, that arise from acquired habits and thought-patterns which condition the senses to experience reality in a very specific way. But the re-configuration of the senses is always facilitated by the configuration of diverse techniques, which are embedded in the practices and disciplines. In the case of Eugen Herrigel, he was provided with an arc and a bow, after which Master Awa Kenzo explained to his pupil that *"the bow encloses the "All" in itself (...) with the upper end of the bow the archer pierces the sky, on the lower end, as though attached by a thread, hangs the earth"* and that this was the reason he had to learn to draw it properly.

In a first approach, the establishment of the connection between the microcosm of the means and the macrocosm that Master Awa Kenzo refers to, is due to the fact that practically all instruments or weapons can be equated to what in physics is referred to as a *spring-damp-mass* system. In this particular case, the elastic feature of the bowstring corresponds to the *spring*, the arrow corresponds to the *mass* and the *"sharp crack mingled with a deep thrumming"* that emerges during the critical moment of release of the arrow can be equated to an expression of the *damp* phenomena. Master Awa Kenzo also made reference to the so-called "spiritual shooting" which corresponded to shots fired by Master Archers, that were visibly much more powerful and accurate, completely independent of the particular type of arc or bow. The important prerequisite for "spiritual shooting" to take place is the channeling of the entropy caused by the friction/damp factor induced by the thumb contact/resistance, so that all entropy production becomes channeled, entailing a negative-entropy process that "pushes" the arrow to one single point, thus converging with the symbol expressed in the arrow. However, in operative terms, this theoretical prerequisite is extremely difficult to accomplish during the moment of maximum physical tension, which takes place when the bow is drawn to its full extent. These are the explanations penned by Herrigel himself:

> "When drawing, the thumb is wrapped around the bowstring immediately below the arrow, and tucked in. The first three fingers are gripped over it firmly, and at the same time give the arrow a secure hold. Loosing, therefore, means opening the fingers that grip the thumb and setting it free. Through the tremendous pull of the string, the thumb is wrenched from its position, stretched out, the string whirs and the arrow flies. When I had loosed hitherto, the shot had never gone off without a powerful jerk, which made itself felt in a visible shaking of my whole body and affected the bow and arrow as well. That there could be no possibility of a smooth and, above all, certain shot goes without saying; it was bound to 'wobble'."

The dispersion of energy that takes place during the sudden release of tension corresponds to the centrifugal dissipation of energy that emerges because of the lack of coordination between the archer and the instrument. By making use of the *mass-spring-damp* model, the initial process of extension and drawing can be explained and calculated accurately. However, Master Awa Kenzo, having already developed the technique of drawing "spiritually," was capable of keeping up maximum tension in a totally relaxed state, which challenges any known law of biophysics. But in the case of a beginner like Eugen Herrigel, his body and movements hadn't yet gained the subtlety and special touch of his Master, so during this phase, his mind was constantly seducing him to understand the mechanical laws supposedly present in actions performed by the Master. This action of release that takes place is what physics considers to be a *non-linear process out of equilibrium* where the laws of causality become totally overthrown by the unexpected phenomena.[246] It is important to recall here that a living organism is characterized by functions and phenomena taking place within non-linear processes out of equilibrium, so it wouldn't be wrong to consider that during those fractions of a second the archer has the chance of relating to an expression of life emerging in an "inorganic" set of objects.[247] If the release of the arrow wasn't a spontaneous release but was a controlled release capable of slowly reversing the extension again in a whole cycle, then it would not be possible to feel any "life" being awakened by the movement of the arc and bow.

Another important aspect is that all living forms is that they correspond to purely energetic dissipation processes which on one hand produce entropy (highly degraded or non-usable forms of energy), and on the other hand this dissipation takes place in living organisms and does not necessarily increase chaos in the system, but instead produces an organization, *being* or self-producing itself (auto-poiesis). In the case of the arc and bow, there is always a dissipation of energy taking place due to the wrenching of the thumb, which in the case of "spiritual" shots is channeled into

---

[246] Le leggi di natura non sono che abitudini; e che la costanza dell'ordine del mondo non è che quella di un carattere trascendentale dell'Io [the laws of nature aren't but habits, and that the constancy of the order of the world is not but that of the transcendental character of the "I"] Julius Evola. *Fenomenologia Dell'individuo Assoluto Iii* ed. corretta: Edizioni Mediterranee, Roma 2007.

[247] La possibilità di una distinzione fra animato e inanimato, fra organico ed inorganico; ma questa distinzione non può essere che relativa: per un riaffermarsi dell'intervallo, ciò che in un dato momento è fissato come inorganico può risultare, così come si è detto, organico rispetto ad un altro termine, e ciò che era organico inorganico rispetto ad un altro termine, il primo di inferiore, il secondo di superiore organizzazione o, in generale, rispetto ad un ulteriore individuarsi del sistema [the possibility of a distinction between what is animate and what is inanimate can not be but relative: by reaffirmation of the interval, that which in a given moment is fixated as inorganic can result, as aforementioned, organic in relation to something else, and that which is organic can be considered inorganic in relation to something else, the first an inferior, and the second as a superior organization, or in general, in relation to a eventual individuation of the system] Julius Evola. *Fenomenologia Dell'individuo Assoluto Iii* ed. corretta: Edizioni Mediterranee, Roma 2007.

the arrow. From a physical viewpoint the process involved in the "spiritual shots" implies the apparition of a set of phenomena that are capable of channeling all the energy dissipation produced by the thumb and centripetally directing it towards the arrow. It is mind-boggling how this phenomenon can occur in situations of non-linearity out of equilibrium, but in the case of Operative Traditions, there is absolutely no concern for aiming to understand the laws that direct the phenomena, but instead to effectively aim at the magic of the phenomena itself.

A specific communication has to emerge between the archer and the arc/bow configuration so that both can express their particular magic. Hence we are referring here to the attainment of a process of *resonance*.

\* \* \*

"You cannot do it," explained the Master, "because you do not *breathe* right."

\* \* \*

The phenomena of resonance takes place whenever a given wave or vibration excites a system in a very specific way, to the point that the system progressively accumulates local energy. A simple example of resonance is the opera singer who is capable of breaking a glass by tuning her voice to a specific frequency; in this case the glass starts accumulating within its crystallographic structure the energy present in the sound wave, and the energy accumulated can easily surpass the energy levels that can be accumulated by crystallographic dilatation. Another simple example of resonance is that of pushing a child on a swing. By choosing to give pushes along with the natural interval of the swing, it causes the swing to attain higher amplitudes, thus accumulating energy. The correspondence between the pushing movements and the natural cycle of the swing corresponds to a *resonance* between both elements. Any mass-spring-damper system is characterized by a specific *resonance frequency* or *natural frequency*.

In the particular case we are addressing the arc and bow, when the arrow is released by a non-experienced archer, the elastic energy formerly accumulated by the drawing movement does not flow totally into the arrow, and to a large extent is dissipated as friction caused by the thumb, causing an imperfect and wobbling shot. The arc and bow, as with any mass-spring-damper system is characterized by a particular resonance frequency, and the only way the archer can impede the friction caused by the thumb, is if somehow if he manages to "couple" the release of his thumb with the initial release of the bowstring, hence ultimately *resonating* with the arc and bow. This resonance between the archer and his weapon is not a resonance that can be accomplished consciously (like when pushing a child in a swing). Because the process takes place in a phase of high tension where the release is a non-linear phenomenon out of equilibrium, very subtle elements arising from the unconscious emerge to the surface hence interfering with the subtle delicacy of the release. It is completely useless to act exclusively on the understanding of the mechanics or physical laws involved. This intellectual understanding is unable to direct the phenomena in its critical points, which means that there is a gap between theory and practice, or rather that the entire notion of physical laws only explains the dynamics under very strict and controlled conditions; in other words, the mechanical laws are essentially *false*. Our categories of thought are unable to dominate the subtlety of the phenomena, and this incapacity also challenges all confidence in oneself, as the supporter of the categories. Nietzsche proposes, however, the antidote to withstand the tragedy: *"the measure of strength is to what extent we can admit to ourselves, without perishing, the merely apparent character, the necessity of lies."*[248] The entire operative discipline aims to configure the predispositions of the practitioner, which correspond to subconscious reactions caused by fear patterns, emotional blockages, atavisms, and all sort of complex or "demonic" instinctive reactions that can overthrow the conscious capacity of the individual, who can only "float" upon the phenomena, and has no chance to dominate

---

[248] Friedrich Nietzsche, Walter Arnold Kaufmann, R. J. Hollingdale. *The Will to Power*, 1968, pg 15.

their direction. It can happen, however, that during the processes the conscious mind intends to repress the irrational reactions, but the effect of this would be even a more intense unconscious reaction, such as punching a rigid wall with twice the power.[249]

\* \* \*

Through all these explanations we are progressively introducing the concept of *technique*. As it can be observed in the example of the arc and bow, the relation of subject-object can be developed in a conscious and functional way which is accessible that doesn't imply the projection of any personal standpoint where subconscious factors are ultimately projected in the phenomena, defining the character, form and development of the phenomena itself. If the production is characterized by harmonic forms or symbols, then we can affirm that the human being who performs the action is accomplishing *poiesis*—a term that shall be more deeply addressed in the third part of the book—implying the activation of a Magical "I." The latter activation implies a complete reconfiguration of the nervous reflexes, and therefore the important aspect of *gesture* in the movement corresponds to another element that allows verifying the existence of the Magical "I."[250] *Gesture* was important to the classical sense of life, as Oswald Spengler explains:

> *"It goes without saying that we, when we turn to look into the Classical life-feeling, must find there some basic element of ethical values that is antithetical to "character" in the same way as the statue is antithetical to the figure, Euclidean geometry to Analysis, and body to space. We find it in the Gesture. It is this that provides the necessary foundation for a spiritual state. The word that stands*

---

[249] Quando si dice «no», evidentemente, si resta pur sempre determinati da ciò che si nega—quale sarà dunque, in genere, il carattere dell'atto? Un tale carattere può solo apparire come quello della spontaneità assoluta. [Whenever it is said "no," obviously one remains determined by that which one denies. Which can be therefore the character of the act? Such character can only appear as that of absolute spontaneity] Julius Evola. *Teoria Dell'individuo Assoluto*. Edizioni Mediterranee, 1988, pg 126.

[250] Il razionale sta al non-razionale come lo stile sta alla libera potenza creatrice di un artista [The rational is to the non-rational as the style is to the free creative power of an artist] Julius Evola. *Teoria Dell'individuo Assoluto*. Edizioni Mediterranee, 1988, pg 136.

*in the Classical vocabulary where "personality" stands in our own is "persona" namely rôle or mask."*[251]

The specific gesture "pilots" the development of the movement with very little need of any assistance by the conscious mind, by allowing the Magical "I" to intervene at all times. Consequently, there are two examples: magical techniques and functional/mechanistic techniques. The former constitute a preparatory base for the latter, like in the case of learning to draw the bowstring of the arc.

In order to approach this, the Master of an Operative Tradition would describe the practices, disciplines and methods required to reach this goal of attaining the Magical "I" which allows the individual to be integrated (in the third part we connect this to Heidegger's notion of *enframing*) into a higher order of reality where everything around him is a coherent manifestation of who he *is*. The presentation of the disciplines by the Master will be developed with simplicity. There is no need to justify the validity of the teachings by resorting to domains of reality which the student has no chance of gaining any experiential access. As Julius Evola points out: *"One can be sure that he who feels the need to show that he is an "initiate," who surrounds the magic methodology with coverage of mystery and occultism, which makes use of the "ineffable" and "higher powers" such as clairvoyance, channeling, etc., to justify their own teachings, he is then necessarily an element of hollow mystical seduction, or just someone who even not realizing it, only to a little or highly confusing degree has not even lived the immanent meaning and the reason for what he has done."*[252]

"Immanent meaning" is the crux of the whole process. There is no need to come up with outlandish explanations of irrational phenomena since it is the character of the operations that have a symbolic language of their own. In an Operative Tradition, the student must embody an extension of the material objects, and the material objects have to correspond to an extension of t he

---

[251] Oswald Spengler. *The Decline of the West*. Alfred A Knopf. 1926. New York, pg 316.

[252] Julius Evola. *Saggi sull'idealismo magico*. Edizione Mediterranee, pg 84.

student, in perfect reciprocity and harmony. Nietzsche made use of the expression "to follow the sense of the earth." These practices promote exactly the same principle. The initial discipline is that of grounding the senses to the earth, and in this sense, a set of basic principles have to be established.

# THE CONCEPT OF VALUE & THE EMPIRICAL STATE OF BEING

*What weapons must we put in the hands of those who strongly aspire to escape from all the deserts based on rationalist and materialist systems, but who are still subject to the coercion of their dialectics?*
Ernst Jünger[253]

*It is not true that the most skillful archer aims the truth.*
Ernst Jünger[254]

Who does not understand by himself, no one will ever be able to make understand, do what he will.[255]

In Julius Evola's TAI-PAI, the issue of certainty we formerly addressed when referring to the domain of phenomena and in terms of what the "I" can affirm to know, is approached by Evola when defining "being" as *"all that which is simply present in terms of internal or external experience"*[256] an *exerting* (Da-sein)[257] which is independent of the degree of development and empowerment of the "I," exerting that *"though unconditional, is unconditionally*

---

[253] "Welches Rüstzeug soll man jenen an die Hand geben, die lebhaft aus der Einöde rationalistischer und materialistischer Systeme hinausstreben, aber noch dem Zwang ihrer Dialektik unterworfen sind?" Ernst Jünger. *Der Waldgang.* Klett-Cotta. Stuttgart 1980, pg 65.

[254] Ernst Jünger. *Eumeswil.* The Eridanos Library. 1980.

[255] B. Trevisano, *De la phil. Nat. des Mét.*, BPC, II, 398.

[256] Julius Evola. *Teoria Dell'individuo Assoluto.* Edizioni Mediterranee, 1988, pg 25.

[257] Heidegger referred to Dasein as the metaphysical comprehension corresponding to the fundamental trait of the human being. This perspective impeded Heidegger to agree with the strictly biological, vitalist and racist implications present in the Nazi regime, despite the fact that he was a political supporter of such movement. For Heidegger, men are impelled to responsibly "make something out of themselves," which implies to care such metaphysical essence or Dasein, as an element implicitly correlated to the active and actual political-cultural forces of a given time. Martin Heidegger, *Being and Time*, tr. by J. Macquarrie and E. Robinson (New York: Harper and Row, 1962).

*meaningful as it is deprived of conditions."* If an individual can define direct inner or external experiences by resorting to a factual description that lacks any subjective mediation (explanation, cause, reason, etc.), then the actions—the *"simplicity of fact"*[258]—correspond to an *"illumination of the spirit."*[259]

Master Awa Kenzo, when providing the arc and bow to Eugen Herrigel, was initially concerned about the way his pupil grasped the bow with his hands and touched it; focused in the form of his movements when nocking the arrow, raising the bow, drawing, remaining at the point of highest tension, and finally releasing the shot ready to start the cycle once again. Ultimately, during this disciplinary cycle, Herrigel was introduced to a context where he had the chance to *be*, to relate to elements of the cosmos, from the mechanistic to the subtle, which in the Hindu Tantras are symbolized by the three gunas (sattva, rajas, and tamas).[260] The whole discipline was that of *focus* and *concentration* on movements. Every single direct action that be considered a fact and experience, though such an experience could not be easily explained. By recalling Evola's definition, because the experiences cannot be explained, this provides the experience with the potential to achieve a value and meaning. In this process the dynamics that take place in the psyche act similar to the chaotic movements of a lake, which is "stagnant" in an impermanent form. Ultimately, the psyche consists of human consciousness that has a great capacity to flow, and to overthrow boundaries, and yet its *form* is provided by its adaptation to a given material or (cosmic) domain. As the philosopher of the Eternal

---

[258] Julius Evola. *Teoria Dell'individuo Assoluto*. Edizioni Mediterranee, 1988, pg 25.

[259] Idem.

[260] "[the gunas] they correspond to various modalities of shakti, which come into play once the ongoing process has led one beyond the metaphysical "point" (…)The three gunas are called sattva, rajas, and tamas. Sattva comes from the word sat, which means "being." The term designates the elements reflecting the stable and luminous nature of being, and it is usually associated with Shiva's nature. Tamas, on the contrary, denotes what is fixed in the opposite sense of a stiffening, or of an automatism (e.g., passive staticity, sheer passivity, the force of inertia, weight, mass, a limiting and obscuring power). Tamas presides over every depleted process and over inactive potentialities. Rajas, conversely, symbolizes dynamism, becoming, transformation, change, and expansion; it corresponds to what we usually designate as energy, life, or activity. Rajas may also be influenced by the other two principles" Julius Evola. *The Yoga of Power*. Inner Traditions, pg 40.

Return, Nietzsche wrote: *"There are no facts, everything is in flux, incomprehensible, elusive; what is relatively most enduring is—our opinions."*[261]

To think that a culture can be reduced to the subjective concepts and structures present in the psyche, and that our subjective views on reality correspond to a reflection of the developments taken place in the cosmic domain is an assumption that can be considered true in exceptional cases. In general, the psyche corresponds to a highly chaotic, unstable, and kaleidoscopic domain present in human nature that is determined by elements beyond it. In the case of the human being, modern psychology has focused on the study of the psyche, but consciousness transcends the psyche and is related to many elements of the cosmos, and to the psychosomatic tendencies or predispositions of the individual. Modern psychology is still unable to address the connections of individual consciousness to specific global operative configurations, as Marshall McLuhan wrote in the 60s:

> *"In the electric age, when our central nervous system is technologically extended to involve us in the whole of mankind and to incorporate the whole: of mankind in us, we necessarily participate, in depth, in the consequences of our every action. It is no longer possible to adopt the aloof and dissociated role of the literate Westerner."*[262]

Hence, despite all efforts emerged in modern psychology, the psyche cannot be grasped by itself,[263] or in other words, it is illusory to "measure" the psyche with the concepts, discursive capacities, or

---

[261] Friedrich Nietzsche, Walter Arnold Kaufmann, R. J. Hollingdale *The Will to Power* 1968, pg 327.

[262] Marshall McLuhan. *Understanding Media. The Extensions of Man.* The MIT Press Cambridge, Massachusetts. London, England, 1994.

[263] L'Io, certo, non lo si «conosce» (chi lo co-noscerebbe?) e non si può dire dire «ciò che» esso sia e ciò che può esser conosciuto e detto non è l'Io; ma rispetto ad esso si può far qualcosa di meglio che non pensarlo o «conoscerlo," cioè si può possederlo, esserlo, in assoluta immanenza [The "I," truthfully can not be known (who could know it?) And one cannot say "what it" is, and what can be actually acknowledged and said is not the "I"; but in relation to it can be done something better than thinking about it or "knowing" it, that is, one can posses it, be it, in an absolutely immanent way] Julius Evola. *Teoria Dell'individuo Assoluto.* Edizioni Mediterranee, 1988, pg 44.

analytical modes implicit in the psyche itself, just like when Baron Munchausen intended to fly by pulling up his hair. In the 19th century, Nietzsche asked the radical question: *"is it likely that a tool is able to criticize its own fitness?"*[264] Kurt Friedrich Gödel (1906-1978) American mathematician and philosopher demonstrated that any formal model, such as those that both Bertrand Russell and Alfred North Whitehead intended for arithmetics in *Principia Matematica*, are always incomplete since they imply the existence of non-demonstrable truths in their interior, that is to say, "self-referenced" types of truths. The latter demonstration on the part of Gödel allows us to conclude a fundamental principle: any mode of self-reference inevitably implies discursive deduction an indetermination. Swiss psychologist Carl Jung addressed this uncertainty in regard to the knowledge of the psyche when writing: *"The object of psychology is the psychic; unfortunately it is also its subject."*[265] By following Jung's assumption on the psyche and the dead-end that inevitably appears when acknowledging the psyche through the psyche itself, Titus Burckhardt, by resorting to the esoteric teachings of Tradition, writes:

> *"According to this opinion [Jung's view formerly stated], every psychological judgment inevitably participates in the essentially subjective, not to say passionate and tendentious, nature of its object; for, according to this logic, no one understands the soul except by means of his own soul, and the latter, for the psychologist, is, precisely, purely psychic, and nothing else. Thus no psychologist, whatever be his claim to objectivity, escapes this dilemma, and the more categorical and general his affirmations in this realm are, the more they are suspect; such is the verdict that modern psychology pronounces in its own cause, when it is being sincere towards itself."*[266]

---

[264] Friedrich Nietzsche, Walter Arnold Kaufmann, R. J. Hollingdale *The Will to Power* 1968, pg 221.

[265] C. G. Jung, *Psychology and Religion* (Yale: New Haven, 1938), pg 62.

[266] William Stoddart. *The Essential Titus Burckhardt—Reflections on Sacred Art, Faiths, and Civilizations.* (Perennial Philosophy) World Wisdom (2003) pg 44

Consequently, the contents, structure, and *state* of the psyche can only be acknowledged by something external to the psyche itself. One thing that has to be acknowledged is that what really matters is not the contents of the psyche (thoughts, world-views, discursive ideas, etc.) but rather the *state* and *structure* in which the contents are formed. Hence, in the same way, water can be found in a solid, liquid, or gaseous state, the contents of the psyche determine individual consciousness to the extent that their *state* has been affected by a given *praxis*. Ideally, at the most advanced stages of an Operative Tradition, the psyche of the individual consciousness must be capable of conserving its basic structures, forms, and states, stabilizing a higher state of freedom beyond that of the determinations caused by diverse psychological contents. Such forms eventually require specific conceptual and scientific frameworks, in a similar manner to some architectural forms require specific materials to be feasibly built. It is extremely worthwhile at this moment to point out the stages indicated by Patanjali. Patanjali refers to:

1. *Pratyâhâra* or the control and dominion over the impressions caused in the senses, and the extremely chaotic associations that appear amidst thought patterns.
2. *Dhâranâ* that entails the focus/concentration on only one object by discarding all others.
3. *Dhyâna* which entails the absolute focus in a pattern produced by the mind and not by the senses.
4. *Samâdhi* or elimination of all mind association and convergence of the mind with its core power.[267]

Ultimately we are faced with the issue of how to *stabilize* the psychological currents within consciousness[268] so that the key issue

---

[267] Pantanjali, B. K. S. Iyengar. *Light on the Yoga Sutras of Patanjali*. Thorsons (2002)

[268] La situazione si presenta (soggetto universale) come quella di un sistema tensionale non sta¬bile, ma suscettibile di stabilizzazione nei termini di una organizzazione, la quale ha sempre il senso di una subordinazione dinamica presupponente un centro che si afferma, si integra, potenzia e sviluppa fino a riprendere nella sua scia tutto il resto [the situation is presented (the universal subject) as that of a system of tensions that is not stable, yet susceptible of stabilization in terms of an organization, which always has the character of a dynamic subordination which presupposes a centre that is affirmed, that

of *certainty* can apply to a psychological domain which has a highly fluid and adaptive nature. Contrary to the operative procedures we are dealing with in this book, most mystical approaches intend to "blend" the unconscious and the conscious mind according to Carl Jung's concept of *individuation* or *mysterium conjuntions*[269] which consist of suggestion techniques that induce a given form in the psyche, which allows one to configure thoughts, ideas, or even mystical experiences. In the case of any mystical experience of this sort, the issue of certainty becomes dissolved, diffuse, and problematic, as there is no absolute reference point, but rather countless ephemeral forms that configure thoughts and concepts. Therefore, the number of different configurations of the psyche is practically infinite, resembling the countless styles that appear in architecture or music in time and space. Yet in TAI-PAI, all these ephemeral modes are considered merely relative, are not absolute. Absolute certainty is provided in Magical Idealism by the specific character of the experiential domain that ultimately sustains the relative character of the psyche, in a similar way to how a given form of the ground provides form to a given amount of water, creating a lake.

When Eugen Herrigel started practicing the art of archery under the guidance of Master Awa Kenzo, his psyche had been formerly forged by the character of the powers that were present in most German institutions during the beginning of the 20th century, Kantian and Hegelian perspectives on Transcendental Idealism, where it is assumed—as formerly explained—that the world is a representation produced—or *placed*—by the thinking, abstract "I." Returning to our lake analogy, this means that human individuality or the self—drowned in a fluid psychological state—perceives the "bottom of the lake" or material domain depends inevitably on the filtering of the light (spirit) through the specific properties of the lake's water, projected upon the matter, and where even the

---

is integrated, empowered and developed aiming to carry in its essence all the rest] Julius Evola. *Teoria Dell'individuo Assoluto*. Edizioni Mediterranee, 1988, pg 176.

[269] *The Secret of the Golden Flower*. Introduction by Carl Jung.

"thing in itself" can be equated with the outer sunlight required for the perception. Hence from the perspective of Transcendental Idealism, the whole issue corresponds to that of "fixing" the basic foundations of the psyche so that afterwards the entire cosmos can be acknowledged scientifically according to translucent foundations or categories. But one problem not devised initially by this position is extremely theoretical and can only be considered as an abstraction, since in any real life experience, the concrete and empirical individual is constantly carried and determined by diverse currents and irrational forces, which Evola referred to as the *"undetermined in the immanent determining of all determination,"*[270] which like in the case of turbulences in fluid mechanics, cannot be grasped qualitatively based on Kant's ideal categories.

Therefore, Eugen Herrigel's initial psychological state was very much analogous to that of a human individuality diving into a sea (the psyche) where all concepts, ideas, languages and thought patterns were "moving around" chaotically like impermanent structures, through which he had the chance to provide impermanent intellectual representation to his perceptions of reality, by allowing the characteristics of objects to filter through the impermanent structures, like a "lens" through which impermanent form can be provided to the objects of reality. So, if for instance, the "lens" was mathematical structures or topologies, and by considering here that the number of mathematical structures or topologies are countless, then perspectives of reality viewed through impermanent "lenses" has to be countless as well. From this perspective the gnoseological problem of certainty and the quest for reference points is definitely not solved, as the relativism of kaleidoscopic approaches to reality based on discursive premises is practically infinite.

Considering the crisis—ultimately a nihilistic crisis—what took place in the West was that the individual focused his knowledge of the cosmos in terms of the discursive "I," and also by discarding or repressing the effects caused by irrational currents on his empirical,

---

[270] "L'indeterminato del «determinare» immanente in ogni determinazione" Julius Evola. *Teoria Del l'individuo Assoluto*. Edizioni Mediterranee, 1988, pg 33.

and real "I." In order to accomplish this, the individual was forced to release from any subconscious influences and to perceive the discursive or secular knowledge modes (scientific, political, and religious) as objects that have no connection to him as a particular individual. Yet adopting this approach is another important issue. Whoever looks at the contents of the lake from above will, first of all, see their reflection on the surface. In other words, our static sense of self corresponds to a highly superficial image that distorts and impoverishes the view of the entire cosmos. If our psyche becomes attached to this image, the effect is a *paralyzing* one, which is contrary to the basic premises of an Operative Tradition, which is devoted to action and experience. The image we have of ourselves as a physical "I" or as a thinking "I" are extremely ephemeral and superficial, floating like a fragile boat upon the world of becoming that ultimately sustains it.[271]

By considering this epistemological context, in any Operative Tradition or in TAI-PAI, and in order to provide the first steps required to surpass any psychological relativism Evola defines the importance of *the empirical state of existence* and the so-called *"naked principle of the I"* which requires the human individual to start focusing on elements of the cosmos, yet with no filtering by the discursive structures embedded in the psyche, thus implying the importance of a *naked* state of perception and openness of the senses, which does not discard any irrational elements, or any repression of passion, that is, passive currents related to a concept Hinduism and Buddhism referred to as *samsara*. In this sense, the discipline implies avoiding as much as possible any rationalization of concrete experience taking place with the outer world as with the inner

---

[271] L'individuale come valore è definito non da qualcosa che è, ma da qualcosa che ha sé - non da un'esistenza, ma da un possesso; il puro possesso, la nuda forma della dominazione, questa è la sua sostanza, ed è propriamente in tali termini che il mondo dell'opzione soggettiva si differenzia da quello del desiderio. Ma un possesso è inconcepibile senza un substrato di cui esso sia pos-sesso. Il concetto di Signore implica necessariamente qualcosa, di cui si sia signore. [What is individual as a value is defined not on whatever it is, but on whatever it possesses –not on existence but on possession; the pure possession, the naked form of domination, such is the substance, and it is in such terms that the domain of the subjective options is different than that of desire. Yet a given possession is unconceivable without a substance it possesses. The concept of Master implies necessarily something that is mastered] Julius Evola. *Teoria Dell'individuo Assoluto*. Edizioni Mediterranee, 1988, pg 79.

world.[272] If, for instance, the student feels tired, all he has to do is to feel tired, without trying to find a cause or explanation for this state. This discipline allows an approach to the concept of *value* applied to the domain of immanent/transcendental experimentalism. Evola defines this *value* as the *"absolute relation between the naked principle of the "I" and whatever is different of such principle in terms of the experience and awareness of the "I" (...) The form by which the relation the "I" and being is presented as unconditional."*[273] In terms of practice and the discipline, the value is characterized by *significance, perfect transparency,* and *simplicity*. Comprehension ("I" comprehend) is always identical to itself in relation to the experience. Hence value implies an act of potency, pure dominion, lack of privation, persuasion, a relation to the absolute, pure will, pure spontaneity, absolute certainty, and a state of justice. Value also refers to a *"living standpoint that faces life in its entirety, experience placed according to all its actual and possible forms."*[274] As we shall see in the third part of this book, such a position is important to address and develop *poiesis* as a form of activity that aims to effectively produce the value. However, in this gnoseological context, to *comprehend/know* does not correspond to the assimilation of intellectual category, but rather *the immediate awareness of a given static or dynamic determination.*

---

[272] Si dovrebbe cioè riconoscere che un razionale che sussista e si svolga per virtù propria è una finzione non esistente né in cielo né in terra, che del razionale stesso la radice è il non-razionale, per cui il rapporto fra i due è non di una opposizione o reciproca esclusione, bensì di una subordinazione gerarchica del secondo al primo. Il non-razionale non deve esser cercato di là dal razionale, ma nel seno di esso, come la sua dimensione interna e come ciò che si esprime ed agisce determinativamente nell'ordine logico: ordine che, disgiunto da esso, crollerebbe nel nulla, perché la sua stessa eventuale costituzione in un sistema di relazioni astratte ha possibilità solo in base ad una volontà d'astrazione. Di massima, può dirsi che tutto è ad un tempo razionale (o razionalizzabile) e irrazionale [It has to be recognized that a rationality that subsists and develops by itself is a fiction that doesn't exist nor in heavens nor earth, that the selfsame roots of the rational is the non-rational, in which the relation between both is not an opposition and reciprocal exclusion, but rather a hierarchical subordination of the latter to the former. The non-rational must not be searched beyond the rational, but within it, as its internal dimension and as that is expressed and acts in a determinant way in the logical order: an order that, separated from it, would fall into nothingness, because its eventual selfsame constitution within a system of abstract relations is possible only to the extent there is a will for abstraction. At best, one can say that everything is at the same time rational (or can be rationalized) and irrational] Julius Evola. *Teoria Dell'individuo Assoluto*. Edizioni Mediterranee, 1988, pg 135.

[273] Julius Evola. *Teoria dell'Individuo Assoluto*. Edizione Mediterranee, pg 25.

[274] Julius Evola. *Teoria dell'Individuo Assoluto*. Edizione Mediterranee, pg 25.

This awareness is provided by the value of the experience itself. It doesn't mean to acknowledge the laws or formal mechanisms by which determination takes place, but just experiencing the determination itself. These determinations and identities merged with consciousness by placing the object of experience, can be considered *objective* and conceiving here *subjective* as knowledge in general,[275] not particularized to the qualitative value of a given experience. For instance, many internal and external experiences can follow strict determinations and as such, they constitute objects of experience. Their awareness is independent of the fact that we might not be able to explain what is occurring. Realists consider that there are objects of experience outside and independent of the act of placing by the "I." From a subjective viewpoint (that of general knowledge) the standpoint can be accepted, yet from an operative viewpoint no objects or determinations can be experienced unless they are produced or placed by the "I."[276] The whole *praxis* requires to *"sustain a limit and eventually overcome it"*[277] referring here to the determination.

Returning to our lake analogy, the individual must first of all *ground the experience*, that is, anchor his experience to an act or series of acts where he is exposed to a limit (determination) that allows him to gain awareness of a specific determination. Evola states that it *"is not possible a deduction of knowledge, but only an exposition of it."*[278] Awareness through direct exposure to experience, can be imagined as an outer light trespassing into the chaotic currents of the psyche and illuminating a solid aspect of reality, an illumination of the spirit accomplished by a simple act of *placing*,

---

[275] Il soggetto universale concepito come un possibile modo d'essere del soggetto stesso, dell'Io [the universal subject, conceived as a possible mode of being of the selfsame subject, of the "I"] Julius Evola. *Teoria Dell'individuo Assoluto*. Edizioni Mediterranee, 1988, pg 179.

[276] "The so called real things are symbols of my non-power, of my privation. The reason I call a thing real comes from my experience of a privation and *not vice versa*." Julius Evola. *L'Individuo E Il Divenire Del Mondo*, Libreria di Scienze e Lettere, 1926, pg 27.

[277] "Si tratta di un mantenere un limite e, nel contempo di superarlo" Julius Evola. *Teoria dell'Individuo Assoluto*. Edizione Mediterranee, pg 33.

[278] "Non sarebbe possibile una deduzione del conoscere, ma soltanto una esposizione di esso" Julius Evola. *Teoria Dell'individuo Assoluto*. Edizioni Mediterranee, 1988, pg 49.

of exerting a *free act*. This qualitative aspect of experience is what provides value to the experience itself; it corresponds to a bridging of the transcendental with the immanent, like a golden thread between both domains, which is the experiential tension given by the determination itself. The further development of this process is what Evola refers to as *experiential transcendentalism,* which establishes a certainty from which it is possible to keep advancing forward towards more integrated states. The more the individual resorts to a subjectivist perception of the given experience, the more powerful the irrational currents of the psyche that interfere with the illumination of the determination become, and this release entails *privation* for the individual, that is, the empowerment of a state of necessity and desire caused by the lack of dominion of a given set of determinations, both internal as external.[279] Such privation goes hand in hand with a basic passivity and *spontaneity* on the part of the individual, who becomes an object of the inner (emotions, passions, greed, thirst, etc.) and external determinations (Heidegger's "enframing" concept, as we shall see.) rather than its subject and producer.

This experiential approach allows certainty to emerge. The individual gains contact with a domain of reality where he is now *anchored*, even though his perceptive faculties and senses might not yet be capable of defining a path of progressive dominion, and many psychological patterns and irrational currents are still impeding the vision of the "solid walls" of the "lake," that eventually reach the surface.[280] The latter corresponds to *the path of the Absolute Individual*, which eventually becomes sovereign of passion, irrationality, and the master of all inner and external drives, as mediator between the two domains.

---

[279] Nel desiderio si ha la correlazione di un elemento con un altro, per la quale il primo si sente come privazione e subordina al riaffermarsi sul secondo la completezza del proprio essere [in the case of desire there is the correlation of one element with another, by which the former is felt as privation and depending on the reaffirmation of the latter the completeness of the own being] Julius Evola. *Fenomenologia Dell'individuo Assoluto Iii* ed. corretta: Edizioni Mediterranee, Roma 2007.

[280] *"Transforming the dark passion of the world into a kind of freedom"* Julius Evola. An Intellectual Autobiography. Path of Cinnabar Arktos 2009, pg 51.

All of the former can be expressed by stating that in an operative discipline it is required to realize one's limitations in terms of gaining certainty in knowledge rather than by compensating for the lack of certainty with any sort of belief, faith, or ideology. In our times, we have been trained to consider that a phenomenon is that which can demonstrate or verify a set of laws or conceptual frameworks, thus discarding a vast spectrum of experiences that do not fit into analytical constructions. In these operative disciplines the standpoint is quite the contrary: what is required is to focus on the multidimensional aspect of experience, releasing it from the repressive and distorting traits of the "laws of nature." The key here is *habit*, not law. Habit refers to a given determination that, in some cases might not even be possible to describe analytically or through language, yet this doesn't mean that the individual is not gaining a substantial knowledge of the phenomena at a level that transcends cerebral functions. In an Operative Tradition what changes in the individual are the deep layers of the subconscious, which finally affects the nervous system, the psycho-somatic configurations, and unconscious processes that take place in the organs of the body, and the conscious mind with all its theories and beliefs corresponds to mere "epidermis." As Nietzsche writes, it is

> *"essential: to start from the body and employ it as a guide. It is the much richer phenomenon, which allows for clearer observation. Belief in the body is better established than belief in the spirit."*[281]

This conscious "epidermis" corresponds to a protective layer and separation between the inside of the individual and the cosmos; a layer that, if hypertrophied, becomes extremely alienating and disempowering, hence favoring what in TAI-PAI is referred to as the *path of the other, the path of alienation (alterity)* and *privation*. As in the case of pharmaceuticals that induce changes in consciousness by altering biochemistry, in the case of an operative discipline, a permanent change of character is induced, but there is, however,

---

[281] Friedrich Nietzsche, Walter Arnold Kaufmann, R. J. Hollingdale *The Will to Power* 1968, pg 289.

a very important difference in comparison to pharmaceuticals or drugs. This is linked to what Jünger wrote: *"a man is able to release from the irruption of pain to the exact extent he can get out of himself."*[282] In the case of an operative discipline, the deep organic functions are integrated in several levels of progressive dominion of the inner and external reality, overcoming the burden of the determinations and reaching states of "hyper-health," referred to in Hinduism as currents of *prâna,* which transcend the physical limitations of the body. The cycles of breathing that characterize *prâna* are to the soul what physical breathing is to the body, and ultimately the physical cycles and patterns are all "projections" of *prâna* as stated in the Upanisad: *"As the spokes of a wheel rest on the nave, so all [in the organism] rests on the prāna."*[283]

Such self-transformation doesn't aim for pleasurable feelings of happiness, but dominion over the inner and external realms and determinisms. Catastrophes or wars at a civilizational level can even take place, but the "wind" that *"bloweth where it listeth (...) You hear the voice of the wind but cannot see where it comes from nor where it goes to"*[284] is beyond the determinations of a given civilization shall always prevail, and the individual who has learnt to incarnate the force can not be destroyed by forces that have been already integrated within. In this sense, we are referring to the idea of "immortality" which does not refer to the Christian idea of everyone having an "immortal soul," but rather the task of *producing* the most integrated "I," that is, of working under the guidance of specific premises towards a state of invulnerability, here and now, that testifies and justifies all former human productions, at an individual, collective, and even historical, level.

In the case of the accounts of Eugen Herrigel with Master Awa Kenzo, it is constantly reiterated that the advice of the Master was to accustom Herrigel to relaxation, to release excessive tension while handling the arc. One of the key requirements, in order to

---

[282] Über den Schmerz: David C. Durst, *On Pain.* New York: Telos Press Publishing (2008).

[283] *Anguttara-nikayo* by Nyanatiloka (Die Reden des Buddhas) Munich, Neubiberg.

[284] John 3, 8.

accomplish relaxation in practice, was to learn to breathe properly. The following are the words of the German author:

> *"The breathing in, the Master once said, binds and combines, by holding your breath you make everything go right, and the breathing out loosens and completes by overcoming all limitations. But we could not understand that yet."*

It can be observed, that even Eugen Herrigel was still intending to understand the words of the Master, but Master Awa Kenzo did not intend to explain anything to his pupil, but only encouraged him to keep practicing and breathing properly. However, as we formerly pointed out, Herrigel was dragging to the spiritual practice the thinking "I" which he had produced while studying in Germany under the influence of Kantian philosophy, which is completely devoted to the idea of a universal and abstract thinking "I." So powerful was the influence of education on the German author that Master Awa Kenzo referred to the tendency as the main trouble. He said:

> *"That's just the trouble, you make an effort to think about it. Concentrate entirely on your breathing, as if you had nothing else to do!"*

Easy to say, but in practice very hard to do for most people, especially when the body is in motion, since motion activates the technical process. Yet the first main purpose of breathing concentration is to allow consciousness to ground from the physical "I" (all bodily functions) to the subtle *pranic* energies, allowing a complete trespassing of the thinking "I," which has to become totally relegated to the "background." To some extent, the difficulty of enduring in this practice is due to the reflexes and habits attained on the part of the individual when living in a civilization determined by a linear conception of time, where all cycles are imposed from the outside. So a considerable gap can emerge between the natural cycles of the body in harmony with the cosmos and the impositions on the part of civilization which constrict and re-adapt cycles to linear activities determined by cause-effect logics. There is inevitably a price to pay when adapting to the external mechanisms. The main issue of relating to external mechanisms corresponds to an instinctive strengthening of the incapacity to "let go," to "release" from mechanistic habits to the point that even security becomes one of the main traits of personal predispositions. So when it comes to breathing, the unconscious physiological cycle is also affected by traits and emotional blockages, repressions, and fears caused by the adaptation of the individual to a given "social I." Ernst Jünger writes in this regard:

> *"I [the Anarch] practice "conscious" inhaling and exhaling. You have to sink the conscious mind down into the diaphragm and suppress any thoughts, which cluster in ravenous throngs. This is difficult—if it succeeds, it also aids normal breathing; it spiritualizes it. It might be better to say it exposes its spiritual strength. That this spiritual strength exists and establishes all Existence was taken for granted in good times and was linguistically attested to by special words like pneuma. Russian pilgrims followed its trail with "perpetual prayer." Prayer becomes breath, and breath prayer.*
>
> *Your breathing is more successful if the mirror is deep, just as, more generally, your vegetative, instinctual disposition increases. To be sure, this disposition must not gain the upper hand in everyday life. In the depth, the animal spark must keep glowing, as on the fuse leading to the powder keg. One can learn that from the samurai: the leap from a motionless state, the deadly thrust with the sword flying from the sheath."*[285]

\* \* \*

> *"All that you have learned hitherto," said the Master one day when he found nothing more to object to in my relaxed manner of drawing the bow, "was only a preparation for loosing the shot."*

---

[285] Ernst Jünger. *Eumeswil*. The Eridanos Library. 1980.

The arc and bow were formerly modeled as a *mass-spring-damper* system, where ultimately the whole issue in terms of handling it corresponds to the establishment of a resonance between the archer and the weapon, especially during the release of the arrow, aiming to accomplish the goal of the so-called "spiritual shot." The discipline implies the progressive creation of a spontaneous "coupling" between the movements of the archer and the selfsame breathing cycle, hence allowing the establishment of harmonic correlations between all elements. It is worth recalling at this point that practically all systems in nature can be modeled energetically as *mass-spring-dampers* where the laws of thermodynamics can be applied. On one hand, the mass-spring components are linked to the amount of energy the system can store while the damper is linked to the specific friction and entropy production on the part of the specific system.

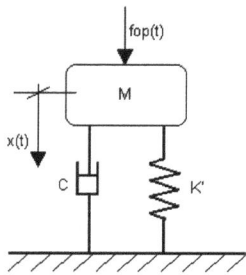

It is important to realize now that in these systems all *breath*, their character is vibratory, and therefore following an oscillatory movement. Furthermore, in nature all these systems are open, therefore causing the nature of their vibration to expand in all directions through the production of wave-patterns characterized by a nature dependent on the medium (sound, electro-magnetic, etc.). Visualizing the process at a larger scale, the set of powers in the cosmos or *physis* (*samskara in Hinduism*) can be conceived of as an *orchestra*, having diverse levels of hierarchy and integration where the ultimate wave-pattern produced by this orchestra directs and integrates the most diverse *mass-spring-damper* "cells" or particular musical instruments. Atoms, molecules, crystals, cells, tissues, organs, and entire organisms are all characterized by ceaseless oscillation and characteristic patterns of vibration and internal rhythm. Through the mediation of men (musicians in this particular example) each instrument has to breath in a way such that it *resonates* with the music. Yet if we analyzed the orchestral system (like any other organism in nature) we would realize that all mechanical energy transmitted by the musicians to the instruments is dissipated in the form of entropy, corresponding clearly to a dissipative system. Contrary to mechanistic systems affected by high entropy production and that finally increase chaos, in the case of the orchestra, dissipation rather produces order, harmonies, and music. It would not be preposterous to hypothesize that the music produced the instruments and not vice versa; to hypothesize that the instruments are "alive" to the extent they serve the music, to the extent their production is integrated within music, and to the extent they *resonate* with it.

In classical or Newtonian physics, energy corresponds to the magnitude that produces mechanical work or heat, following the direction of the process determined by the 1st and 2nd law of thermodynamics. Yet in the case of *physis*, which corresponds to a more integral idea of the cosmos than the way classical physics or cybernetics conceives of nature, there are forms of energy that are much more subtle and "capricious" than the measurable energy forms, and are decisively powerful. The way to access those subtle energy forms is by the establishment of a resonance between the individual and the subtle patterns. For instance, by applying this core idea to a materialistic domain, a system as apparently chaotic as the sea can be transformed into order and energy when a given collector (a tide-wave station) manages to resonate with the patterns of the tides. Another example is the classical Atmos clock, which has no need to ever be wound since it's source of energy emerges from the resonance it establishes with the temperature cycle-patterns on a daily basis. These two examples are based on a materialistic framework where every single part of the process follows classical laws of physics. These devices comply perfectly well with the two first laws of thermodynamics; they correspond to sources of negative entropy, hence transferring chaos into order. Yet the external chaos where the devices create order is only an apparent external chaos. These devices are only capable of resonating with the overall patterns if they have "awareness," being formerly designed according to the *form* present in the external patterns, and hence being capable of *tuning* into the subtle form of energy. Therefore, any deep awareness of an overall pattern embedded mysteriously in the currents of samsara can produce local energy, that from the viewpoint of classical physics apparently emerges "out of nowhere." Specific breathing techniques are required to "tune" the bodily reflexes in order to link to the forces at a level that becomes instinctive and permanent. The Eastern *prânâyama* techniques aim to create the powerful energetic resonance, aiming to converge the patterns of *prâna* with those of the "I" which were symbolically represented in the Eastern Traditions as *kundalinî*. The physical outburst of this subtle energy, referred to as "Chi" in Eastern traditions can be considered "free" energy in the sense that from a mechanistic viewpoint it appears to arise "out of nowhere," yet the techniques required to access the energy are not "free" at all, as they require the absolute combustion of any sense of individuality, ego, and the dominion of all intrusive and alienating thought-patterns.

So when we are referring to the capacity breathing techniques have in order to reconnect to the subtle force of *prâna*, we are actually dealing with the process of *tuning* both the cycles of the body as the instinctive reflexes with a non-corporeal entity or overall pattern which guides the subtle currents of entropy production into one single point, that is, it acts as a centripetal guidance. This is the aim, yet this discipline not only requires hard practice and discipline, but also the cultivation of *virtue*, which we shall describe in the following pages when explaining the phenomenology and development of the practice.

# THE PHENOMENOLOGY OF TAI-PAI

*Don't think of what you have to do, don't consider how to carry it out! " he exclaimed. 'The shot will only go smoothly when it takes the archer himself by surprise. It must be as if the bowstring suddenly cut through the thumb that held it. You mustn't open the right hand on purpose."*
*Master Awa Kenzo*

All the former indications and premises explained about operative practices are rather simple, hence facilitating the student or initiate to start properly grounding the development task from the foundations of a method that is experiential, practical and based on action. Having an operative formation based on the military traditions of the old Prussian empire, Ernst Jünger also affirms that in order to attain a higher sense of being:

> *"There are methods to strengthen individuals in that direction and at first it is irrelevant that they are exercised mechanically. They resemble exercises to restore life to the drowned, which also are mechanically executed at first. Breathing and heartbeats come later..."*[286]

Or as another warrior spirit, Yukio Mishima, shows: *"no technique of action can become effective until repeated practice has drummed it into the unconscious areas of the mind."*[287] Yet the focus on constant

---

[286] "Ihr Leiden kündet ihnen einen höheren Zustand an. Es gibt Methoden, sie in dieser Richtung zu kräftigen, und es ist unerheblich, daß diese zunächst mechanisch geübt werden. Das gleicht den Wiederbelebungsübungen an Ertrunkenen, die auch zunächst exerziert werden. Dann treten Atmung und Herzschlag hinzu." Ernst Jünger. *Der Waldgang*. Klett-Cotta. Stuttgart 1980, pg 65.

[287] Yukio Mishima, John Bester. *Sun and Steel*. Kodansha USA (2003), pg 35.

repetition and habit, being apparently easy in practice, ends up putting the patience and faith of the practitioner in the whole process to the test, which is ultimately a test of the faith he has in himself. There is nothing better than surrendering to these operative practices for years, without seeing any visible result, in order to verify to what extent an individual is aiming to reach his own invisible centre, that is, to encounter an unconditional meaning for his life, actions, commitments, and loyalties. The mandatory character of a doctrine required to overcome this challenge is also presented in the Bhagavad-Gita, where it is stated: "*In liberty from the bonds of attachment, do thou therefore the work to be done: for the man whose work is pure attains indeed the Supreme.*"[288] Whenever we encounter in our lives a teacher or Master who is instructing an operative discipline like martial arts, for instance, and the issue of distilling individual virtue is not present, then it can be stated that the whole discipline has already become an end in itself, and this shall necessarily empower the attachment of the individual to the physical "I." A case of this has proliferated in countries like the US or UK during the last century, where martial arts such as Karate, Aikido, or Kung-Fu that were imported from the East countries surrendered to a highly physical identification of the "I." These arts eventually became sports, where the importance conceded to virtue was progressively eroded and the need to surrender to social acceptance and competition was hypertrophied. However, in the case of all imperial operative arts, it was never the public who had to be satisfied, not even the family, but the Master who incarnated legitimacy and the spirit of the Tradition. This intrusion demonstrates clearly the surrendering of the spirit to the chaotic and irrational determinations caused by social passion. This tendency is widespread in our times, especially in those domains such as sports where competition enhances the importance granted to the "individual," when actually in any authentic Operative Tradition the character of separation of the "individual" is totally burnt and the practitioner can no longer be defined in terms of an artificial ego, but rather by the way he defines his life.

---

[288] Bhagavad-Gita, III, pg. 19.

In the alchemical/hermetic traditions of the West these operations are considered as a "mortification" and "putrefaction" of the individual, ultimately aiming to separate the "subtle elements from the dense elements." In order to better understand the former, let's consider we have a variety of grapes and we want to produce wine, that is, we aim to extract the subtle and spiritual essences of the grapes. The grapes cannot autonomously produce this essence but must comply with a set of production standards and fermentation procedures that allow the transubstantiation to eventually take place. Once a given variety goes through the process and becomes wine. What is the name of the wine? Is it that of the grape variety? Is it that of the origin of the grapes? No. The wine has become a qualitatively different substance than the original grapes. The particular wine identification is that of the particular process of fermentation (namely, the wine producers tradition) induced on the grape variety, in order to optimally extract its essence at a given level. If the grapes never went through the fermentation process, their essence would have never been known, so in this sense even in the case of humans, it is highly speculative and abstract for any individual to refer to "identity" without having formerly gone through a specific operative process that "forges" and "distils" the essence at sacrificial temperatures, that produces an *individuation*. There is still a remnant of this idea when referring to someone as "Dr." or "Sir," etc., but the human essences that are produced in today's highly functional, administrative, and bureaucratic institutions have become to a large sense sterile and highly artificial, like plastic flowers or artificially-sweetened wine.

In the same way that grapes eventually release qualitative substances that are linked to a particular origin, the individual who goes through an operative discipline will eventually have to face predispositions, traits, feelings of guilt, emotional blockages, and fears that were naturally assimilated from the family, lineage, social contexts, and other forms of suggestion. Passions such as anger, hatred, envy, and resentment all become eventually released during the practice,[289] like in the case of toxic alcohols that emerge in any distillation process of the spirit. Such "toxic" releases emerge during moments of fear, that is, during moments when the student is unable to find the "center" of his being which allows the practice to converge properly and the breathing techniques to develop smoothly. However, none of these unconscious toxic predispositions are obstacles to the process, instead to a certain extent, they are necessary "substances" to allow the spiritual combustion to take place. Nietzsche once wrote:

> *"I know the hatred and envy of your hearts. Ye are not great enough not to know of hatred and envy. Then be great enough not to be ashamed of them!"*[290]

The main issue caused by all these toxic burdens is that they cause the practitioner to identify with them, and this attachment fortifies the sense of ephemeral identity which compromises the important capacity to *release*, especially when it comes to breathing and resonating with *prâna*, as formerly explained. These toxic passions, if not dominated by the progressive development of a Magical "I," shall favor a state of *privation/alienation* where the individual is still not capable of "anchoring" in fixed experiential modes, causing deep psychological instability during the moments of the release of inner tension.

From the viewpoint of an Operative Tradition, in order for the Master to impart a teaching which maximizes the potentials of all pupils, it appears obvious that it would be a waste of time and effort to teach the discipline and methods to large groups, as the number of individuals who will sacrifice for the attainment of a spiritual goal are extremely few in number. The recruitment processes which took place in the Shao Lin Temple in China, the cradle of Eastern Martial arts, where the admission in the teachings was extremely strict, were also open to individuals who did not strictly belong

---

[289] *"One needs to affirm the act which consumes deficiency."* Julius Evola. *An Intellectual Autobiography. Path of Cinnabar* Arktos 2009, pg 49.

[290] Friedrich Nietzsche. *Thus Spoke Zarathustra*. War and Warriors.

to the lineages and castes who formerly were the depositories of Tradition. The monks of Shao Lin Temple often traveled all over the countryside, in order to encounter orphans they could godfather. Yet the admittance procedures in the case of those who did not belong to the lineages demanded the aspirants to remain outside the Temple for many years, suffering all sorts of material deprivations and personal conflicts. The monks were entitled to observe the specific attitude and character of the aspirants when they faced such conflictive situations. Finally, those who demonstrated the best attitude were admitted, going through an initial discipline that implied extreme physical training for many years. Hence some initial processes of selection had to take place, in order to allow the best teachers to contact the best students, which corresponds to a principle of optimization in any teaching procedure. As expressed formerly, the key difficulty in the practice is for the student to encounter *a pure center of being* which he can resort to during the moments when all ephemeral individualistic identifications fall apart. The issue is therefore the following: *Do I know who I am, independently of all physical, cultural, social or economic elements I resort to as forms of identity?* Only those individuals who have been capable of living authentic lives beyond the demands of their society have the chance to answer this question. Gaining a set of absolutely genuine and unrepeatable experiences on the part of the individual thus constitutes the "raw material" required to begin with; the experiences are like particles moving through sidereal space, and the issue then becomes that of finding the *star*[291] they are revolving around. The importance of gaining a genuine experience that can't be framed by any standard cultural assumption or ideology is the source of pure creativity, as was stated by Ernst Jünger when writing:

> *"These are the things that constitute the pride and sorrow of a life. All the individual's great moments, the ardent dreams of youth, the intoxication of love, the fire of battle, all of which coincides with a*

---

[291] "*I as the 'pure and immovable centre of light'*" Julius Evola. *An Intellectual Autobiography. Path of Cinnabar.* Arktos 2009, pg 45.

*deeper awareness of the figure; and the memory itself is the magical return of the figure."*[292]

In many cases, it is experiences of unconditional love and the proximity of death that grant the individual the chance to expand beyond rigid and ephemeral identifications that, if not perceived as highly artificial and alienating, control one's entire life . By referring to Max Stirner, Evola describes the issue as following:

> *"He (the student) must generate the power to provide live to himself through the fire and catastrophe of his selfsame life. Hence he must attribute negative potency to all form: he must deny every faith, violate all law, moral or social, despise any feeling of humanity, all love, generosity, all passion, he must reaffirm himself before all science and speculation, thus finally advancing towards a conscious and reasonable madness, this is the Stirnerian "Ich habe meine Sache auf nichts gestellt" (I have not placed my foundations in anything external to myself)."*[293]

This vital trait has always been linked to those groups of individuals who are devoted towards *freedom*; not freedom conceived in terms of the pressure of social, cultural, and economic constriction— thus deeply aiming to promote subversive forms of anarchy and nihilism—but rather freedom conceived in the devotion to living a life of an absolutely genuine, spontaneous, and bold attitude within the margins that social and economic constrictions allow. As we shall further explore in the sections devoted to the idea of

---

[292] "Dies macht den Stolz und die Trauer eines Lebens aus. Alle großen Augenblicke des Lebens, die glühenden Träume der Jugend, der Rausch der Liebe, das Feuer der Schlacht, fallen zusammen mit einem tieferen Bewußtsein der Gestalt, und die Erinnerung ist die zauberhafte Rückkehr der Gestalt. Ernst Jünger. *Der Arbeiter*. Ernst Klett, Stuttgart 1981. Klett-Cotta, pg 17.

[293] "deve cioè generare in sé la forza di darsi la vita mediante l'incendio e la catastrofe di tutta la sua stessa vita, in quanto vita connessa ad un esterno o «altro." Così deve investire in potenza negativa ogni forma: negare ogni fede, violare ogni legge sia morale che sociale, di sprezzare ogni sentimento di umanità, ogni amore e generosità, ogni passione, riaffermarsi di contro alla scienza e alla speculazione in uno scetticismo attivo implacabile e ogni pervadente, spingersi infine sino ad una follia cosciente e ragionata, in una parola: egli deve a sé stesso farsi l'estrema ragione - lo Stirneriano «ich habe meine Sache auf nichtsgesiellt» [«io ho posto la mia casa sul nulla»] gli deve divenire una realtà vivente" Julius Evola. *Saggi Sull'Idealismo Magico*. Edizione Mediterranee. 2006, pg 87.

work in *Operative Traditions Volume II*, corresponds to the pristine essence of *otium, sacrum otium*, which today is distorted when its essence is veiled by the word *leisure*, which is related to idleness and entertainment. The same core idea embedded in *otium* is expressed by Master Awa Kenzo when advising Herrigel that:

> *"The right art is purposeless, aimless! The more obstinately you try to learn how to shoot the arrow for the sake of hitting the goal, the less you will succeed in the one and the further the other will recede. What stands in your way is that you have a much too willful will. You think that what you do not do yourself does not happen."*

The capacity on the part of the individual to create a margin of absolute freedom in life that can eventually constitute the main capital required in order to withstand the discipline imposed by an Operative Tradition corresponds to a capacity that was assigned to the aristocratic and noble castes of societies. So, in order to facilitate things and optimize the long teaching process, the Operative Traditions were taught to the higher castes of society such as kings, queens, princes, and warriors, etc., since these groups were prone to have a natural-born association with freedom. As we shall see in further sections, what effectively determines the belonging to a given *caste* are always operative factors, that is, the instinctive way a being relates to territorial/material dominion. This criteria itself, which overcomes the idea of economic *class*—where the external objects are no longer dominated but rather determine the individual's freedom as economic factors—allows us to realize that today we can no longer refer to the existence of *active* castes in society beyond that constituted by the horizontally dispersed and highly networked, decentralized groups formed by technocrats, technicians, bureaucrats and business developers, who despite not constituting a strict political caste, actually engage in practice operative/technical power. All old political castes of Europe have necessarily surrendered to the power of technocracy and business, hence demonstrating the decline of the political element they formerly embraced and therefore verifying their incapacity to incarnate a State, whose ultimate substance is freedom. Thus it would be unrealistic to apply the same criteria of prioritizing individuals of the old aristocracies to receive the teachings of an operative discipline. And yet the process of selection shall occur as it has always occurred: the individual (no matter what the social, cultural, economic, and ethnic conditionings are) if selecting a way of life that is purely authentic, unconditional, and unrepeatable, shall eventually have greater chances to directly or indirectly access teachings. These teachings are themselves a constant school of self-discipline and self-sacrifice, where ultimately an individual will have no other alternative to survive through the whole process except by relying on a natural sense of non-egotistic pride based on a conscience that is well aware that through all former life's victories and defeats there was always implicitly present a core aim to act based on *justice*. Such justice, which demands to find the perfect balance of forces in a situation, eventually becomes a huge inner capital to resort to whenever all the ephemeral identifications are challenged by an operative process. Such an individual candidate will have to learn to suffer, learn to loose, learn to win, learn to be economically rich and learn to be economically poor; the candidate will have to accept all fair/unfair social criticisms, all social rejections, and he should not ever lose the awareness of his core being, his natural sense of dignity and justice, even when faced with the most terrible injustices or indignities. Such an individual will have to learn to obey, and learn to command; thus natural dignity shall be shown by the fact that such an individual prefers to remain at the margins of society, remain marginalized, and even despised by family and acquaintances instead of serving a cause that does not encourage sovereignty, freedom, dominion and responsibility. The candidate must naturally realize that every time he serves a corrupt institution his core being also risks becoming corrupted. He must have a pure, sincere, unconditional and loyal heart, always acting spontaneously and honestly, without expecting anything in return. Hence we are not defining here someone who rigidly follows moral, religious, or ethical standards, but someone who already has a natural sense of *what is right and what is wrong*, not in terms of any social conformist criteria but in terms of what has to be done fairly, independent of the consequences for one's individuality, or the respect others can then grant.

All of the latter implies that we are not referring to those individuals who get along well with everybody, and who are always intending to be accepted by society. A natural sense of justice and dignity in an individual corresponds to a strictly political idea about values where societal configurations are but a product of such values. The tendency to aim for societal acceptance without questioning the values of the acceptance degrades the value of freedom and responsibility of the candidate who will be challenged by the operative practice. So one of the attributes revealed when an individual is potentially capable of surviving the hardship of an Operative Tradition is the mix of sensitivity, delicacy, and love, combined with virile traits that value commitment, loyalty, combat, and sacrifice. The capacity to recognize one's limitations and mistakes is absolutely crucial, as well as a deeply-hearted respect for the Master and for the Tradition that a Master incarnates.

The Master of an Operative Tradition must always cherish the wisdom (*Arcanum magisterium*) capable of discerning the precious noble attributes of a student, that is to say, those subtle elements of character and temperament that shall are crucial in order to withstand the hardest moments of the practice. If the Master rejects teaching a given student, this decision must not be conceived in terms of discrimination or that of a formal "selection process," instead it is because the Master does not want to cause unnecessary psychological trauma. As formerly explained, *the whole aim of an Operative Tradition is that of distilling the genuine essence of a given individual, that is, to spiritualize such essence based on the objective, transparent, and strict methods defined by the doctrine.* In order to attain a perfect harmony of breath that allows resonance of the currents of the body with those of the cosmos, the *prâna* energy will have to release it from all the psychological and physical determinations where it is confined, and this can only be accomplished if the student has a natural predisposition towards freedom. This is absolutely crucial, though Jünger wrote that:

> "Today is especially difficult to sustain freedom. Opposition demands great sacrifices and this explains the huge number of human beings who prefer coercion."[294]

Deep down the practice of an Operative Tradition will require the student to struggle against himself, along a path where his ego, physical "I," and thinking "I" shall constantly burden the emergence of the Magical "I" where he aims to devote his entire being and essence. Hence those individuals who, despite behaving socially in correct and decent ways, have a natural predisposition to feel happiness when alone—prone to reflect about themselves and their relation to the world—shall be those who have considerable motivation and "fire" to go through the process and accept what the discipline entails. Yet this predisposition towards solitude must be aimed freely, and responsibly. This solitude is not to be confused with that of those individuals who end up alone against their will because they can't maintain a normal and healthy relation with people in general. In other words, we are referring here to *a responsible form of solitude* and not a solitude caused by guilt, or incapacity to relate to other individuals.

In the case of Operative Traditions, there is always a stage during the advanced training when the student has to learn to let go of all control emerging from his will and psychological patterns. In this extremely delicate stage there is no way the Magical "I" can take over the movements and gestures during this release if the student's sense of "I" is still captured within the ego, the physical "I," or the thinking "I." As Oswald Spengler writes: *"The secret of motion awakens in man the apprehension of death (...)It is the thinking man who is perplexed by movement, for the contemplative it is self-*

---

[294] "Es muß nun zugegeben werden, daß die Behauptung der Freiheit heute besonders schwierig ist. Der Widerstand erfordert große Opfer; daraus erklärt sich die Überzahl derjenigen, die den Zwang vorziehen" Ernst Jünger. *Der Waldgang*. Klett-Cotta. Stuttgart 1980, pg 45.

*evident."*[295] One of the main contradictions taking place today is that it is very common to see Karate black-belts who are attached to the specific results of their businesses or private matters. Others are still attached to feelings of anger and hatred. This circumstance reveals that no matter how sophisticated and skillful the results achieved by these students, and no matter how many competitions this person might have won, the Magical "I" has not been released. It must be recalled that once the Magical "I" is released it affects all levels of the life of an individual (social, economic, and work.) but in order to release the Magical "I" *it is precisely all elements of attachment in one's life that must be considered in the spiritual practice*. As Master Awa Kenzo affirms: *"leaving yourself and everything yours behind you so decisively that nothing more is left of you but a purposeless tension."* So as a corollary, no Master of an authentic Operative Tradition would ever impart teachings to a student caught in financial debt or rigid social commitments, since under this context only the physical "I" can be worked out. The "grapes" can become "grape-juice," but are not yet "wine"…

In an Operative Tradition, it is, therefore, illusory to think that one can separate the practice from one's specific life conditions. Yet this gap is something that was widened during the last century as a very bodily based conception of these traditions was imported into the West, developing in parallel to the cult of physical fitness, bodybuilding, etc. As said earlier, these are "sports" that hypertrophy the physical "I," the ego, and some character traits, but are completely insubstantial when aiming to awaken the Magical "I," which requires focus at a social, economic, and cultural level. As a premise, it can be considered that *the freedom of the individual in relation to the world corresponds macroscopically to the freedom obtained by a free and unburdened action microscopically developed in the selfsame hand in the form of a qualitative gesture.*

However, it would be an error to consider the non-attachment trait as that of fearful behavior that emphatically rejects all contacts and dependencies caused by interacting with the circumstances of

---

[295] Oswald Spengler. *The Decline of the West*. Alfred A Knopf. 1926. New York, pg 389.

life. This attitude easily aborts any chance of relating to the powers of the world and to the determinations induced by the powers. Freedom—strictly speaking—only makes sense when referred to the harmonization and integration of a given set of determinations. What really matters in this regard, is the capacity to gain a specific experience (based on the premises of the empirical state of being) then to acknowledge the determination, and finally release from the determination. This release can never be accomplished by escaping the determination, *only by integrating it*. Evola presents this development in the following way:

> *"a sort of heroic sympathy that is in practice a worthy cornerstone for measuring the extent the "I" has truly surpassed the limits of the individual, thus also in order to measure its power. Whatever is the appearance that the "other" presents, happy or unhappy, dark or enlightening, mean or noble, must, however, remain evident the following principle: "Also this is what I am," or better: "I can be, I can also want this," not in the sense of the already debased, promiscuous and pantheistic identities, but rather in the sense of the acknowledgment of one's own transcendental possibilities, as a challenge of responsibility, resistance. It is the feeling of the infinite—as truly unconditional and non-graspable—which is taken for being able of intrepidly assuming the person of whatever other being, a feeling as being able of absolutely wanting such. The latter is, deep down, one of the paths, after which the irrational is enlightened in absolute rationality, the attachments are removed, and the experience of the many and the world becomes exactly that of freedom."*[296]

---

[296] Una specie di simpatici eroica, la quale praticamente può valere come pietra di paragone per la misura in cui l'Io ha veramente superato il limite della persona, quindi anche per misurare la sua forza. Qualunque sia il volto che l'«altro» presenta, felice o infelice, oscuro o luminoso, abietto o nobile, dovrebbe rivestire evidenza il principio: «Anche questo sono io» o, meglio: «Io posso essere, io posso volere anche questo," non nel senso della già condannata identità promiscua e panteistica, ma appunto nel senso del riconoscimento di una propria possibilità trascendentale, epperò anche di una specie di prova di carico, di resistenza. È il sentimento di infinito—come vera incondizionalità e inafferrabilità—che si coglie nel poter assumere intrepidamente la persona di un qualsiasi altro essere, nel sentire di poterla assolutamente volere. Questa è, in fondo, una delle vie, lungo le quali l'irrazionale si rischiara nell'assoluta ra-gione, il vincolo è reciso, l'esperienza del molto e del mondo diviene quella

The latter entails the overcoming of *desire*. *Desire* emerges whenever any external object or determination has power over the individual, that is, over the individual's ego, the physical "I," or the thinking "I." Desire can be then visualized as the attracting force caused by all external aspects effecting the individual, who has no chance but to passively surrender to the determinations that the external force causes. Let's provide some easy examples of the different levels. In the case of the ego—or artificial self-image that an individual creates in a state of privation and lack of actual dominion/persuasion—all external elements that feed and strengthen the ego shall determine the individual responses. This can be observed very easily in the case of advertising techniques, where given individual identifications are constantly produced aiming to *induce the desire* of consumption on the part of the individual who lacks a consistent sense of Magical "I" and has to compensate it with an artificially created "I," often by inducing mind associations between a given product and a happy-looking famous individual or celebrity. As Robert Lane points out:

> *"It is not true that the function of advertising is to maximize satisfaction; rather, its function is to increase people's dissatisfaction with any current state of affairs, to create wants, and to exploit the dissatisfaction of the present. Advertising must use dissatisfaction to achieve its purpose."*[297]

Hence fear drives this process. Jünger writes in this regard: *"The easiest men to scare are indeed who believe that it all ends when it is extinguished their ephemeral appearance. The new slave traders know this and the importance such people grant to materialistic doctrines is based on this fact."*[298] Once this fear/desire couple is activated, such passion completely takes over the thinking "I," which can obviously

---

stessa della libertà. Julius Evola. *Teoria Dell'individuo Assoluto*. Edizioni Mediterranee, 1988, pg 193.

[297] Lane, R. E. 2001. *The Loss of Happiness in Market Democracies*. New Haven, CT: Yale University Press, pg 179.

[298] "Der freilich ist am leichtesten einzuschüchtern, der glaubt, daß, wenn man seine flüchtige Erscheinung auslöscht, alles zu Ende sei. Das wissen die neuen Sklavenhalter, und darauf gründet sich die Bedeutung der materialistischen Lehren für sie." Ernst Jünger. *Der Waldgang*. Klett-Cotta. Stuttgart 1980, pg 92.

analyze the behavior as irrational or inconvenient, yet ultimately the thinking "I" has no power over the individual's actions, since the "I" corresponds to a very thin layer of human consciousness. In the case of the physical "I," desire can easily emerge in the case of the need for comfort, food, medicines, and sex. It is obvious that any individual requires minimum conditions of physical care and nurturing from the outside, but when the individual's consciousness is completely overwhelmed by the need of more comfort, food and sex, this actually signifies that there is a gap between the unconscious realm of the instincts and that of the Magical "I." It must be acknowledged that the Magical "I" has the power to "metabolize" all essences and qualitative values potentially emerging from any experience that affects the physical "I," such as eating, working or sexual experiences. The Magical "I" is also the source of a higher form of physical health, and therefore the taking of medicines and pharmaceuticals required to compensate at a mere physical/biological level, the physical "I" is no longer required in the case that the Magical "I" has become activated. So those individuals overwhelmed by the demands of their instinctive drives, shall be propelled by irrational forces that desire the satisfaction of the instincts, and nor thought, nor morals, nor any ego shall be capable of easily withstanding the pressure, and if the elements do actually repress the urges, then the instinctual nature shall progressively die-out, causing physical illnesses and psychological pathologies, as for instance, Wilhelm Reich[299] pointed out many decades ago. As the operative efforts and disciplines required to activate the Magical "I" have not been promoted in our modern societies, what took place, instead was the inevitable transposition and sublimation of the instincts through entertainment, diversion, social activism and hyper-consumerism, which are all activities that evidence modes of individual privation, and correspond to the degraded forms of *otium* (leisure) that we'll soon deal with in further sections.

The thinking "I," or the abstract faculty of thought was overemphasized in the idealist philosophy of the West since the times

---

[299] See Wilhelm Reich. *Orgone*, volume II: The Cancer Biopathy. 1948.

of Immanuel Kant. The thinking "I," when hypertrophied in the consciousness of the individual, also corresponds to a symptom of desire and need, which nonetheless, often claims to approach reality in a way Nietzsche defined as intellectually "immaculate." The German philosopher writes:

> *"This would be the highest to my mind"—thus says your lying spirit to itself—"to look at life without desire and not, like a dog, with my tongue hanging out. To be happy in looking, with a will that has died and without the grasping and greed of selfishness, the whole body cold and ashen, but with drunken moon eyes (...) O you sentimental hypocrites, you lechers! You lack innocence in your desire and therefore you slander all desire."*[300]

This occurs in the case of many intellectuals, scholars, scientists or dilettantes who have not managed to operatively cope with their instinctive nature—or their instinctive nature is weak—and it wouldn't be misleading at all to consider that their intellectual activity corresponds to a very pale mode of repression and instinctual sublimation, which socially justifies their privation, as our society is much more prone to accept to live with a sophisticated thinker or philosopher than with a criminal. Evola refers to all of this as "an interplay of prestigious dialectics, as transposed phenomenalism."[301]

But deep down, there is not so much of a difference between these two types, and it's hard to tell who of both to trust in situations of danger of life conflict.[302] Both are determined by their passions, yet in the former case the passions are sublimated and transposed, and in the second case they are affirmed in a raw, direct manner. And what's more, if we look closer at the ideological foundations existent within the intellectual products that a high percentage of

---

[300] Friedrich Nietzsche, Walter Kaufmann. *The Portable Nietzsche*. Penguin Books (1977), pg 234.

[301] "I giuochi di prestigio dialettici, in un fenomenismo trasposto" Julius Evola. *Teoria Dell'individuo Assoluto*. Edizioni Mediterranee, 1988, pg 41.

[302] Friedrich Nietzsche's words are highly meaningful in this regard in The Pale Criminal. *Thus Spoke Zarathustra*: "Many things in your good people cause me disgust, and verily, not their evil. I would like that they had a madness by which they succumbed, like this pale criminal!"

scholars and thinkers promote in society through journalism, we can easily perceive Darwinian, Marxist, Liberal, and even Freudian standpoints very much present, which all correspond to intellectual justifications of the most primitive, irrational, and animalistic elements in men. A nihilistic view of life is enhanced whenever the ideological constructs are promoted, where it is provided with a justification to given tendencies. As Nietzsche writes in this regard: "one does not want to fight weakness with a système fortifiant [a method that strengthens] but rather with a kind of justification and moralization; i.e., with an interpretation."[303] Hence deep down, all these intellectual constructs correspond to transposed modes of human primitiveness and privation, seductively and sophisticatedly projected into the intellectual domain, inducing a very severe and dangerous psychological contagion, especially in the case of those youngsters who have not gained any genuine experience that can ultimately serve as the "fuel" for disciplining themselves in operative practices. The lack of adequate institutions in our time required in order to configure and forge the spirit of the individuals and release them from the fear of themselves is expressed by Jünger when writing:

> *"The great loneliness of the individual person is one of the characteristic signs of our time. The unique person is fenced, surrounded by a fear that pushes like a wall."*[304]

The attachment on the part of the individual to all justifications and rationalizations induced in consciousness by the thinking "I" is something that the student, first of all, has to be aware of, conceiving the attachments as transitional modes of privation that necessarily exist along the entire process. By mere mental effort, it is not possible to just "forget" the influence of the constructs nor release from desire since the privation of the "I" transcends the faculty determined by thought. What's more, a mere denial or frontal confrontation against

---

[303] Friedrich Nietzsche, Walter Arnold Kaufmann, R. J. Hollingdale. *The Will to Power*, 1968, pg 29-30.

[304] "Die große Einsamkeit des Einzelnen zählt zu den Kennzeichen der Zeit. Er ist umringt, ist eingeschlossen von der Furcht, die sich gleich Mauern anschiebt gegen ihn." Ernst Jünger. *Der Waldgang.*. Klett-Cotta. Stuttgart 1980, pg 56.

the pervasive influence of the thinking "I" will not be effective either, and will make the contents of the thinking "I" even more reactive and subversive. By pointing out this conflict, Yukio Mishima writes:

> *"Words are a medium that reduces reality to abstraction for transmission to our reason, and in their power to corrode reality inevitably lurks the danger that the words themselves will be corroded too."*[305]

Hence what is required is to keep working along the process of dominion and persuasion, connecting with the body, and directly relating to the external objects as the only framework for criticizing discursive constructs/morals/assumptions,[306] and all this discipline is practiced in combination with the focus in correct breathing. The most difficult transition, which actually constitutes a qualitative or "quantum leap" in consciousness corresponds to the capacity to practice all these operative exercises for their own sake, that is, unconditionally and by not expecting anything in return. Whenever the student expects results from the practice, the expectation corresponds to a sign of privation, lack of release capacity, and lack of inner freedom, not necessarily during the specific practice, but during life on a daily basis whenever involved in other activities and affairs. The whole aim of this discipline is to develop what in Eastern traditions was referred to as wei-wu-wei, the art of acting without acting or what Master Awa Kenzo referred to as the "gentle art." This is the development of activities that are free from desire and that are released from all determinations. It is at this stage of the practice where the core nature of the student shall be tested in all its magnanimous generosity and nobility since the student will have to be capable of focusing his entire concentration in the core essences of all things.[307] These essences are presented to our

---

[305] Yukio Mishima, John Bester. *Sun and Steel*. Kodansha USA (2003), pp 6.

[306] "Those who have first not known how to doubt *everything*, can have absolute certainty of *nothing*" Julius Evola. *L'individuo e il divenire del mondo*, Libreria di Scienze e Lettere, 1926, pg 12.

[307] "Not until he has suffered the unreality of every reality, the uncertainty of every fact, the darkness of every light" Julius Evola. *L'individuo e il divenire del mondo*, Libreria di Scienze e Lettere, 1926, pg 17.

senses only in terms of direct experience, with no mediation. "The essential changes are the most difficult to grasp since they occur in what is obvious"[308]—as Ernst Jünger points out. The concept of quality embraced by Robert M. Pirsig also points to the essences, and as the American author remarks: "intellectuals usually have the greatest trouble seeing this Quality, precisely because they are so swift and absolute about snapping everything into intellectual form. The ones who have the easiest time seeing this Quality are small children, uneducated people and culturally "deprived" people."[309] In the same way some individuals can love the beauty of a flower without expecting anything of it in return—that is, without wanting to pick it up, grab it, without wanting to release sentimental feelings, take it away, or even pay for it—this is the same unconditional love required by the practitioner of an Operative Tradition who aims to release from desire and eventually reach the Magical essence. Julius Evola expresses in *Phenomenology of the Absolute Individual* that: "Love here is therefore not need, it is not abandonment, it is not "altruism"—nothing sentimental, nothing "moral," "beautiful," "noble," or anything else, nothing that basically men or rather women often call it—it is something pure, frightening, transcendent."[310] Therefore no need of emotional comforting, in addition to the capability of withstanding solitude are at this point key elements of the practice. Many issues of conscience and guilt can therefore easily arise. Let's provide a simple example: Let's say someone has a pet (dog or cat), and if we ask the person to tell us why they feel they love their pet; we will soon identify that in most cases love corresponds to a sense of emotional attachment, and a compensation of the fear of being alone, that what Abraham Maslow referred to as "deficiency

---

[308] "Die wesentlichen Veränderungen am schwersten zu erfassen sind, eben deshalb, weil sie im Selbstverständlichen vor sich gehen." Ernst Jünger. *Der Arbeiter*. Ernst Klett, Stuttgart 1981. Klett-Cotta, pg 117.

[309] Robert M. Pirsig. *Zen and the Art of Motorcycle Maintenance: An Inquiry into Values*. 1974 (William Morrow & Company).

[310] Amore qui non è dunque bisogno, nor è abbandono, non è «altruismo»—niente di sentimentale, niente di «morale,» «bello,» «nobile,» o altro, niente, in fondo, di tutto ciò che gli uomini o, per meglio dire le donne, chiamano tale—è qualcosa di puro, di spaventoso, di trascendente. Julius Evola. *Fenomenologia Dell'individuo Assoluto Iii* ed. corretta: Edizioni Mediterranee, Roma 2007.

motivation" and "deficiency love" neurosis inducers.³¹¹ Very few would say that they love the pet because of its unique essence, but these few will be able however to accept the final days of the pet with calm and sovereignty since their love impels them to conclude that this essence is always present, that this essence doesn't die, even if the physical body inevitably perishes.

At this stage in the practice, it is still far away from the ultimate goal. However the inevitable consequences and limitations imposed—though "imposition" was not initially the intention—by the type of spiritual development proposed by one of the most prestigious and influential gurus of the last century Jiddu Krishnamurti (1895-1986), whose spiritualist doctrines aimed to liberate the psyche of mankind from all the restrictive barriers of superstition and thought—even the limitations of the theosophist schools he was raised in—thus proposing a path for self-autonomy without the need of any Masters or any institutional support, in order to attain the incorruptible "I."³¹² Krishnamurti was clearly aware of the false identifications caused by the ego and the thinking "I," thus conceiving the existence of a higher form of life that ought to be liberated from the limitations and barriers, in order to reach what he claims to be "individual uniqueness," which is life conceived in a formless and pure cosmic state,³¹³ also considered to be an indescribable and untamable mode of life. Such proposals on the part of Krishnamurti constitute a very healthy reaction and the overcoming of the limitations caused by many aspects of modern culture, and even a positive superseding of the speculative limitations promoted by modern theosophical schools. The main danger that arises with Krishnamurti's proposals is that the release from all identifications of the "I" and submerging into the force of "liberated life," and "lawless perfection"³¹⁴ in all its purity can be like leaping into the sea without knowing "how to swim." What's more, Krishnamurti

---

[311] "Most neuroses involved, along with other complex determinants, ungratified wishes for safety for belongingness and identification, for close love relationships and for respect and prestige" Abraham Maslow. Towards a Psychology of Being. Van Nostrand Reinhold Company, New York, pg 21.

[312] *La Dissolution de l'Ordre de l'Etoile.* J. Krishnamurti, Ommen, 1929.

[313] J. Krishnamurti, *La Vita Liberata*, Trieste, 1931.

[314] J. Krishnamurti, *La Vita Liberata*, Trieste, 1931.

considers all forms of discipline, methods and techniques as a fixed pathway that also must be discarded as modes of authoritarianism and "patriarchy" (a term very much used by those who directly or indirectly follow his teachings). He writes:

> *"I maintain that truth is a pathless land, and you cannot approach it by any path whatsoever, by any religion, by any sect. That is my point of view, and I adhere to that absolutely and unconditionally."*

However, in the case of an Operative Tradition, strong discipline is as equally important as the "letting go" of egotistic mechanisms, as proposed by Krishnamurti. An Operative Tradition embraces the core requirement of release from any psychologically fixed patterns, yet the discipline is extremely focused on perfecting the body movements, breathing, attitude, and the constant awareness of the action exerted upon the subtle powers of physis, which correspond to the particular domain that Krishnamurti's constantly referred to as "liberated life." Contrary to most spiritual gurus of the last century, an Operative Tradition does not propose liberation as a path attainable by everyone in the entire world, since release from the ego can be extremely traumatic for individuals burdened by deep trauma, guilt, or fear, which risks strengthening the ego or neurotic predisposition seven further. A strict Operative Tradition always provides an extremely rigorous and time-consuming discipline for learning to "swim" during the key moment of release, by always following the rule that the number of individuals who can cope with the rigors of the practice are always very few in number.

One of the apparent paradoxes emerged in the case of spiritual doctrines like those proposed by Krishnamurti is that in the case of the hippie and countercultural movements of the 1960s, the teachings reinforced a very strong sense of "self-cocooning" and higher levels of alienation in sectors of societies formerly influenced by Krishnamurti's views. Inevitably, "spirituality"—converted into a commodity—descended to the level of aiming to solve private matters related to the attainment of happiness. However, the a paradox is solved by recalling the inevitable "boomerang effect" of the ego when consciousness is weak at an operative and magical level. Hence, during the last decades, the figure of Krishnamurti has been necessarily followed by those individuals who aim to "free their mind" and become "spiritual," believing that the issues of technique, technology, or economics correspond to a "non-spiritual" area that has to be outsourced to experts, scientists, engineers, economists and politicians. This trait clearly expresses the gap between the Magical "I" which is intended to be awakened by an Operative Tradition, and the actual conditions of existence that affect the individual.

Hence by burning and exhausting highly speculative ideas such as those proposed by Krishnamurti, and by keeping up the discipline, tests the capacity for self-sacrifice, and the capacity to transcend all thought identification fears. As Eugen Herrigel writes: *"his conquests and spiritual transformations, so long as they still remain 'his', must be conquered and transformed again and again until everything 'his is annihilated,'"* In effect, fear is always related to the feeling of facing death and of release from life's conditions, entering into the realm of the unknown. The kind of knowledge that perceives the determinisms of the world as forces embedded with specific attributes that have to be addressed, is a kind of knowledge that not only allows relating to the needs of our time, but also allows us to overcome the implicit nihilistic character present in such needs. As Ernst Jünger writes in *Over the Line*, (1951): *"Who has not experienced upon himself the enormous power of the nothingness and has not suffered its temptation knows very little our age."* And yet if the practitioner learns to devote his life to relate to things *as they are*, to imagine them as they are, and not as they are presented to his mind, needs or desires, he will eventually have the capacity to relate to all essences, and to those timeless figures that never die at the level of the empiric state of being. The symbol of a rose never dies, the symbol of a hawk never dies, a musical composition never dies, the fragrance of a purely hearted and fair woman never dies, the beauty of a Tradition never dies.

To release the senses and progressively *resonate* with such essences is at this point of the practice the key fundamental factor. And the ultimate proof of such accomplishment, in addition to the integration of consciousness into the Magical "I" shall be demonstrated empirically by the modification of gesture and of the natural reflexes of the body when developing any action that causes the individual to "let go." As Nietzsche wrote: *"the phenomenon of the body is the richer, clearer, more tangible phenomenon: to be discussed first, methodologically, without coming to any decision about its ultimate significance."*[315] A very specific form of balance and stability shall emerge in the physical body that challenges any bio-mechanical law, resembling the extraordinary and fascinating way that a snake manages to effortlessly crawl around the ground. Eventually, the practitioner shall notice that a specific force or subtle energy shall take over his consciousness and body reactions when letting go of all conscience control. These forces or energies correspond to subtle forms of entropic currents that "pull" the body into given modes of movement that aim towards a center of overall balance, but that can only be acknowledged *when the body is actually in movement*. The idea of *chreode*, which is related to that of an attractor, can come here "to the rescue" of our explanations, in order to visualize the character of the experience. Conrad Hal Waddington suggested the concept *chreode* (from the Greek *chre*, it is necessary, and *hodos*, route or path), to express the determinism and paths of development of given non-linear processes out of equilibrium, and illustrated it by means of a simple three-dimensional "epigenetic landscape."

When it comes to an Operative Tradition, the whole idea of the practice is to configure the "chreodes" that the practitioner progressively learns to feel as determining the development of his movements when his conscious mind releases control. This idea, which can be thought initially as inconceivable on the part of the student during the initial stages of teaching, can be approached by understanding that, as it occurs in handwriting where the unconscious patterns that impel the movement of the hand are

---

[315] Friedrich Nietzsche, Walter Arnold Kaufmann, R. J. Hollingdale. *The Will to Power* 1968, pg 270.

different in each person independent of their strict intellectual/discursive formation, morphogenetic fields that are active in a given individual are not the same, depending on the operative training of the individual. Contrarily to gravitational, electric, or magnetic fields, the morphogenetic fields do not act on all elements/particles, but rather "pull" those that have been forged by a qualitative aspect capable of integrating and synthesizing all previous forms and figures that emerge in the operative type of training.

The whole idea of the practice is for each individual to progressively configure, integrate and harmonize in his action the morphogenetic fields he becomes attached to; hence, this process corresponds strictly to a series of temporary spiritual attachments, not in the sense of mystical or ecstatic connections, but rather, in terms of magical and operative ones, which can be verified in practice, whenever the body moves. An Operative Tradition is the only discipline capable of verifying the existence of the morphogenetic fields, which have constituted in modern physics the controversial idea, since hypothetical fields have not been measured by any technology, such as in the case of other fields like the gravitational or the electromagnetic. However, from an epistemological viewpoint, it shall always be impossible to "measure" the fields through devices based on cybernetic techniques, because nature conceived in strictly materialistic terms, or even the same fields that are registered in such devices, are a "mirror" of the internal functional characteristics of the same devices. So this is why in an Operative Tradition all mediations with nature/physis have to be minimized, and it is the whole set of senses of the individual that are forged and configured, following the phenomenological assumption that the experiential domain is a function of the "I" the individual is related to.

In spite of the popularity gained by quantum mechanics or Einstein's relativity, none of these domains of physics have any specific application in an Operative Tradition, because, especially in the case of quantum mechanics, there is a probabilistic/statistical approach to nature which runs countercurrent to the focus required on the part of the student in terms of their unique and qualitative character. In an Operative Tradition, there is no need to "predict" the development of a given phenomena since any attempt of prediction already constitutes a sign of conscious control on the part of the mind which affects the character of "breathing." The specific developments impelled by morphogenetic fields overcome the restriction of physical law, and the "breathing" itself is intended to resonate with the ultimate attractor of the field along a path—the "chreode"—that becomes "charged" through a resonance process, determining the character of the movement once the conscious mind has released attempts to control the non-linear phenomena.

If there is some resemblance between quantum mechanics and the process described here, it refers to the fact that in quantum mechanics there is always a given wave function that indicates the probability of the presence of the particle along a specific coordinate or set of coordinates. We can merely use this quantum analogy to describe that also in the case of an individual who releases from conscious mind coercion within a given body/operative action, a "wave function" can be visualized, which corresponds to a development in time and space. This wave function is potentially capable of developing a resonance with a given layer or state present within a morphogenetic field (morphic resonance) and through a centripetal/negative entropy process. This resonance induces in t the body specific gradients of development that "pull" the movements, like the ocean lifting a surfer. In order to achieve the qualitative movement, the focus on conscious breathing on the part of the student is necessary yet not sufficient. We are referring here to a more "subtle" form of "breathing" that implies an a harmonization of all life experiences (past and present), and the creation of a "music" one carries within, and that expresses its character precisely in those moments where the conscious mind is absent and all physical laws lose their legitimacy. So, if there is also another adequate analogy within the domain of quantum physics, it is that the influence of the "wave function" of the student's "breathing" is not only present during the specific training sessions, but is "spread" in time and space (like in a "multiverse"), beyond the body, and qualitatively configured depending on the way life experiences were rendered into "art/music" as a mode of redemption. All forms of personal guilt or material attachment impede correct "breathing," in spite of the fact that a conscious discipline of physical breathing during the activity can be performed perfectly well by the student. The difficulty of attaining harmonic levels of unconscious breathing entails the presence of an underlying fear when it comes to "let go" of the conscious mind, and the temporary impossibility of experiencing the world beyond the conscious mind, as a centripetal configuration of symbols and figures that resemble a musical composition. As Julius Evola points out in his *Phenomenology*:

> *"The object of the new experience is nothing but a "rhythm," just like the shape of the gesture caught in flight, before it precipitates into a thing: it is not a new kind of phenomena, but rather the resolution of what in general is actually phenomenal in the sense of a particular kind of spiritual movement—almost a synthetic "sensation" of such process. As such, it corresponds to the object of a productive perception, as the individuality of a gesture in which the I itself is expressed, almost expressing a deeper and buried mode of life. The I is therefore really the creator of such a world, which is only it to the extent it becomes it. On the other hand this reality is also objective: it is experienced as the resolution of a real object, not as an "other world."*[316]

Let's present now to the reader a simile which allows us to visualize the extremely subtle process of action-reation that takes place during these phenomenon, since at this stage of the practice, imagination of the symbolic aspects of reality is a crucial factor, which helps a lot to transcend typical cause-effect thought configurations:

---

[316] L'oggetto della nuova esperienza non è null'altro che un «ritmo," quasi è da dirsi la forma del gesto colta a volo, prima del suo precipitarsi in cosa: non è una nuova specie di fenomeni, sibbene la risoluzione di ciò che è fenomenico in generale nel senso di una particolare qualità di movimento spirituale—quasi una «sensazione» sintetica immateriale del suo processo. Come tale, esso non è che come oggetto di una percezione produttiva, come l'individualità di un gesto in cui l'Io stesso si esprima, quasi esprimendo una sua vita più profonda e sepolta. L'Io è dunque realmente il demiurgo di un tale mondo, che è soltanto in quanto egli divenga lui. D'altra parte questa realtà è altresì oggettiva: essa è sperimentata come la risoluzione di un oggetto reale, non come un «altro mondo." Julius Evola. *Fenomenologia Dell'individuo Assoluto Iii* ed. corretta: Edizioni Mediterranee, Roma 2007.

Let's suppose a charged particle is present within an electric field. This particle shall only be capable of generating a magnetic field when it is in motion. In addition to this scenario, let's suppose that the particle "breathes" and that it performs a particular vibration/oscillation. Due to this vibration, an electromagnetic field shall be generated, which corresponds to the projection of mechanical energy present in the charged particle. Because of the irreversible centrifugal character of generated wave-patterns, from a thermodynamic viewpoint the electromagnetic radiation produced by the particle can be conceived as the production of entropy, or in other words, it can be affirmed that the particle is dissipating its mechanical energy in all directions. However, the particle shall also be affected by the electromagnetic wave it is producing according to phenomena called *induction* (in this case electromagnetic induction). To express this phenomenon with other words, *the particle shall be partially affected by the character of its actions/movements*. ("What goes around comes around.") Such induction inevitably interferes with the vibratory "breathing" of the charged particle, and what's more, it can even alter the particular charge distribution of the particle. This is exactly the main issue that appears during the moments of "release" of conscience during the disciplining of physical motion taking place when following an Operative Tradition. What is required at this crucial release stage of practice is, therefore, the capacity to "surf" the self-induction phenomena, that is to say, to create a constructive interference and resonance between the motion and the irreversible dissipative effects caused by the motion. If this is accomplished successfully, then a given proportion of the dissipated energy release is re-channeled centripetally, and the action becomes *perfect*. The movement "feeds on its own waste," so to speak, causing extreme power releases that apparently defy the laws of physics, yet only apparently, since what is really taking place is that the physical laws (for instance those of thermodynamics) are superseded by a higher harmonic principle that integrates q physical determinations and laws.

Therefore, we can conclude by stating that in no case are the physical movements intended to be *efficient* (as occurs in all sports that are devoted to performance) but rather dissipation and waste are here the secrets to perfection, ultimately linked to the dissipation and sacrifice of all self-identifications. Jünger writes:

> *"The sacrifices are important. With them is also associated, however, an immediate increase of sovereignty."*[317]

The flight of an eagle in the skies is also an adequate image of this process, as the apparent effortlessness of the bird when flying high would be impossible in practice without a perfect connection to the implicit patterns of heat transfer present in the winds.

---

[317] "Die Opfer werden bedeutend sein. Jedoch verbindet sich mit ihnen auch ein unmittelbarer Gewinn an Souveränität." Ernst Jünger. *Der Waldgang*. Klett-Cotta. Stuttgart 1980, pg 81.

# IT:

## THE SPIRITUAL "FLASH," THE LIGHTNING OF ZEUS

This sudden self-lighting is the lightning-flash. It brings itself into its own brightness, which it itself both brings along and brings in. When, in the turning of the danger, the truth of Being flashes, the essence of Being clears and lights itself up. Then the truth of the essence, the coming to presence, of Being turns and enters in. To flash *[blitzen]*, in terms both of its derivation and of what it designates, is "to glance" *[blicken]*. In the flashing glance and as that glance, the essence, the coming to presence, of Being enters into its own emitting of light. Moving through the element of its own shining, the flashing glance retrieves that which it catches sight of and brings it back into the brightness of its own looking. And yet that glancing, in its giving of light, simultaneously keeps safe the concealed darkness of its origin as the unlighted.

-Heidegger[318]

In the case of the specific archery practice exerted by Eugen Herrigel, we formerly described how the arc and bow could be well modeled as a *spring-mass-damper* system. These physical systems are also operatively equivalent to the LCR electric circuits (*inductance-capacitor-resistance*), which constitute the main electrical configuration that photoflashes and countless other electronic devices require for causing a high tension *pulse*. Such *pulse* can be considered equivalent to the sudden phenomena taking place during the moment of release of the arrow during the instance of highest tension on the part of the archer. But, as an electric projection of the strict metaphysical character of the phenomena, let us first analyze what happens in the case of the flash technique in a camera.

Though the battery set of a photoflash might only be characterized by a rather modest 4.8 V nominal tension and a maximum power

---

[318] Martin Heidegger. *The Question Concerning Technology*. Garland Publishing, Inc. New York & London 1977, pg 66.

output of 25 Watts when the batteries are fully charged, the inner capacitors of the photoflash can easily attain mortal tensions of 380 Volts, which is the basic requirement for instant power release in the form of light. Technically, the high voltage in the capacitor is attained when producing a very specific electrical oscillatory signal in a specifically designed RLC circuit.

*RLC series circuit*

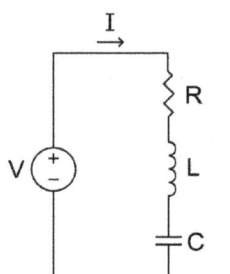

- V, the voltage source powering the circuit
- I, the current admitted through the circuit
- R, the effective resistance of the combined load, source, and components
- L, the inductance of the inductor component
- C, the capacitance of the capacitor component

Any RLC circuit (just like any mass-damper-spring system) is characterized by a typical, natural frequency or resonance frequency, which in this case depends on the specific RLC configuration and values.[319] The resonance frequency is applied as an electric signal excitation to the circuit, the tension in the capacitor shall progressively increase up to a given maximum tension value. During the charge of the capacitor that takes place in parallel to the progressive increase of its voltage (the intensity of the electric field), what occurs is that the perfect combination of electromagnetic induction, dissipation and the particular characteristics of the capacitor when increasing the intensity of the electric field, all in combination, allow considerable amount of energy to be stored in the electric field. Supposing we have a 1,500 µF capacitor charged with 380 V, the energy contained in the field is then 108 joules.

---

[319] In the case of a series RLC circuit the frequency of resonance $w_o$ is given by the formula:
$$\omega_0 = \frac{1}{\sqrt{LC}}$$

Apparently not much, but because the flash takes place as a sudden discharge of the energy contained in the electrostatic field during 1/1000 of a second, the power release is 108000 Watts (108 KW). Compare this value to the nominal power output of the batteries (about 25 Watts), or even the average 3-8 KW max power. Besides, touching a 4.8 V source is not dangerous, yet touching a 380 V source can be pretty "shocking."

So, if we make an effort of trying to induce the metaphysical aspects of the process Master Awa Kenzo dominated, by visualizing and meditating on this flash example, such visualization will allow us to relate more and more to the crucial and subtle elements present in the case of the cycle of the release of the arrow on the part of Herrigel during his long apprenticeship. In the same way that in a flash device the capacitors can not be charged to their maximum tension without the application of a very specific electrical signal, so in the case of Eugen Herrigel the specific breathing cycle is the absolute key to understanding the process of effortlessly attaining the maximum tension just before the release. The spiritual flash or *numina* can be visualized as the ethereal correspondence of a physical sexual orgasm, which firstly requires a progressive process of perfect coupling and resonance between the polarity of the bodies. In this case, however, the "coupling" is not taking place at a physical level, but at a ethereal one, and the maximum tension attained is not perceptible in the muscles of the arms, shoulders, etc. of the Master when grabbing the arc at maximum amplitude, since these physical elements are all in a state of extremely high relaxation. All the energy is rather stored in the "fields" surrounding the archer, which corresponds to the "body made mind" or "apparent body" (mâyâvî-rûpa) in the Hindu tradition, or the "body of transformation," "immortal," "magical" or "free" (nirmâna-kâya) of the Trikâya doctrine in the Mahayana Buddhist teachings.

In the case of the flash device, the maximum tension attainable in the electrostatic field confined in the capacitor is not dependent on the user (in the case of the photographer who pushes the button) but on the specific characteristics of the RLC electric circuit. In the same way, the archer is not free to choose the maximum tension of the bow nor the moment of release of the arrow. Such elements are determined beyond the will of the archer; all the archer is encouraged to do through the entire teaching is to learn to release from his individual will and learn to resonate with all elements through attaining breathing resonance, allowing the surrounding fields to accumulate subtle forms of energy.

\* \* \*

*"The right shot at the right moment does not come because you do not let go of yourself. You do not wait for fulfillment, but brace yourself for failure. So long as that is so, you have no choice but to call forth something yourself that ought to happen independently of you, and so long as you call it forth your hand will not open in the right way."*

\* \* \*

We can recall at this point a remarkable experiment developed by a researcher in the physiology department of the University of California, San Francisco, Benjamin Libet (1916-2007) who demonstrated the existence of a time-gap between the conscious decision of a given individual when intending to perform an action, and the action itself, which is impelled directly by unconscious factors. In other words, the thinking "I" or "free will" is overthrown by irrational and unconscious factors.

\* \* \*

*"I myself was not breathing but—strange as this may sound—was being breathed."*

\* \* \*

If we understand all the former, or at least visualize the parallelisms exposed, it shall also be easier to understand why during the critical moment of release, any wobbling or disruptive "noise" on the part of the archer-arc couple is practically eliminated. Any disharmonic reaction on the part of the archer or arc would reveal that energy was still stored at a physical level in both elements.

In many different areas of physics, fields act as attractors of phenomena, they determine the development and gradients of a dissipative system, especially in non-linear processes out of equilibrium, imposing patterned restrictions on the outcomes. We are mostly acquainted with electromagnetic fields (since our modern technical devices mostly depend on these) and we are also naturally familiar with fields such as the gravitational. Considered as a release from all interferences arising from the discursive mind and as a release from all subconscious blockages, in the case of an Operative Tradition, the term "spirituality" can only be conceived in terms of fields, of operative attractors, integrating factors; as *entelechies,* which is a Greek word (*en-telos*) indicating "something" that contains a goal in itself, an aim that determines an entire organization, macro and microscopically. All identifications and attachments are necessarily burnt along the whole operative discipline in order to energize these fields.[320]

Such fields condition the subtle developments taking place in the physical plane, whenever the physical plane is impelled to attain non-linear conditions out of equilibrium and high levels of energy dissipation and entropy production where the capacity for predictability is extremely diminished. In physics, the equations that express the intrinsic behavior of a given system are called *differential equations,* since in the case of the behavior of any element in the physical plane, the variation of factors is as crucial as the factors themselves. For instance, an acceleration of a given object (like that of Herrigel's arrow during its motion) corresponds to a variation

---

[320] Un implacabile ed onnipervadente scetticismo corrodente ogni certezza filosofica e scientifica, sino alla violazione deliberata di ogni legge morale o sociale; dalla riaffermazione di là da ogni valore riconosciuto e da ogni autorità, fino alla negazione di ogni fede, ideale o entusiasmo e al disprezzo di ogni sentimento di umanità, di amore o rispetto [a relentless and all-pervasive skepticism eroding all philosophical and scientific certainty, the deliberate violation of all moral or social law; a reaffirmation beyond any recognized value and all authority, towards the denial of any faith, ideal or enthusiasm and the contempt of every feeling of humanity, love or respect] Julius Evola. *Fenomenologia Dell'individuo Assoluto Iii* ed. corretta: Edizioni Mediterranee, Roma, 2007.

in its velocity, and the variation has crucial effects in terms of the inertial forces that an object is exposed to. As long as these variations can be framed in space/time/mass magnitudes, the variations can be measured. And yet what happens in the real world of our actual experience can only be scientifically predicted by integrating the differential equations when we deal with extremely mechanistic configurations. Yet when dealing with the domain of organisms that develop their attributes based on *poiesis*, integrating figures or symbols, the differential equations show—in all fields of physics—their highly limited capacity for prediction. And what's more, when friction and opening to the whole occur in these systems, a disturbing phenomena takes place: the so-called "sensitivity to the initial conditions," which means that any microscopic perturbation induced by unknown subtle and chaotic factors can radically alter its predictability and behavior. Such sensitivity to the initial conditions was expressed as the "butterfly effect" decades ago by Edward Lorenz when affirming that *"one flap of a sea gull's wings would be enough to alter the course of the weather forever."*[321]

In his last work, Nietzsche wrote: *"Whatever Christians and idealists have devised has neither rhyme nor reason: we are more radical. We have discovered the "smallest world" as that which is decisive everywhere."*[322] Extremely subtle forces can have a huge impact in the physical domain, and the whole practice of an Operative Tradition is that of channeling the forces into very specific modes of motion and gesture. The way to accomplish this aim is by configuring and developing the Operative "I," so that the subtle forces are assimilated harmonically, as occurs with the case of a typical "domino configuration" where the physical work exerted by the entire domino system when it is falling can be vastly more significant than the first tiny impulse that triggered the motion in the first domino piece. Ultimately the symbols and figures embedded in a given technique developed by the operative practice act as specific

---

[321] Lorenz, Edward N. (1963). "The Predictability of Hydrodynamic Flow." *Transactions of the New York Academy of Sciences*. 25 (4): pp 409–432. Retrieved 1 September 2014.

[322] Friedrich Nietzsche, Walter Arnold Kaufmann, R. J. Hollingdale *The Will to Power* 1968, pg 525.

"receptacles" of subtle and microscopic influences, in a similar way to how a telescope or a microscope amplify the connection with the elements present in the macrocosmos and the microcosmos. And yet this little analogy of the microscope/telescope also presents us with an important dilemma. The act of developing a firm discipline and practice based on formal teachings is no guarantee whatsoever that the subtle domain of forces shall be channeled adequately, in the same way that knowing how to use a microscope is not sufficient to be able to perceive a specific bacteria amidst given conditions. Based on the former explanations on the limits of transcendental idealism and the new approach to Magical Idealism presented by Evola, the "I" can only perceive in experiential terms *that which it already is*. Any of us can have a photographic camera or set of instruments or techniques, and yet what determines our specific *focus* in a given domain of reality or existential issue is not only a function of the technical characteristics of the medium but also a function of *who we are*. So, as Master Awa Kenzo pointed out to Eugen Herrigel, the whole idea behind developing the technique of handling the arc and bow corresponds to the means required in order *to point to oneself*, to focus *the operative "I" into oneself*, into one's *state of Being*. This state of Being has a transcendent, metaphysical character; it transcends any psychological or anthropomorphic conception of a given personality, and yet it ultimately determines the *focus* that is addressed by technical means, when operatively linked to the individual. In other words, such a state of being provides a very specific aim for integrating the means.

So, in this sense, in any Operative Tradition, the Master encourages the student to focus both in the immanent and the transcendent domains. The immanent domain corresponds to that domain where the student relates directly to the forces that determine motion in the physical plane, and the transcendent domain corresponds to the domain where the student relates to the subtle forces that determine the influences he shall be capturing in practice, and that shall determine the character of his gesture. Ultimately the student is encouraged to aim for the center, to the heart, to *his heart*, thus allowing the convergence of efforts in a single direction. A deep sense of respect towards the Master, and towards the Tradition he incarnates is therefore absolutely crucial, in order to be capable of aiming for the regal center. It is then the "I" not only perceives reality according to specific configurations but what's more, *produces* this reality, especially in terms of the subtle influences or "butterfly effects." Evola referred to this domain as the "dynamic super-sensitive domain" and he affirms that this domain,

> *"appears from an expressive gesture emerging from the inside, directly related to the "I." Now, when in the experienced "body of activity," such centrifugal direction is then suddenly overturned in the opposite, he will feel certainly different in terms of the former distinctive power. And afterwards, in the further illuminative flashing of a more intense and decisive gesture of evocation-distinction."*[323]

At a personal level, what allows the centripetal direction of motion in the body towards one single point—assisted by the subtle/entropic "winds"—is the sacrificial and heroic capacity on the part of the student to release from all attachments (psychological, material, etc) that strengthen the ego that disrupts the powerful influence of the subtle "butterflies." This metaphysical domain is *primordial*, that is, it corresponds to a domain that once rendered active, centripetally configures all material objects and influences in such a way that it can only be considered as a magical process, and thus all the developments taking place in the physical plane are then conceived of as *secondary* or merely *derived* from the primordial, center of potency and light. As Evola writes: *"compared to the "subtle" dynamism, the sensible is felt as mere shadow and passion, an even, compared to it, it is the selfsame subtle dynamism what*

---

[323] Il sovrasensibile dinamico procedeva da un gesto espressivo dall'interno, direttamente riferito all'Io. Ora quando nel corpo di attività, che egli sperimenta, una tale direzione centrifuga risulti subito travolta e capovolta nell'opposta, egli si sentirà certamente distinto nella sua anteriore potenza distintiva. E dunque, nell'ulteriore lampeggiamento illuminativo di un più intenso e deciso gesto di evocazione-distinzione Julius Evola. *Fenomenologia Dell'individuo Assoluto Iii* ed. corretta: Edizioni Mediterranee, Roma, 2007.

*becomes a symbol, an abstract embodiment of expression.*"[324] In myth, the connection established between the primordial metaphysical domain and the immanent is generally symbolized by lightning and thunderbolts, a natural phenomena that can only take place when clouds in the atmosphere are electrically charged, to the point that the electric field created disrupts the dielectric character of the atmosphere, causing massive amounts of charged particles to flow from the clouds to the ground as lightning. Lightening is also a very powerful symbolic expression of a centripetal form of energy, as it concentrates on an extremely local point the potential energy contained in the electric field present in a vast region of the atmosphere.

In an Operative Tradition, the Magical "I" can also be symbolized as a "lightning rod" that connects the immanent conditions with the extremely subtle conditions that emerge in any non-linear physical phenomena out of equilibrium. These highly entropic conditions, which are mostly registered with today's devices as pure noise and can only be treated statistically when approached with electronic devices, constitutes for modern science one of the most mysterious domains of *physis*. And yet this domain was approached by Belgian physical chemist and Nobel Laureate, Ilya Prigogine (1917-2003), who explained that whenever any physical domain reaches such state, it appears as if new configurations appeared in such highly dynamic conditions. Long-range patterns and wave configurations appear in extremely subtle domains where any sensitivity to the initial conditions, or the openness to "butterfly effects" becomes extremely high. Probably the best visual example of the former—just for didactic purposes—is the Belousov-Zhabotinsky chemical reaction, where a given chemical composition (generally a mix of potassium bromate, cerium, sulfate, malonic acid, and citric acid in dilute sulfuric acid) reaches a state out of equilibrium and entropy production where colored spot configurations appear as the harmonic outcome of a type of non-linear process out of equilibrium that decades ago was

---

[324] Rispetto al «sottile» dinamico il sensibile fu sentito come mera ombra e scorza, così, rispetto a queste, è il dinamico stesso che diviene simbolo, astratto corpo espressivo. Julius Evola. *Fenomenologia Dell'individuo Assoluto Iii* ed. corretta: Edizioni Mediterranee, Roma, 2007.

conceived in physics as typically turbulent and chaotic, and where no overall forms or figures could be perceived. Even Nobel Prize winner Richard Feynmann denominated turbulence as *"the most important unsolved problem of classical physics."*[325] However, what the Belousov–Zhabotinsky reaction shows is that when matter is forced to reach a dissipative state, it then enters into a subtle physical domain, where it is then exposed to overall configurations that ultimately determine its specific organization and material patterns. So the Belousov–Zhabotinsky reaction allows us to discard any separation between what is generally conceived as inert/inorganic, from what is intuitively conceived as "alive"/organic. In effect, when exposed to energy dissipation, all material objects appear to "surf" upon subtle waves and patterns that determine their development. The higher or lesser integration of material elements with the patterns depends on the higher or lesser resonance that exists between the individual particle and the whole, in other words, the ability of the "surfer" particle, to "surf" the waves.

The Belousov–Zhabotinsky reaction also shows us that any potential material configuration that is forced to attain a high dissipative state has the chance of "singing," so to speak, and to connect to an overall, yet extremely subtle and delicate pattern that transcends any time-space determination. Besides, the Belousov–Zhabotinsky reaction is also a relatively simple chemical experiment that gives us the chance to perceive the emergence of overall patterns, and inevitably this process causes us to suggest that the patterns might not only be restricted to this particular BZ chemical reaction, but rather is present in every single development that takes place in nature (*physis*). As the B-Z reaction shows, these overall figures that emerge in a state out of equilibrium are directly linked

---

[325] Besides, the mathematician and nobel prize winner J. Nash wrote: *"The open problems in the area of non-linear partial differential equations are very relevant to applied mathematics and science as a whole, perhaps more so than the open problems in any other area of mathematics, and this field seems poised for rapid development. Little is known about the existence, uniqueness and smoothness of solutions of the general equations of flow for a viscous, compressible, and heat conducting fluid. Also, the relationship between this continuum description of a fluid and the more physically valid statistical mechanical description is not well understood. Probably one should first try to prove existence, smoothness, and unique continuation (in time) of flows, conditional on the non-appearance of certain gross types of singularity, such as infinities of temperature or density. A result of this kind would clarify the turbulence problem."*

to the self-organization of the same system, providing the forms, figures, and patterns that it shall evolve towards. It was this kind of phenomena among others what suggested to the English biologist, Rupert Sheldrake, the idea of *morphic fields*, as a set of timeless configurations, characterized by endless layers of patterns which are constantly present in an apparently turbulent domain. And we write here the term "apparently" because anything that appears as chaotic or turbulent to the "I" only reveals an impotence of the "I" to discern the symbolic character of the configurations. For instance, if our senses were not developed to perceive the patterns present in a given musical piece we would be forced to consider the piece as a chaotic set of sound frequencies.

We can finally visualize the Magical "I" that Eugen Herrigel has to configure—which at the maximum moment of tension "pulls" the movements of the body in an harmonious and powerful way—as linked to a *morphic field* that develops the configuration of the "I," and the whole process by which Herrigel establishes contact with that field and energizes it is called *morphic resonance*. The adjective *morphic* (Greek *morphe* = form) refers to *structure*, order out of chaos: *neguentropy*. Contrarily to electromagnetic fields—which can be generated at will by the vibration of a given charged particle—morphic fields, like the Magical "I" are always potentially present, in the whole practice of an operative tradition to eventually *tune* and *resonate* with them.

\* \* \*

*One day I asked the Master: " How can the shot be loosed if "I" do not do it?" "It shoots," he replied.*

\* \* \*

In his philosophical work, Julius Evola resorts to very suggestive and rather provocative expressions that assist extremely well in order to visualize the operative effects caused by the activation of the Magical "I," once the student has managed to transcend the constrictions of thought and all former "ideas." The Italian author writes that the "ideas" are *"weak forms of reality, while realities are Ideas that are more powerful"* [The Capital I of "Ideas" doesn't appear in the original text, and is presented this way in order to distinguish Magical Ideas from discursive and abstract "ideas"]. When the student of an Operative Tradition learns to see with a new sense objects and elements surrounding him, he no longer intends to "understand" the objects, but rather perceives all directly related objects as progressively hierarchical dispositions of elements that converge in one single point, in one single source, the *Idea* that is the more powerful. The external disposition of elements, objects, and beings that allow direct interaction all correspond to *actual Ideas*. Evola writes: *"ideas are potential realities, realities are actual ideas."*[326] The statement that *"ideas are potential realities"* allows us to visualize and imagine that anything beyond the elements present *here and now* which are assumed to be ideas on the "supernatural," the "transcendent," the "metaphysical," the "political," or the "religious" correspond to realities that can only be considered potential and not actual, that is, not *Ideas*, if the conditions of reality are not dominated and persuaded by a higher configuring principle. Hence, in a state of privation on the part of the individual, all discursive ideas are symptoms of a reality that is alien to the individual, a reality that however conditions and determines the individual, and not vice versa. Yet a state of dominion and persuasion of reality on the part of the Magical "I" makes it is extremely to perceive the existence of rational and ordered elements separated from irrational elements or chaos. Herrigel writes:

> *"This state, in which nothing definite is thought, planned, striven for, desired or expected, which aims in no particular direction and yet knows itself capable alike of the possible and the impossible, so unswerving is its power—this state, which is at bottom purposeless and egoless, was called by the Master truly 'spiritual' (...) This*

---

[326] Julius Evola. *An Intellectual Autobiography. Path of Cinnabar* Arktos, 2009, pg 53.

*means that the mind or spirit is present everywhere because it is nowhere attached to any particular place."*

From the perspective of the Magical "I" both the rational and the irrational are well integrated, as occurs in the case of the geometry of the circumference, which in effect can be one of the most suggestive symbols of the Magical "I." Hence, this leads us to another expression employed by Julius Evola, apparently paradoxical from a conceptual or discursive perspective, yet absolutely true when not read from a consciousness in state of privation: *"truth is a powerful error; error is a weak truth,"*[327] which is equivalent to the one we present here: *"truth is a powerful chaos; error is a weak chaos."* So when all apparently irrational elements are dominated and configured by a powerful principle, irrationality itself becomes powerful, that is, "error/chaos" is powerful; yet when the contingency of phenomena and all apparently chaotic elements are highly determining factors that lack integrating principles, this means that truth—by conceiving the term *truth* here in the sense of dominion and persuasion—is necessarily weak. In order to provide a "pointer" example of what we are referring to, let's suppose I have a set of chaotic numbers like 35 8979323846264338327950288419716939937 5. If I intended to express mathematically this chaotic series of numbers (this "reality") in order to conceptually dominate it, I would not initially be capable of encountering any mathematical function that could analytically integrate the series without the function constituting an extremely complex mathematical expression which would require more bits to be expressed than that amount of bits (63) required to define the series itself.[328] So the "idea" or "truth" that would express the series of numbers in mathematical terms would be weak because it would require more information—it is *"heavier"* than the series itself, thus demonstrating the burdening weakness of any analytical and rational mathematical function. However, if I awaken a Magical "I" capable

---

[327] Julius Evola. *An Intellectual Autobiography. Path of Cinnabar* Arktos 2009, pg 54.

[328] The Binary equivalence of such series of numbers would be: 111110001110001010001001001011111 10 11000100011110010111110011 (63 bits)

of perfectly integrating the rational with the irrational, I would end up realizing that such a series corresponds to the 40 decimals that appear after the 8th decimal of pi, which can be expressed like π *(8-40)*, which is an extremely simple, "light," and yet powerful truth that integrates perfectly well the apparent contingency that appears in the series. Therefore π (8-40), having a symbolic nature, is a *powerful truth*. So ultimately all abstractions correspond to weak forms of reality and truths correspond to stronger forms of reality and observed phenomena, just as the great developer of "General Systems Theory," Ludwig Von Bertalanffy, points out— *"the choice is not whether to remain in the field of data or to theorize; the choice is only between models that are more or less abstract, generalized, near or more remote from direct observation, more or less suitable to represent observed phenomena."*[329]

Eugen Herrigel was constantly encouraged by Master Awa Kenzo to discipline a perceptive state that could allow the transcendental view of all things,[330] the convergence with the Idea.

> *"Focus your minds on what happens in the practice-hall. Walk past everything without noticing it, as if there were only one thing in the world that is important and real, and that is archery!"*

All the former explanations, which all inevitably constitute mere conceptual "pointers" towards the centre of the Idea we are aiming at, allows us to realize that in terms of visualizing the phenomena—once experienced—the most alluring hypothesis of what the Magical "I" is relates extremely well to the hypothesis of the existence of morphic fields.

*  *  *

> *"The child plays with the things, yet it is equally true that things are playing with the child."* Master Awa Kenzo.

---

[329] Ludwig Von Bertalanffy. *General Systems Theory.* George Braziller. New York. 1968, pg 162.

[330] "These things, which depend in fact entirely on the power of the 'I'" Julius Evola. *L'individuo e il divenire del mondo*, Libreria di Scienze e Lettere, 1926, pg 21.

The idea of *morphic fields* was brilliantly and extensively elaborated by British biologist Rupert Sheldrake, and refers to the existence of fields that *"interact directly or indirectly with electromagnetic and quantum fields, imposing patterns on their otherwise indeterminate activities."*[331] The hypothesis proposed by Sheldrake and David Bohm on the existence of *quantum vacuum* fields is something that any student like Eugen Herrigel who succeeds in an operative practice can feel in his senses; a *vacuum* where "quantas" or "pulses" emerge practically out of nowhere.

One of the controversies which arise from the idea of the morphic fields is that the hypothesis was received by the scientific community in general as developing on the thresholds of pseudo-science and spiritualism since the measurements attempting to demonstrate the fields are not yet totally conclusive. But at this point, if we recall the key stages of TAI-PAI and the problem of epistemological certainty established at the beginning of the practice, we'll realize that *in order to know an object, one has to be that same object*. Hence, the hypothesis of causative formation by morphogenetic fields or the idea of morphic fields corresponds to ideas as equally speculatively acceptable as "God," "Historicity of the spirit," the "thing-in-itself."[332] The actual validity of the idea of morphic fields thus only resides in the capacity of the individual *to be the same field*, with no mediation, no need for any explanation, and no justification. In terms of the development of TAI-PAI, the visualization of morphic fields is much more effective in practice than any vague spiritualist, theosophist, or religious ideal, since most of the latter are highly characterized by moral, historicist, evolutionist and ethical norms, which in the case of an Operative Tradition also has to be superseded, to the point that the practitioner has to eventually be "beyond good and evil," yet without embracing the idea as Nietzsche had conceived it when

---

[331] Rupert Sheldrake. *Morphic Resonance. The Nature of Formative Causation*. Park Street Press.

[332] *"The imperative is not to call 'being' what is non-being"* Julius Evola. *An Intellectual Autobiography. Path of Cinnabar* Arktos 2009, pg 49.

referring to the "will to power," since in the case of an Operative Tradition the way of gaining power is by incinerating the individual "will." When referring to the powerlessness of human will on it's own, Friedrick George Jünger wrote:

> "Success, for instance, does not depend exclusively on will, even the greatest effort of will cannot force it. Rather, accomplished and perfect motion is distinguished by the fact that in it willed effort recedes. Great works of art, for example, always appear effortless; in an excellent painting, in a superb statue, the artistic effort and painstaking workmanship vanish in the perfection of the whole. Will and success are not identical, and therefore the will to power by itself does not accomplish anything."[333]

\* \* \*

"Why try to anticipate in thought what only experience can teach?"
- Master Awa Kenzo

\* \* \*

"All perfect acts are unconscious and no longer subject to will; conscious action is the expression of an imperfect and often morbid state in a person." - Friedrich Nietzsche[334]

\* \* \*

As the methods of modern science necessarily have to be founded in space, time and strict causality foundations in order to gain universal validity and the chance of peer-reviews and repetition of the experiments in laboratories around the planet. The existence of bridging between the experiences provided by an Operative Tradition on morphic fields and a "demonstration" of the fields for

---

[333] Friedrick George Jünger. *The Failure of Technology*.
[334] Friedrich Nietzsche, Walter Arnold Kaufmann, R. J. Hollingdale. *The Will to Power* 1968, pg 163.

the scientific community seems extremely unlikely. For instance, how could individuals like Eugen Herrigel explain the field—the so-called "It"—by means of a language based on cause-effect premises, if causality itself had become one of the most difficult and hard barriers for him to attain the truth? As Bernard D'Espagnat[335] stated, *"the principles of physics cannot even be formulated without referring (though in some versions only implicitly) to the impressions—and thus to the minds of the observers."* As all mind patterns constitute the most important barrier to the production of the Magical "I" in TAI-PAI, it is obvious that the morphic field character of the Magical "I" cannot be formulated dialectically, it has to be *operatively experienced*. A Master in these operative arts must be capable of maintaining a sense of pure objectivity, an absolute presence of the external and internal configurations, without staining the experience with any discursive interpretation, that is, accepting a completely "mind-boggling" situation with absolute naturalness. Once Eugen Herrigel reached this state of spirit, these were his words:

> *"I'm afraid I don't understand anything anymore at all, "I answered, " even the simplest things have got in a muddle. Is it " I " who draws the bow, or is it the bow that draws me into the state of highest tension? Do " I " hit the goal, or does the goal hit me? Is " It "spiritual when seen by the eyes of the body, and corporeal when seen by the eyes of the spirit—or both or neither? Bow, arrow, goal and ego, all melt into one another, so that I can no longer separate them."*

*Sub specie aeternitatis:* it is eternal that which is contemplated through eternal eyes. In *Essays on Magical Idealism*, Evola resorts to a very alluring image to express this "eagles" long-range state of perception, referring to it as the perception of a *"submarine mass of a continent that causes continuity and unifies the multiplicity of islands that emerge from the waters,"*[336] and in *Phenomenology of the*

---

[335] B. D'Espagnat, *The Conceptual Foundations of Quantum Mechanics* (Reading, Mass.:Benjamin, 1976), pg 286.

[336] Julius Evola. *Saggi Sull'Idealismo Magico*. Edizione Mediterranee, pg 100.

*Absolute Individual* he writes that *"all multiple experiences can not be but particular determinations of a single consciousness."*[337] The eagle consciousness.

It must be clear that having formerly resorted to visualization of electromagnetic or "wind" similes in no case entails that we are dealing with strictly electromagnetic or aerodynamic issues. The nature of the subtle currents integrated within a morphogenetic field can be visualized as a complex of multileveled field gradients having a qualitative character, affecting all living things, but to affirm that the fields are electromagnetic or gravitational would be extremely reductionist. None of our modern techniques can measure the morphic fields, but nonetheless through the practice of an Operative Tradition, our hands and senses have the chance of "touching" the fields and "feeling" them, even if our mind is incapable of grasping concepts or metaphors. Resonating adequately with the fields has an *inductive* capacity, something Nietzsche also referred to as,

> *"a part of spirituality (induction psycho-motrice, Charles Féré thinks). One never communicates thoughts: one communicates movements, mimic signs, which we then trace back to thoughts."*[338]

Most scientists would claim that, since these energies cannot be measured, these energies don't exist either. Yet the Master of an Operative Tradition would always remain silent before such claims, since they say are just "sayings," and the cosmos always expresses itself through other means.

When the accomplishment of the spiritual goals embedded in an Operative Tradition is ultimately attained, the issue of *communication* between beings also changes its nature. Language or even symbols can only correspond to mere shadows of the metaphysical energy contained in the spiritual "flashes" or "pulses" we formerly described through a couple of electromagnetic field

---

[337] Le molteplici esperienze non possono essere allora che particolari determinazioni di un'unica coscienza. Julius Evola. *Fenomenologia Dell'individuo Assoluto Iii* ed. corretta: Edizioni Mediterranee, Roma 2007.

[338] Friedrich Nietzsche, Walter Arnold Kaufmann, R. J. Hollingdale *The Will to Power* 1968, pg 428.

analogies. In the direct presence of any Master accomplishing the ritual actions that characterize an Operative Tradition —thus causing the brilliance of magic to appear qualitatively present in physical gesture—a very "shocking," psychologically overwhelming and highly mystical experience can take place for those spectators who are relatively close to the center, to the "bridge" or "lightening" between "heaven" and "earth"; between the morphic fields and the operative conditions of reality. As Evola writes in *Phenomenology of the Absolute Individual*: *"this selfsame gesture does not produce, but rather is an immediate bright force, a spiritual dynamic of the ethereal, which is the substance in which such is recreated, where it is integrated, reborn and purified in action the whole realm of the phenomenon— namely, not even the physical determinations but also the world of thoughts, feelings and forces; though from such the person can only capture abstract and particular sensations."*[339] On one hand, the social exposition of the ritual arts has to be configured and developed with extreme care, allowing all spectators to assist the sacred act like configuring a Hindu *mandala*, by the establishment of hierarchical layers surrounding the ritual, where the higher initiated and more experienced individuals are placed close to the center, and the least experienced or general public are placed in the periphery. In the Chinese traditions, these rites constituted the *"channels by which we can apprehend the ways of Heaven"*[340] and the *"manifestations of the heavenly law."*[341] Operative Traditions are extremely rigorous with these severe norms that aim for developing transcendent immanence. As during the extraordinary "flash" or "lightening," a powerful wave pattern of unknown nature (not electromagnetic or gravitational) is generated in all directions, and like all wave irradiations—like the electromagnetic—have decisive induction

---

[339] Questo stesso gesto non produce, ma e immediatamente una forza luminosa, un etere dinamico spirituale, che è la sostanza in cui si ricrea, in cui rinasce integrato e purificato in attività l'intero regno del fenomeno—e cioè: non pure le determinazioni fisiche, ma anche lo stesso mondo dei pensieri, dei sentimenti, delle forze; ché anche di questi—come si disse—la persona coglie solamente astratte e particolari sensibilizzazioni. Julius Evola. *Fenomenologia Dell'individuo Assoluto Iii* ed. corretta: Edizioni Mediterranee, Roma 2007.

[340] Li-ki, VII, IV, 6.

[341] Tsun-shung, XXVII, 6.

properties, almost *"as if in the warrior there was a force similar to a fluid that was capable of creating new knights by direct transmission,"*[342] an average individual in the presence of these "pulses" can become psychologically shocked, and permanently disturbed at a deep unconscious and psycho-somatic level, and can easily become severely ill, hence "blinded" (the perceptive faculty of the senses) by the "light." Conversely, an individual who is more experienced can be very positively stimulated by the phenomena. It is the main responsibility of the Master to rigorously determine, classify and establish the different layers of the students and the general public. These rituals were not aimed for entertainment but instead were characterized by political aims and purposes. During the past, it was quite usual that Kings and Emperors used to have a privileged place in these operative rituals, as these castes were characterized by a natural relation to freedom. However, just placing all the ritual elements together is no guarantee that the ultimate purpose can be accomplished. The difference between these specific rituals we are referring to and the mere ceremony or ritualism typical of our times is equivalent to the difference between an electric motor that is connected to the "grid" and another one that is not... The former *works* (channels energy) and the latter doesn't...

This entire ritual, when not constituting mere ceremony or ritualism, and when actually *working*, that is, when the Magical "I" is effectively activated and a bridge is then established between the transcendent realm and the operative domain, constitutes that which Rupert Sheldrake defines as a *morphogenetic germ*, though the British biologist applied this idea much more in the domain of atomic, molecular, and biological formation. This morphogenetic germ generates a surrounding morphogenetic field where "pulses" of wave patterns are emitted in all directions, integrating and providing a form to all the physical developments taking place in the closest physical dissipative systems out of equilibrium. The formative influence on the part of the center of irradiation depends on the capacity of the surrounding elements to establish morphic

---

[342] Julius Evola. *Revolt Against the Modern World*. Inner Traditions, pg 65.

resonance with the core ritual operative elements, depending on the diverse levels of initiation in the arts. So we are referring here to a very powerful *suggestive* capacity on the part of these spiritual acts, which are capable of forging deep predispositions of consciousness in those individuals who are present during the ritual.

\* \* \*

*"You've heard of the importance of eye contact in the classroom? Every educationist emphasizes it. No educationist explains it."* - Robert M. Pirsig. Zen and the Art of Motorcycle Maintenance: An Inquiry into Values.

\* \* \*

In Eugen Herrigel's account, the former phenomena emerge in what Master Awa Kenzo referred to as the *"immediate transference of the spirit,"* which corresponded to the act performed by the Master of giving some shots whenever those students with a high degree of advanced practice exposed unbalance during their own movements. The mere closeness to the act performed by the Master, "magnetized" and "fixed" the movements of the student in a more stable and harmonic way, allowing progressive contact with the field generated or that which Master Awa Kenzo referred to as the *Buddha* or "It," that is, the Evolian *Absolute Individual* or *bodhisattva* whose essence (sattva) is not made of matter, but of "illumination" (bodhi).[343]

---

[343] (Egli-gli-Dèi) di apparire, di prender forma nei varî «salvatori," nei bodhisattva, esseri la cui sostanza (sattva) non è fatta di materia, ma di «illuminazione» (bodhi). Julius Evola. *Fenomenologia Dell'individuo Assoluto Iii* ed. corretta: Edizioni Mediterranee, Roma, 2007.

# CONCLUSION

*"Making from a finite being an infinite being, rendering "earth" into a "Sun," a God, an authority of power, glory and dominion is not fantasy, it is not a myth, it is not an object of faith and hope, it is rather an actual possibility for a will who knows how to say to itself: or I succeed or I fail in the attempt"*
Julius Evola. Phenomenology of the Absolute Individual.[344]

This part of *Operative Traditions* has intended to present the assumptions, practices, and developments expressed in philosophical terms by Julius Evola in the two main volumes: *"Theory of the Absolute Individual"* and *"Phenomenology of the Absolute Individual,"* in addition to *"Essays on Magical Idealism"* where the two first volumes constitute the most extensive and coherent philosophical work of the 20th century. In addition, *"Essays on Magical Idealism,"* by presenting influences from the Eastern esoteric traditions can be considered as a linking text for those who intend to establish extremely powerful connections between the esotericism of the West and East, while *"Theory of the Absolute Individual"* and *"Phenomenology of the Absolute Individual"* are built upon the ultimate developments and idealist conclusions drawn by strict Western philosophical thought, even though some elements of the Eastern traditions are also spread in the works, having more presence in the second volume: *"Phenomenology of the Absolute Individual."*

---

[344] Fare di un essere finito un essere infinito, trarre da una «terra» un «Sole," un Dio, un ente di potenza, di gloria, di dominazione, non è fantasia, non è mito, non è oggetto di fede e di speranza, è invece una possibilità concreta per una volontà che sappia dire a sé stessa: Riuscirò o mi spezzerò nell'impeto. Julius Evola. *Fenomenologia Dell'individuo Assoluto Iii* ed. corretta: Edizioni Mediterranee, Roma 2007.

This study of the extraordinary philosophical works of Julius Evola intends to introduce to English readers the *applications*[345] of his work. In effect, Evola's philosophical works amount to more than 600 pages in Ediciones Mediterranee Italian editions, and his philosophical writings not only require a lot of publishing space, but are also highly multidisciplinary, and not easy to cope with at a conceptual level. Consequently, this section of Operative Traditions constitutes a summary of Evola's works.

Because of the character of TAI-PAI and the gnoseological foundations it presents, it is very unlikely that TAI-PAI will be reviewed or presented to the readers through academic publishing or universities, since in our times the character of the institutions (especially in the philosophical domain) do not rely on operative teaching factors as an experiential basis, and emphasize a separation between the concrete and empiric "I" of the students and the teaching of a "knowledge in general," to be distributed to the public in general, and mostly based on humanist and scientific foundations. Ernst Jünger considered that, in order to be released from the "umbilical cords" of a bourgeoisie era characterized by apparent modes of dominion, the less "culture" gained by the individual—in the sense of all the ideological and discursive ties to the bourgeoisie world—the better.[346] This corresponds well with the position presented by Jacques Ellul in his book *Propaganda. The Formation of Men's Attitudes*, where the French author shows that intellectual groups correspond to those more affected by propaganda techniques, which promote the individual to be more prone to maintain a polemic world-view than to encourage an individual to have a specific experience of

---

[345] "If a further development beyond Magical Idealism is to be imagined, this will be not a philosophical development, but a kind of action... (...) what matters is for individuals to understand the meaning of and the need for the final philosophical step which I have outlined, and to move ahead"Julius Evola. *An Intellectual Autobiography. Path of Cinnabar* Arktos 2009, pg 61-62.

[346] "Ein Einschnitt, der tief genug ist, um uns der alten Nabelstränge zu entledigen, kann in der nötigen Schärfe nur gezogen werden durch ein starkes Selbstbewußtsein, das in einer jungen und rücksichtslosen Führerschaft verkörpert ist. Je weniger Bildung im üblichen Sinne diese Schicht besitzt, desto besser wird es sein" Ernst Jünger. *Der Arbeiter*. Ernst Klett, Stuttgart, 1981. Klett-Cotta, pg 105.

the world.³⁴⁷ So the inevitable consequence is that, as Jünger writes: *"today we can easily hear a thousand of clever people discussing the Church, yet we would vainly encounter the ancient saints who lived in the rocks and woods."*³⁴⁸ Hence, these formative/academic contexts do not have conditions required to *resonate* with Evola's philosophy and are necessarily alien to it. However, any individual who develops an activity within environments where particular arts, techniques, traditions, and crafts still allow a relation to the operative factors of the world might attain a state of consciousness that shall allow, firstly the ability to resonate with *Operative Traditiona*, and secondly perceiving something in TAI-PAI that makes sense to their lives.

Any individual in our times who is developing an activity where a sense of responsibility combines with a sense of virtue and personal fulfillment through direct experience will naturally attract this book with no need for any academic mediation. This is the magic implicit in TAI-PAI: the individual, by strongly focusing on the empiric state of being can easily attain new traits of the "I" which *produce* a given reality, or what is more often expressed as the *attraction* of a specific reality (a specific "layer" of *physis*). As is shown in TAI-PAI there is no reality or objective world beyond that produced by the empiric and concrete "I," and the domain of reality can eventually constitute a dominated and persuaded experiential framework where the "I" is situated in the center, determining and producing all facts. This idea, which corresponds to that of the fulfillment of the *Absolute Individual*, is however challenging for any scholar who works in a given university as a teacher of philosophy, science, or history etc., since TAI-PAI would require the teacher to determine, beyond the classrooms, which are the concrete elements of his life he presumes

---

³⁴⁷ "Intellectuals are more sensitive than peasants to integration propaganda. In fact, they share the stereotypes of a society even when they are political opponents of the society. (…) Intellectuals are most easily reached by propaganda, particularly if it employs ambiguity. The reader of a number of newspapers expressing diverse attitudes—just became is better informed—is more subjected than anyone else to a propaganda that he cannot perceive, even though he claims to retain free choice in the mastery of all this information. Actually, he being conditioned to absorb all the propaganda" Jacques Ellul. *Propaganda. The Formation of Men's Attitudes* (1962) Vintage Books, 1973, pp 76, pp 113.

³⁴⁸ "Wie man heute mit Leichtigkeit tausend gescheite Leute über die Kirche räsonieren hören kann, während man die alten Fels- und Waldheiligen vergeblich sucht" Ernst Jünger. *Der Arbeiter*. Ernst Klett, Stuttgart, 1981. Klett-Cotta, pg 105-106.

to be mastering. Can he master his relationships? Can he master the economic conditions of his life? Can he master his health with no need for pharmaceuticals? Can he master something as simple and as annoying as a smartphone device and leave it silent for a while?

Depending on how the teacher answers these questions of conscience based on the premises of TAI-PAI, he will have or not have access to the extremely powerful TAI-PAI teachings. Hence working as a philosophy teacher or scholar does not guarantee factors that contribute to accessing TAI-PAI, and what's more, the intellectual activity can constitute a barrier for reaching TAI-PAI, in the same way the abstract "I" of Transcendental Idealism in Western philosophy (Immanuel Kant, Hegel, Schelling, Fichte) favored human consciousness to fly astray in the most chaotic and ungrounded abstract domains, like a tree having no trunk nor roots. Jünger's figure of the *Anarch* is also characterized by overcoming Idealism, and he writes:

> "*idealism is far from my mind, even though I have made sacrifices to it. I, in turn, feel it does not suffice to grasp facts according to their weight but not their eros. Matter is concentrated in eros; the world becomes exciting.*"[349]

Even though this idealistic standpoint is not conceivable from the viewpoint of TAI-PAI, it has constituted a strong trait in modern Western culture. As in the case of Eugen Herrigel, any Operative Tradition is extremely demanding in terms of the student aiming to reach the core of his life and truth, symbolizing the attainment at an empirical level with the spiritual shot, where the arrow finally flows softly along fields of energy that appear as miraculous, in terms of how they supersede and *play* with physical laws.

TAI-PAI, which corresponds to the most appropriate framework in order to visualize the operative aspects of traditions such as Zen, shows that the accomplishment of the egos' self-sacrifice and that of all mind identifications—hence allowing afterwards something

---

[349] Ernst Jünger. *Eumeswil*. The Eridanos Library, 1980.

deeper and more powerful to emerge—is something that can *be proven empirically*, by the specific character and "wave-sound" (in the case of Eugen Herrigel's account, the unforgettable "deep thrumming" of the bow), caused by the movements of the student when releasing from conscious control. The main issue of certainty embraced by many philosophical schools of the past, and all the epistemological and gnoseological approaches that aimed to find an ultimate "firm grounding" for any form of knowledge in the cultural domain has in the case of an Operative Tradition a starting point for establishing certainty, connected to what Evola refers to in TAI-PAI as the *empiric state of being*. When the subtle gestures of the practitioner are ultimately transmuted, this is proof of *transcendent immanenc*. Let's recall that Eugen Herrigel, after many years of practice, finally managed to exert a spiritual shot. Master Awa Kenzo, immediately noticing "IT," bowed before the phenomena, as a divine act. "Divine act": This is an expression that encompasses perfectly well the idea of *transcendent (divine) immanence (act)*.

Any student who succeeds in the discipline and who manages to empirically demonstrate what formerly was referred to as the "spiritual flash," as a mind-boggling energy form, might afterwards appear before our eyes as a normal and average man, yet his actions and purposes are devoted to another realm, that of the transcendent and Magical "I," that is, a sacred domain. And this is so, because he is no longer following human goals, but superhuman ones, incarnating in human form the highest form of spiritual initiation. Aware of the power he *embraces* in his hands, he shall also be characterized by a cautious wisdom that shall induce him to decide if the powers must be exposed or not in most human contexts, whenever these contexts are not properly prepared at a ritual level.

For the *Absolute Individual*, all former speculation on gods, religions, and philosophy shall afterwards constitute mere "pointers" for all those individuals who still haven't encountered the most important foundational truth, the "philosophers *stone*," that is, that the truth can only be attained by relating *to what is immediate, to what is concrete and to what is specific experience*. And from this standpoint everything follows. As Evola states in *Phenomenology of the Absolute Individual*:

> "The real question then is this: to understand that what matters now is only what is real, the concrete, a stripped relationship of the self with all things and beings; relations that for men are extrinsic and contingent in terms of the physical state of existence and the space-time categories that govern it."[350]

So the realization of the "philosophers *stone*" as the key that opens the doorway of consciousness, allowing it to penetrate like a sharp knife into new and unimagined realms of power and dominion—which in myth, legend, and saga is always equated with *magic*—an absolute requirement in our times for recovering the ultimate purpose is to acknowledge the empirical conditions of existence, since otherwise the path of the Absolute Individual can't be properly *grounded* in the immanent "soil." So, this requirement implies an absolute focus in the empirical conditions of existence, in combination with an adequate acknowledgment of the specific objects we *handle* on a daily basis. This *handling* factor shall provide us with most of the required insights, since in TAI-PAI the experiential domain to be configured by the "I" must always be *non-mediated*, but *direct, raw*, experience. This is one of the main purposes of the second part of *Operative Traditions*. Having allowed the reader to visualize the path of Absoluteness and its core requirements and disciplines, such Absoluteness must now be projected into the domain of the relative, particular and specific, connected to the actual conditions and experiences of the particular individual who in our times lives in a highly complex techno-industrial society.

We might not be aware of this factor but we live in the most operative and technical world ever, to the point that several authors like Jacques Ellul and Langdon Wienner have referred

---

[350] Il problema vero è dunque questo: capire che ciò che importa ormai è soltanto il reale, concreto, denudato rapporto di sé con le cose e con gli esseri; rapporto che per gli uomini è quello estrinseco e contingente proprio allo stato fisico di esistenza alle categorie spazio-temporali che lo reggono. Julius Evola. *Fenomenologia Dell'individuo Assoluto Iii* ed. corretta: Edizioni Mediterranee, Roma, 2007.

to the concept of *technological determinism*, and also that the actual problems in our times are all of a technical nature. Yet here a question emerges: How is it that we do not have a *synthetic* knowledge that embraces this operative domain, thus establishing the different levels of homeostatic dominion of all fields amidst it? Jünger himself posed this question in the 1930s when outlining that it is *"surprising that technicians are unable to even inscribe their own definition within an image that captures life in all its dimensions."*[351] As a vast set of multidisciplinary domains, both scientific and engineering knowledge do not have a synthetic character, but are rather fragmented in countless areas and specializations, that could only become connected—yet not *integrated*—through cybernetic paradigms and mind-sets, which as we referred to in the first part, is a paradigm that only modifies the cause-effect mechanistic approach of former times into a circular "feedback" approach, where homeostasis only exists within the technical devices (servo-systems, etc.) but no longer as a relation subject-object, (I/ non-I) which is the aim of the synthetic Magical "I."

If we have properly assimilated the core foundations of TAI-PAI we shall then ultimately conclude that we can't access the synthetic knowledge to the extent we have not generated the knowledge in ourselves, amidst our specific conditions of empirical existence. This itself would reveal that we are still in a state of privation where, as planets passively revolving around the sun, it is still unreachable as the capacity for transmutation, causing our actions to be determined by a "centre" which is still alien to us. This is what Evola refers to as "alterity" and lack of centrality of the "I." For instance, let's suppose a space-ship pilot, in order to conquer a new territorial dominion, aims to reach a planet that is in an orbit closer to the sun. There are two alternatives available for him: the first alternative is TAI-PAI's "path of the other," which means that the pilot of the ship can decide to *hibernate* (allowing all automatic pilots to take over) and then being provided with entertainment to withstand the boredom of

---

[351] "Auffällig ist es im besonderen, daß der Techniker selbst seine Bestimmung nicht in ein Bild einzuzeichnen vermag, das das Leben in der Gesamtheit seiner Dimensionen erfaßt" Ernst Jünger. *Der Arbeiter*. Ernst Klett, Stuttgart 1981. Klett-Cotta, pg 77.

the journey. The second alternative is the "path of dominion," which would require the pilot to first of all be aware of all the immediate determinisms that condition the ship (speed, position, energy reserves, gravitational variables, etc.) and once properly informed of all these aspects, decides to *pilot* the ship towards the destination, where territorial dominion shall constitute the material projection of the capacity on the part of the pilot to create a homeostatic balance among all technical determinisms, or in other words, the projection of his capacity to *provide an aim to all determinations and functions*. Such is the "arrow" that the pilot aims for. This piloting would correspond to the exertion of technical mastery accomplished by the activation of a Magical "I" where all individualistic traits of the pilot were formerly burnt and sacrificed, allowing something deeper and more powerful to emerge. This kind of pilot devotes all his action to the fulfillment of the "I," as a symbol of power, a dominion ultimately expressed in the territorial conquest accomplished.

Like the pilot of the aforementioned example, in our times every single one of us has the chance to go for the path of alienation, or the path of dominion; that is: the path of the "individual," or the path of the "Individual." If, after having adequately assimilated TAI-PAI we realize that the Magical "I" corresponds to an *impersonal* idea that integrates and transcends any anthropomorphic and humanist conception of the cosmos, it is also easy to deduce that all developments and creations produced and induced by the "I" can take place regardless of traits that are considered to be typically human. For instance, all the Christian/Protestant ethics and morals that are still used to justify certain actions—thus providing the actions with a pseudo-religious character—can easily be discarded by operative processes, though morals are often considered as genuinely human attributes.[352]

We all probably recall the first moments in our lives when we knew we actually existed. It could have been at the age of 2, 3, or 4 years old. It was precisely at that moment that a singularity in

---

[352] In realtà, la morale come categoria autonoma è un non-senso; la morale non è che religione secolarizzata [In truth, morals as an autonomous category is meaningless; morals aren't but secularized religion] Julius Evola. *Teoria Dell'individuo Assoluto*. Edizioni Mediterranee, 1988, pg 103.

our consciousness took place, and for the first time in our lives we referred to ourselves as "I," and the world around us suddenly became illuminated. Before this singularity had taken place, a complex process had developed inside our bodies where massive processes of cellular destruction and creation aimed to produce the enlightening singularity of the "I," and all the highly unconscious and mysterious developments taking place were suddenly affirmed and synthesized into a new being, a new awareness. As with a baby, an equivalent singularity takes place with cultures, and also with the emergence of a Magical "I."

No human attribute can determine the emergence or not of a given Magical "I" nor the correspondent operative processes that the "I" determines in the cosmos. It is as impossible for humans to determine the traits of a given Magical "I" as it is to determine the pure essence of a rose or to determine the specific development of the seasons of the year. The latter cosmic forms and cycles are *given* conditions. The latter is a key idea present in the societies of the past, where according to Alan Drengson, in ancient China and Greece:

> *"Humans were seen as ignorant and finite; nature as mysterious and infinitely powerful. It was believed that it is futile for humans to try to control the natural world, or even to drastically alter it."*[353]

The human condition is potentially capable of bridging the forms and cycles to the particular conditions of life in a balanced way, but in no way is the human condition *privileged* for attaining the purpose. What's more, if we look at the ecological disasters, extreme resource depletion, and the destruction of native cultures accomplished by human action during the last two centuries, it isn't hard to admit that the human species has demonstrated itself as the least adequate species in order to properly establish a coherent and harmonic relation with the cosmic cycles. The planetary disease and lack of homeostatic harmonies we are clearly testifying—

---

[353] Drengson, A. 1995. *The Practice of Technology—Exploring Technology, Ecophilosophy, and Spiritual Disciplines for Vital Links.* Albany, NY: State University of New York Press, pg 197.

which necessarily implies that the highly synthetic "I" at a cosmic level is still in a *transitional* phase of creative development—symbolizes a vast process of exhaustion of what are considered typical anthropomorphic elements (discursive/analytical mind-sets, emotional, romanticist and sentimental traits, etc.) which are all insatiably materialistic in operative/technical terms, hence lacking any goal beyond their own pathological self-reinforcement. And ultimately it is only the development of economics that can satisfy this self-reinforcement, regardless of the overall issue of dominion. *"The virtuous, rational and ideal image of the world corresponds here to an economic utopia of the world where all approaches have as a reference point economic demands(...) this is one of those games where apparently all players involved can win, but where only banks can ultimately win"*—writes Ernst Jünger.[354]

Yet, as we explained formerly in terms of the methods and disciplines required for gaining access to the Magical "I" through negative entropy processes, we shall recall that the morphic fields "feed upon their own waste," so to speak, and this inevitably causes us to realize that the destructive traits of our age are still serving a higher purpose, unknown for most—incognito for many—like in Ancient Egypt when thousands of those individuals who actively participated in the building of the pyramids never had the chance of seeing the "I" of Horus, as the ultimate divine "attractor" of the entire process took place over different generations.

*"We must in fact seek perfect life where it has become least conscious (i.e., least aware of its logic, its reasons, its means and intentions, its utility). The return to the facts of bon sens, of the bon homme, of the humble people of all kinds. The stored-up integrity and shrewdness of generations which are never conscious of their principles and are even a little afraid of principles. The*

---

[354] "Das vernünftig-tugendhafte Idealbild der Welt fällt hier zusammen mit einer wirtschaftlichen Utopie der Welt, und wirtschaftliche Ansprüche sind es, auf die jede Fragestellung sich bezieht.." Ernst Jünger. *Der Arbeiter*. Ernst Klett, Stuttgart 1981. Klett-Cotta, pg 13.
"Es gehört dieses Spiel zu jenen, bei denen scheinbar jeder Mitspieler, in Wirklichkeit aber nur die Bank gewinnen kann." Ernst Jünger. *Der Arbeiter*. Ernst Klett, Stuttgart 1981. Klett-Cotta, pg 37.

*demand for a virtue that reasons is not reasonable—a philosopher is compromised by such a demand."* - Friedrich Nietzsche[355]

---

[355] Friedrich Nietzsche, Walter Arnold Kaufmann, R. J. Hollingdale *The Will to Power* 1968, pg 242.

# PART III

# SOCIAL & POLITICAL IMPLICATIONS OF TAI-PAI

> "Mine was a philosophical introduction to a *non-philosophical world*"
> Julius Evola[356]

> As theoretical knowledge frees itself from confiding acceptance, it is marching to self-destruction,
> after which what remains is simply and solely technical experience.
> Oswald Spengler[357]

The first part of *Operative Traditions* has mainly presented us with the actual operative and technical conditions of our times, as the basic "grounds" or "philosophical *stone*" on which any operative analysis and synthesis can be considered. Therefore, this part of the book was intended to introduce the reader to the *immanent* conditions of existence, where a given individual is faced with problems that require an active link to reality in order to cope with the circumstances, on a practical level. Ultimately, it was concluded in the first part of the book that the operative links we establish today with the world have a highly systemic and cybernetic character.

In the second part of *Operative Traditions* we presented Julius Evola's philosophical work, as the required counterpart or idealist approach—not in terms of Transcendental Idealism but in terms of overcoming this—for learning how to intellectually approach immanent conditions in general, by establishment of a series of methods and disciplines that root their foundations in the core traditional/operative arts of the West and the East. These traditional

---

[356] Julius Evola. *An Intellectual Autobiography. Path of Cinnabar* Arktos, 2009, pg 64.

[357] Oswald Spengler. *The Decline of the West*. Alfred A Knopf. 1926. New York, pg 269, 270.

teachings do not aim for aesthetic or contemplative delight, nor aim to arouse feelings and sentiments in the spectator, but rather correspond to active processes of initiation, aiming to release the power of the spirit from immanent conditions, transforming the deep layers of consciousness in the advanced practitioner. It is a mutation that can be empirically proven when the artist experiences a considerable change of character when released from conscious mind limitations.

The third part of *Operative Traditions* can be considered as the application of the teachings expressed in the second part (TAI-PAI) amidst the conditions defined in the first part.

The challenge presented to us by such application is that, while in the case of most traditional arts (*techné*) the steps and procedures were normative, doctrinal, fixed and harmonically linked with stable conditions of existence—symbolized perfectly well in the tools and instruments employed by the artist in the traditions—in our times we no longer live amidst "fixed" conditions, but rather amidst the most dynamic operative times in History. The changes taking place in the technological, technical, and economical realm at a global level develop at exponential speeds, and even those high-tech corporations that aim to master the conditions and profit from technical dominion are more and more stressed by the apparently never-ending need to be technologically up-to-date, in order to not lose presence in the consumer and stock markets. Not only corporations, but billions of global-wide individual workers are also constantly forced to stay updated in terms of skills and know-how; circumstances that can easily make them change their professional activity several times in a lifetime, or as Ernst Jünger puts it, by resorting to some notions of Total War: *"men no longer fall fighting, but suffer a glitch and are then out of service,"*[358] since *"the front of war and the front of work are identical."*[359] Such dynamic changes can induce psychological unbalances, anxiety, fear about the future, risk

---

[358] "Mman fällt nicht mehr, sondern man fällt aus" Ernst Jünger. *Der Arbeiter*. Ernst Klett, Stuttgart 1981. Klett-Cotta, pg 54.

[359] "Daß Kriegsfront und Arbeitsfront identisch sind" Ernst Jünger. *Der Arbeiter*. Ernst Klett, Stuttgart 1981. Klett-Cotta, pg 55.

of unemployment, fear about oneself, one's identity, etc. which are all clear symptoms of a lack of homeostatic factors at an individual and long-term level.

Now, assisted by TAI-PAI, we can realize that global dynamics are extremely chaotic to the extent that our eyes—the "I" we are individually linked to—can only perceive chaos present in the internal world. If this "I" could mutate into the *Absolute*, then the perception of these dynamics would be substantially modified, and a immutable centre of power would be perceived—and what's more, *lived*—where all processes and dynamics that appear as chaotic and highly complex would merely correspond to the impermanent currents of becoming or *samsara* that feed that centre through subtle neguentropic currents, while the center or "pole" would justify and synthesize the whole development in a purely creative way, like a flower justifying the former plant development. This potential path is implicitly present in TAI-PAI, where it is shown how the individual, by embracing a deep sense of dominion and operative power, can ultimately attain the center, like stepping on a hill that shall eventually become a mountain, where the Absolute Individual, in complete solitude, not only acknowledges the conditions of the former path, but perceives the entire cosmos in a completely different way.[360] This *centripetal* ascent towards the top places the individual in the realm of heavens and directly exposed to their influence; he then incarnates the bridge between the transcendent domain and immanent reality, capable of causing "thunderbolts" at will. As formerly shown, the "thunderbolts" correspond to an image which is one of the signs that express the consecution of the goal intended by an operative tradition such as Zen Buddhism, which is that of bridging the transcendent and immanent through the individual. "It" can be rigorously equated as a *super*-power, as its source of energy does not discharge from the physical domain, but from the subtle elements (morphic fields) that the student is encouraged to learn to "charge," so to speak, through adequate breathing techniques and a

---

[360] "With the highest point of the process being that of the pure act, 'monad of monads' or 'God' Julius Evola. *An Intellectual Autobiography. Path of Cinnabar* Arktos 2009, pg 51-52.

deep sense of self-sacrifice. This Individual, this *Master*, conscious of having attained the Absolute *with his own hands* and living in the "*pole*" will be extremely cautious not to expose this super-power, as he is aware of the psychological "shock" the phenomena can entail at a subconscious level for those who haven't received proper guidance and discipline. And yet, silently and humbly, the enlightenment of a new world shall slowly grow from this magical seed, a genesis that is necessarily invisible for almost all contemporaries, yet that as always, shall flow into myths, legends, and sagas. Freedom was the main capital of these men. Jünger states: "

*Only free men can make real history. History is the imprint that free man gives to destiny.*"[361]

*Are there in our times any specific profiles or modern individual traits that are the most appropriate for withstanding the requirements of this path towards the absolute?* Surprisingly enough, the traits were present in one of the most powerful books of the last century. We are referring here to Ernst Jünger's "*Der Arbeiter: Herrschaft und Gestalt*" (1932).

---

[361] "Dennoch kann echte Geschichte nur durch Freie gemacht werden. Geschichte ist die Prägung, die der Freie dem Schicksal gibt." Ernst Jünger. Der Waldgang. Klett-Cotta. Stuttgart 1980, pg 45.

# DER ARBEITER

## THE WORKER:
### THE MODERN OPERATIVE FIGURE FORESEEN BY ERNST JÜNGER

> *"You I advise not to work, but to fight. You I advise not to peace, but to victory. Let your work be a fight, let your peace be a victory!"*
> Friedrich Nietzsche, *Thus Spoke Zarathustra*

> *"Der Arbeiter is more than a philosophy; it is a poetic creation"*
> Marcel Decombis[362]

*Der Arbeiter: Herrschaft und Gestalt* was translated into English as *"The Worker: Figure and Dominion."* This misleading translation was very likely one of the main reasons why the book was not properly grasped in terms of its core transformative message, and why it was often misinterpreted from political standpoints. In order to not further confuse the readers, in *Operative Traditions* it has been chosen to make use of a new expression: *"The Operator"* (instead of "The Worker") intending to approach in a new renovated way a *figure* which is naturally predisposed towards *operative dominion*.

Julius Evola, aware of the powerful ideas embedded in Jünger's book, published an introductory book on "The Operator" for Italian readers called *"L'Operaio nel pensiero de Ernst Jünger"* (*The Worker in the thought of Ernst Jünger*) which is a summarized approach to the essay of the German author, and which was elaborated—

---
[362] Marcel Decombis, *Ernst Jünger et la "Konservative Revolution,"* GRECE, 1975, p. 8.

especially in the last chapter "Conclusions"—from the perspective on Tradition that Evola relied on during his entire life. Evola also agreed that by referring to the term "Arbeiter/Worker" Jünger was not referring to anything that could resemble a proletarian ideal or the idea of the worker as reduced to fulfilling an exclusive economic/ production aim.[363] These were the impressions of Evola on the work of Jünger: *"The merit of Jünger in this first phase of his thought, is to have recognized the fatal mistake of all those who think that everything can be reordered, that this new threatening world, always in progress, can be tamed or interrupted on the basis of vision of life and the values of the previous era; that is, of the bourgeois civilization."*[364]

The most substantial aspect of Ernst Jünger's work corresponds to presenting to Western culture the issue of *technique,* as an *operative factor* that has become extremely formative in terms of the actual conditions of power and dominion. Presenting this idea for the first time corresponded in the case of Ernst Jünger, to an extremely revolutionary accomplishment, since one of the traits of Western culture since the Enlightenment was to grant power, relevance, and dominion to scientific thought and to the discursive and rationalistic mind (eventually becoming philosophically defended the tendency by Transcendental Idealism). Yet it was precisely Jünger's heroic experiences in warfare and extremely dangerous situations what allowed him to perceive *"the power of the Zeitgeist, which degrades*

---

[363] These are some ideas of Julius Evola on Jünger's "*Der Arbeiter*": "A book that on the controversial plan is opposed to economic materialism, the ideals of "cattle" prosperity, the gentrification of the selfsame groups that formerly claimed anti-bourgeoisie standpoints, while on the constructive plan such book aims to affirm, albeit with sometimes unacceptable undertones, the need for education, in order to form a new type of man, who is willing to give much more than to demand, and in order to overcome the crisis which has shaken the modern world (...) Because by allowing the penetration in all revolutionary movements of economic terms and civilization ideals of the Third State, it indicates the merely apparent revolutionary character of such movements, ultimately becoming irrelevant movements, in addition to the impotency of all revolutionary social dialectics designed by the same left-wing factions, postulating contrarily a space for a different figure that, according to Jünger, will characterize the new age (...) It illustrates that the reduction of every revolutionary movement to the economic sphere, by becoming exhausted in it, is one of the techniques required to keep alive the principle of "society," that is, to reconfirm, in spite of everything, the world of the Third Estate" Julius Evola. *L'Operaio Nel Pensiero de Ernst Jünger*. Edizione Mediterranee. 1998.

[364] Este y Oeste, *Archè*, Milán, 1982, pg 69.

*the ideal into an illusion*[365] and visualize the dynamic mobilization based on the power provided by technical and operative factors. Amidst this new configuration of *physis*, all former ideals, religious frameworks and secular ideals were becoming mere subsidiary factors, "shadows," and in most cases justifications *a posteriori* of all phenomena taking place within a *novel reality* that was no longer accessible to the bourgeoisie speculative and idealistic approaches to the cosmos, but rather to a figure characterized by an intimate aim for dominion: *the Operator*, the *"Lord and dominator of the world, an imperious type who is in possession of a full power only dimly glimpsed so far"*[366]—as Ernst Jünger writes.

Having described the core philosophy of Julius Evola as a framework that assists the individual to relate to the empirical conditions of the world, that Jünger considered as *"the only possible heir of Prussianism,"*[367] the *Operator*. This corresponds well to the Individual who has formerly burnt away all sense of artificial and abstract individualism, and who develops an activity—technique—which awakens an "I" potentially capable of homeostatic dominion over the effective conditions of reality. Such an Individual would have to necessarily arrive at the third stage defined by Evola, which is the development phase of Magical Idealism, or what in his phenomenology he refers to as the Stage of Dominion ("era della dominazione"). This stage is not determined by economic power, but rather the modes of power and dominion over the economic conditions. Therefore, whenever referring to the *Operator* it is crucial to draw a distinction line between the *Operator* and the idea of the *worker*. The Operator is a figure that is more primordial than the worker, and *"beyond dialectics,"*[368] as Jünger remarks; the *Operator* sets the territorial dominion on which the worker can afterwards develop

---

[365] Ernst Jünger. *Eumeswil*. The Eridanos Library. 1980.

[366] "Der Herr und Ordner der Welt, als gebietender Typus im Besitze einer bisher nur dunkel geahnten Machtvollkommenheit" Ernst Jünger. *Der Arbeiter*. Ernst Klett, Stuttgart 1981. Klett-Cotta, pg 20.

[367] "Der einzig mögliche Erbe des Preußentums" Ernst Jünger. *Der Arbeiter*. Ernst Klett, Stuttgart 1981. Klett-Cotta, pg 34.

[368] "Jenseits der Dialektik" Ernst Jünger. *Der Arbeiter*. Ernst Klett, Stuttgart 1981. Klett-Cotta, pg 39.

a production/economic activity.[369] The former distinction is crucial, since most of the misconceptions on Jünger's essay arose from the assimilation of the *Operator* figure to that of any individual developing an activity related to production/economic purposes. This distinction carries Jünger's approach to the Operative Traditions of the West and East, where territorial dominion and the establishment of strong homeostatic links between men and the environment constituted the legitimizing factor for developing economic activity. The Operative Freemasons cherished this idea during medieval times, and as we've seen in the account of Eugen Herrigel with Master Awa Kenzo, the entire archery practice is also intended to be uncoupled and released from any economic *need*. As we've already seen, in the case of an Operative Tradition such as the "Great Doctrine," Eugen Herrigel finally managed to properly assimilate and *be one with* it. What is intended to be produced is not a specific form of matter, but rather a specific form of energy which directs the developments of matter and its particular character of mobilization; hence the archer is intended to constitute an operative "bridge" capable of serving as a "channel" or "funnel" between both realms. By accomplishing this task successfully, the economic domain—strictly linked to the conditions of material production—serves a purpose that, for instance, in the case of Zen Buddhism was often assigned a ritual and regal function, which was related to the *State*. The same idea emerged during Medieval Europe with the Operative arts being integrated in Gothic architecture, which all dispersed across Europe before the emergence of the State-nations, and constituted clear imperial developments at a continental level.

The latter entails another key characteristic of the *Operator*: its intimate relation to the State. However, once again another key differential nuance has to be pointed out. Jünger did not refer to the State-nations as the territory the *Operator* aims to dominate or politically link with. Ernst Jünger's idea of the State is the *imperialer*

---

[369] An insightful cinematographic portrait of this primordial character of the type over the individual worker is Philip Kaufman's film *"The Right Stuff"* (1983) showing how the piloting capacity of Chick Yeager in the American Air-Force constitutes the main factor of the development of military techniques and consequently economic power.

*Räume*[370] where hierarchy—like in the case of the Prussian Empire—was determined by a sense of duty, self-discipline, self-sacrifice, and devotion towards military activities. These character traits constituted valuable elements to Eugen Herrigel when he was under the guidance of Master Awa Kenzo, a man who incarnated imperial values. In fact, the "distillation" of Herrigel's personal virtue and essence would have been impossible if not for having embraced those crucial traits during the years of training. Yet these imperial character traits can easily cool when the individual is placed in a society where economic individualism (liberalism) or the cult of economic production as a socio-political factor (Marxism) constitute ideologies that no longer aim to "distill" the needs and desires, but instead tend to increase and hypertrophy all needs and desires. In this context—which has proliferated in the West since World War I—the *Operator* would have to search for territories where technique can be developed as the main configuring factor of reality. Yet in order to attain this release from the intoxicating effects of ideologies such as liberalism and Marxism, the *Operator* would have to first be free inside—in terms of conscience—and this entails freedom from any *bourgeoisie* world-view.

There have been many descriptions of the bourgeoisie class by historians, sociologists and philosophers, and to some extent most of the descriptions have gained a stereotypical character. It is remarkable that today the word *bourgeois* has practically disappeared from the mainstream, as if the *bourgeois* individuals no longer existed.[371] However, when Jünger focused on the importance of inner feeling released from what Marx referred to as the bourgeoisie "class consciousness," the German war hero was not even referring to a "class" or social strata as Marxism did, but rather to a *style*, a way of doing things and a way of being. If we recall Julius Evola's stages of development of the "I" in TAI-PAI, the first and second stage corresponds to the individual finding a sense of self and identity through identification with the contents of the thinking "I"

---

[370] Ernst Jünger. *Der Arbeiter.* Klett-Cotta. 1981, pg 154.
[371] Jacques Ellul. *Métamorphose du bourgeois.* La petite vermillon, pg 71-72.

(conceptual structures, world-views, ideologies, beliefs), the physical "I" (properties, emotions, sensations, feelings, self-image), the *self* of modern psychology (character traits, psychological predispositions, traumas, fears) and the *ego*. As a class that historically existed as an intermediary social layer between the old aristocracies and the developing working classes, the "I" or class consciousness of the bourgeoisie was mostly located in TAI-PAI's second stage. When Jünger refers to the traits of the bourgeoisie, he mostly focuses on how the individuals who belong to this class establish a barrier of conceptual structures, world-views, ideologies, religions and beliefs between their deep inner predispositions and reality. This is a "prophylactic" symptom of evasion and alienation from reality, and whenever such an individual operates on the external reality he always feels the need to justify his actions. Let's recall that the first and second stage of TAI-PAI are characterized by *privation*, which means that the lack of operative dominion and persuasion is necessarily compensated for with the need for speculative justification. So when Jünger points to the main weakness of the bourgeoisie, he is pointing to their consubstantial and intimate *privation*, which inevitably triggers needs and desires, which if not satisfied, have to be repressed, sublimated, or transposed.

When Jünger states that the *Operator* has to be free from the bourgeoisie predispositions, this is practically the same as saying that the *Operator* is not in a state of privation, and hence the *Operator* develops a close relation with given structures of power without need for justification. Privation is the main trait of those individuals who are determined by emotions, anger, sentiments, feelings, and intuitions (what Jünger constantly refers to as "elementary powers") who have not managed to dominate them or *affirm* them, by channeling them operatively in the physical world. It is therefore inaccurate to consider that the traits of natural sovereignty emerge as substantial traits in those individuals who have managed to fully dominate themselves in accord arid with "stoic" forms of rigid and insensitive characters, but quite the contrary. These individuals— like the *Operator*—who manage to dominate the "elementary powers" and attain a natural and spontaneous balance, could only have achieved this state by profound feelings of love, especially towards the beauty of all the essences of *physis*. So we are referring here to an *aristocratic* sense of love that intimately resorts to the power of beauty as the homeostatic affirmation of all raw powers of the cosmos.[372] Ernst Jünger captured this idea in very few words: *"Like all things of this world, plants also speak to us, men; but to understand their language a lucid spirit is required."*[373] A good example of this way of perceiving the cosmos was expressed in the 1930s by a group of young Germans who referred to this as *New Essenciality* (In German: neue Sachlichkeit).

Franz Matzke, one of the main leaders of the juvenile movement, when referring to this transparent view on *physis* wrote:

> *"The way we see reality is beyond the thoughts of men, the reality of things is great, infinite and whatever is human is little, conditioned, impregnated by feeling (...) Only what is expressed in terms of reality is what we are interested in (...) In every field we dismiss the vanity of the author and whatever is objective and universal we perceive as beyond any private psychology (...) It is not that we are insensitive, only others might think that a soul that is silent is no longer a soul; we too have a sensitivity, yet such sensitivity is no longer triggered by the feelings of others, it is only triggered when faced with real things and when faced with that which in men is real, elementary (...) Nature for us is the great kingdom of all things, the realm of all those things that do not demand anything from us, that are not*

---

[372] Ora la persona in quanto ama perfettamente, epperò genera amore, riesce veramente a riconoscersi nell'altro, a risolverne la resistenza, l'oscurità e l'eterogeneità inerenti alle precedenti apparizioni di questo, epperò a conciliarsi con sé stessa in una circolare mediazione che rende adeguato l'antecedente al conseguente della categoria. Tale punto si esprime nel trapasso dalla donna al figlio. [Hence, the person, to the extent he perfectly loves, consequently generates love, succeeds in recognizing itself in the other, by solving all resistances, all darkness and the heterogeneity that are inherent to the former appear from this, thus in a self-conciliation through a circular mediation that renders adequate the cause to the effect of the category] Julius Evola. *Fenomenologia Dell'individuo Assoluto Iii* ed. corretta: Edizioni Mediterranee, Roma 2007.

[373] "Como todas las cosas de este mundo, también las plantas nos hablan a nosotros, los hombres; pero para entender su lenguaje es preciso poseer un espíritu lúcido." Ernst Jünger. *Sobre los Acantilados de Mármol.* Ediciones Destino. [Auf Den Marmol-Klippen], pg 22.

*imposed on us nor expect a given attitude on the part of our soul; which are before us as silent as a world eternally strange on its own. And it is precisely the latter what today we require, a grandeur and farness, an entity that relies on itself, both in the highest of the little joys as in the tiny sufferings from men. A close kingdom of objects, where we ourselves can feel ourselves as objects –absolute freedom regard to all that is merely subjective, of all vanity, of the nullity caused by all narcissism- such is that which we conceive as nature. All predispositions towards any cult or devotion is by us denied (…) We comprehend the grandeur of anonymity during the Middle-Ages, free from any personal vanity, where nobody even cared about transmitting to others the pains or joys of their own heart (…) No work any longer refers to the author; such works remain before us closed and independent as "things" in the highest sense. (…) Do we affirm the existence of God? Do we deny it? Nor one thing nor the other. These problems have lost for us any meaning. "Culture" for us is the form of an inner attitude, where grandeur is measured by its unitary and conclusive character: nothing to do with paintings, poems or speculative research."*[374]

Hence, Jünger's idea of the *Operator* is not characterized by insensitivity, but rather a higher and more subtle sensitivity aiming for the artistic configuration of the material domain, as an expression and materialization of the power. It is difficult not to perceive this in Nietzsche—who Ernst Jünger considered to be an important influence—in these sentences:

> "'What is good?' Ye ask. To be brave is good. Let the little girls say: 'To be good is what is pretty, and at the same time touching.'"[375]

Another example also equates spiritual power as dominion over the irrational elements of one's psyche: "The multitude and desegregation of impulses and the lack of any systematic order among them result in a "weak will"; their coordination under a single predominant impulse

---

[374] Franz Matzke, Jugend bekennt: so sin dwir! Leipzig 1930 (The Youth, that's how we are).

[375] Friedrich Nietzsche. X. *Thus Spoke Zarathustra*. War and Warriors.

*results in a "strong will": in the first case it is the oscillation and the lack of gravity; in the latter, the precision and clarity of the direction."*[376] Hence the *Operator* corresponds to a strong sovereign figure when it incarnates boldness and equilibrium. These heroic predispositions allow the relation to technique to be clear, concise, transparent, and also as Jünger expresses: *"beyond utility in itself, beauty in itself and that which is obvious."*[377] Power that emerges from a domain of reality that transcends the determinations present in the economic domain, corresponds to a layer of *physis* that is intrinsically formless and dependent on the specific techniques applied on a given territory.[378] This economic domain allows keeping up the material conditions required in order to allow the *Operator* to configure territory. It must be recalled that even Aristotle considered that wealth was based on the operative idea of caste, and therefore based on the specific set of techniques a given caste could apply to a given territory, ultimately being the highest form of richness that constituted the king,[379] characterized by imperial and State traits, which in the German language is linked to the term *reich* "rich" and *Reich* "realm," which are related to the terms *regal* and *noble*. *Reich* is the same as the Latin *rex*, and the Sanskrit *rajan* (king). Hence as Jünger points out: *"the figure of the Operator is related to an element of abundance and not of poverty,"*[380] yet this abundance is not to be

---

[376] Friedrich Nietzsche, Walter Arnold Kaufmann, R. J. Hollingdale *The Will to Power* 1968, pg 29.

[377] "Die typische Bildung kennt daher nicht das an sich Zweckmäßige, das an sich Schöne oder an sich Einleuchtende." Ernst Jünger. *Der Arbeiter*. Ernst Klett, Stuttgart 1981. Klett-Cotta, pg 120.

[378] "It was a classical idea, for instance, that perfection cannot be measured with a material criterion, but that it rather consists in realizing one's nature in a thorough way. The ancients also believed that materiality only represents the inability to actualize one's form, since matter (úle) was depicted in Plato and Aristotle's writings as the foundation of undifferentiation and of an evasive instability that causes a thing or being to be incomplete in itself and not to correspond to its norm and "idea," (that is, to its dharma)" Julius Evola. *Revolt Against the Modern World*. Inner Traditions, pg 74.

[379] Aristotle: "there are four kinds of economy, that of the king, that of the provincial governor, that of the city, and that of the individual. This is a broad method of division; and we shall find that the other forms of economy fall within it"
Jonathan Barnes. *The Complete Works of Aristotle*. The Revised Oxford Translation, Vol. 2. Princeton University Press (1991) Economics Book 2, pg 8.

[380] "Der Gestalt des Arbeiters nicht ein Element der Armut, sondern ein Element der Fülle zugeordnet ist." Ernst Jünger. *Der Arbeiter*. Ernst Klett, Stuttgart 1981. Klett-Cotta Vorwort Zur Ersten Auflage, pg 4.

understood in terms of material abundance, but in terms of power and the capacity for configuration and dominion. What's more, this power can also be compatible without economic abundance. Jünger writes: *"Institutions such as the Teutonic Order, the Prussian Army and the Society of Jesus are perfect examples of this, and it is worth not forgetting that soldiers, priests, wise men, and artists are characterized by a natural relationship with poverty."*[381] Hence the main focus of the *Operator* corresponds to that of a purely ascetic focus on *technique* and that of furthering the technical development, as a natural focus on the specific territorial conditions that ultimately drive the aims of economic capital, resources, and the workforce.

At this point it is therefore important to clarify any confusion which emerged when equating Jünger's figure of the *Operator* with that of the *worker*. The Operator is he who is devoted to *opus*, a rather primordial conception of human activity related to the aim of inner transformation or *metanoia* by means of *opus*.[382] The latter term *"metanoia"* entails a change of inner perception of *physis*, whilst *opus* corresponds to a term recovered from medieval accounts, corresponding to the idea of taking to perfection any form of activity, as an unconditional discipline that is good in itself, characterised by a *magical* character (*opus magnum, opus magicum*) and which is equivalent to the Hindu ideal presented in the *karmayoga* teachings and—as we already have exposed—in the operative aspects of Zen Buddhism. Julius Evola captured this idea of inner transformation by disciplining external activity in his *Phenomenology of the Absolute Individual*, where the Italian author writes: *"Every externalization of the interior awakens a deeper interiority."*[383] *Opus* corresponds to a mode of activity that aims to bridge the sacred with the profane, thus providing a justification for developments taking place within the profane domains. It must be stressed here that when referring

---

[381] "Erscheinungen wie der deutsche Ritterorden, die preußische Armee, die Societas Jesu sind Vorbilder, und es ist zu beachten, daß Soldaten, Priestern, Gelehrten und Künstlern. Ernst Jünger. Der Arbeiter. Ernst Klett, Stuttgart 1981. Klett-Cotta, pg 104-105.

[382] Peter Senge. *La quinta disciplina*. Barcelona: Granica. 1995.

[383] Ogni esteriorazione dell'interiore eccita una interiorità più profonda. Julius Evola. *Fenomenologia Dell'individuo Assoluto Iii* ed. corretta: Edizioni Mediterranee, Roma 2007.

to "profane" domains we conceive this term as related to all those aspects of life which are fragmented, and thus lacking integrating and synthetic factors. Yet when intending to establish an adequate conception of the term "profane," it is crucial to point out that there are no such things as "profane things," but rather a *profane view on things*. From the perspective of the Magical "I" developed in this book, nothing in the cosmos is separate but instead converges into a single point, like Eugen Herrigel's "spiritual" arrow. A profane view of things emerges whenever the cosmos is filtered through the thinking "I" or through the complex set of chaotic emotional components embedded within the physical "I." Hence this distinction allows one to finally supersede all apparent dichotomies. As Ernst Jünger writes: *"Our age has exhausted its forces in most antithesis, in a very similar way as how it has exhausted also another antithesis, that of idea and matter, that of blood and spirit, that of power and right; but the only outcome of such antithesis are perspectivistic interpretations that sheds light over this or that partial claim."*[384]

Ultimately, there is one sacred way of dealing with profane things, and countless profane ways of dealing with the sacred. Due to the stereotyped and deceptive filtering of the cosmos caused by the thinking "I," vast amount of individuals consider that devoting their lives to the "religious," "spiritual" or "transcendent" areas of knowledge corresponds to non-profane activities. However, *opus*—as a sacred mode of activity—can be encountered whenever the transmutation of extremely precious "gems" occurs in the most unexpected and supposedly "profane" domains of life: factories, farms, hospitals, commerce, etc. Charles Peguy,[385] refers to this *way of doing things* that characterised *opus* during medieval times:

*"To work (ouvrier/opus) is joy itself, it is the deep root of being [...] There is an extraordinary honour in such activity [...] A shirt's*

---

[384] "Unsere Zeit hat sich in dieser Gegenüberstellung ganz ähnlich wie in jenen anderen Gegenüberstellungen von Idee und Materie, Blut und Geist, Macht und Recht erschöpft, aus denen sich nur perspektivische Deutungen ergeben, durch welche dieser oder jener Teilanspruch belichtet wird" Ernst Jünger. *Der Arbeiter*. Ernst Klett, Stuttgart 1981. Klett-Cotta, pg 21.

[385] Charles Peguy. *L'Argent*. Librairie Gallimard, Paris (1932).

*button was required to even to be well done[...]It was not required to be well done because of a salary [...] the master's approval, or the acceptance on the part of the connoisseurs [...]The activity was required to be well done for its own sake [...] This is the principle itself of the Cathedrals...*"

We can therefore conclude that whenever a given activity is devoted towards *opus*, the economic factors derived from the activity become secondary elements, and do not constitute the main aim of the mode of activity. *Opus* constitutes a typically transmuting activity that provides a meaning to all means, a harmonic stable/homeostatic configuration to elements present, a synthetic perspective on diverse modes of knowledge, and a justification of all productive/economic efforts; a form of art that as Jünger states *"allows to conceive life in its highest sense as totality."*[386] *Opus* aims to synthesise the "cherry at the top" of a series of economic activities and production layers that typically have their integrating principle elsewhere. In other words, *opus* corresponds to an ultimate justification, integration, and synthesis of all activities characterised by a necessary sense of *privation*. So *opus* is to totality as *work* is to division. McLuhan writes:

> *"Where the whole man is involved there is no work. Work begins with the division of labor and the specialization of functions and tasks."*[387]

These division of labor activities—characterised by fragmentary and conditional modes of action—correspond to what in Greece was referred to as *ponos* and *ergon* (which as we shall develop in *Operative Traditions Volume II*) can be considered as closely linked to the concept of *work*, which is directly linked to material production aiming for the satisfaction of needs. Hence, it is misleading to

---

[386] "Die Kunst hat zu erweisen, daß das Leben unter hohen Aspekten als Totalität begriffen wird" Ernst Jünger. *Der Arbeiter.* Ernst Klett, Stuttgart 1981. Klett-Cotta, pg 109.

[387] Marshall McLuhan. *Understanding Media. The Extensions of Man.* The MIT Press Cambridge, Massachusetts. London, England, 1994.

characterise Ernst Jünger's most mysterious figure (Der Arbeiter) as "The Worker" (as it has been translated into English) but rather as what is proposed in this book: the *Oper-ator*, that is, the renaissance of the alchemical nature of *opus* activities within the complex and highly systematic conditions of our urban-industrial societies.

The main requirement for developing an activity based on *opus* is that of being released from all identifications and attachments on the part of the individual, something extremely close to being released from the fear of death. As Jünger sees it: *"And so it is that human beings, by discovering their figure, discover their own mission, their destiny; such discovery enables them to sacrifice, which reaches its most significant expression in the offering of blood."*[388] The ephemeral attachments that surrender to such figures, which constantly trigger desire—desires that according to TAI-PAI correspond to modes of privation—are conceived of in Buddhism to be the root cause of *suffering*. In Buddhism this state is called *dukkha,* which refers to the feelings of despair and non-fulfilment caused by the attachment to external elements, and thus lacking homeostatic integration. Suffering arises by the inner tension or painful "splitting" of consciousness when intending to attach to elements that are produced by different "I's." This difficulty causes a constant anxiety of being "in the middle of nowhere," and therefore the temporary satisfaction of a need to attach to external elements relieves the tension temporarily. Hence, the only way to be free from this state of agitation is the of development of a Magical "I," which would allow all external elements to "orbit" around one's life. In this case, all determinisms orbiting around the "sun"—would all be acknowledged in a harmonic and integral way, corresponding to a view of the cosmos gained by what the medieval scholars referred to as *intuitio intellectualis* or νοῦς. However, any individual who has not yet embraced a Magical "I" and still perceives the world through the "lens" of a given cultural standpoint or belief-system, shall *necessarily* perceive all external phenomena as complex, that is, the

---

[388] "So kommt es, daß der Mensch mit der Gestalt zugleich seine Bestimmung, sein Schicksal entdeckt, und diese Entdeckung ist es, die ihn des Opfers fähig macht, das im Blutopfer seinen bedeutendsten Ausdruck gewinnt." Ernst Jünger. *Der Arbeiter.* Ernst Klett, Stuttgart 1981. Klett-Cotta, pg 18.

individual perceives the world as lacking a synthetic and hierarchical integration. In mathematics, whenever one aims to find the equation that defines a given geometrical trajectory, it is absolutely crucial to correctly define the type of coordinates involved in the problem (Cartesian, polar, etc.) and then establish its *centre*; this way, what might appear to be complex geometrical objects—like spiral-type plotting—can be acknowledged by quite simple expressions. In other words: when an individual perceives the world as complex or non-integrated, and the perceptions induce nihilist feelings of despair and suffering, it is due to the non-activation of a Magical "I" that, situated in the correct centre, perceives all life-processes as easily integrated and hierarchical. Jünger writes:

> *"It is true of course that it is more difficult than ever to achieve security in the midst of a situation that is purely dynamic and which one can not see any axis."*[389]

So the *Operator*'s activity is situated in a position where suffering is highly minimised, and all developments are therefore not determined by pain or pleasure. It is a paradisaical state of being beyond pain and pleasure, beyond good and evil. Yet contrary to this particular style of activity, most religious belief-systems that directly or indirectly derive from the Judeo-Christian predisposition consider the world as a "veil of tears," and what's more, they propose that the only valid way of redemption is that of becoming alienated/separated from the world (in the form of mystical evasion, body-mortification, etc.). In the case of the *Operator*, the standpoint is the contrary: the *Operator* is deeply related to the developments taking place, and does not repress any predisposition of his being through moral/religious or social modes of self-coercion. The specific view of the *Operator*, situated in that of the Magical "I," perceives all social, moral, and religious concepts as being *derived* from the power of the ultimate State, which the *Operator* constantly aims to incarnate and relate to. This State—as formerly pointed out—has nothing to

---

[389] "Daß freilich die Sicherheit inmitten eines scheinbar rein dynamischen Zustandes, in dem keine Achsen zu erkennen sind" Ernst Jünger. *Der Arbeiter*. Ernst Klett, Stuttgart 1981. Klett-Cotta, pg 30.

do with the modern State-nations, and corresponds more to a centre of justification and synthetic integration of all contradictory or ephemeral processes taking place. The *Operator* intimately knows that it is the flower what determines the material development of a plant, and not the opposite; amidst the interplay of forces of all objective and worldly powers there is an invisible "centre of coordinates" that most traditions point to,[390] where everything makes sense, where everything ultimately encounters its destiny. Situated amidst these dynamics, whatever is referred to as "human, all to human"—in Nietzschean terms—that is, all the moral, religious, economic, cultural interpretations of humanity correspond to mere shadows, pale reflections of a cosmic process that transcends conceptual denominations,. What ultimately symbolises human processes are not words, nor utopian visions of History, but *figures*, and the more the *Operator* can artistically conceive the emergence of such, the more his activity transcends and integrates all worldly powers through dominion and persuasion. The *Operator* does not intend to "control" such a process—since this is as unrealistic as intending to control the seasons of the year—but he intends to follow the entire development up to its highest peak, like a mountain and aims to climb to the top. Hence freedom does not correspond to a final or utopian state of achievement, but rather a *process* of constant pursuit, and if there is a mode of activity that reveals the pursuit of the path of freedom it is that of *opus*.

*Opus* corresponds to the highest mode of activity; an activity that is free from any speculative and economic concern. It has an operative character which is present in alchemical lore. Yet before delving into *opus* as the main operative style of the *Operator*, it is convenient to introduce the issue of *technique*, since as an operative

---

[390] René Guénon: "In the current period of our earthly cycle, i.e., in the Kali Yuga, this "Holy Land" defended by "gatekeepers" that conceal it from profane eyes ensuring nonetheless some external relations, is in effect invisible, inaccessible, but only for those who do not have the qualifications required to enter it. However, its location in a particular region, should it be considered as literally effective, or just symbolic, or both? We shall answer this question, by simply stating that for us, geographical facts themselves, and also historical facts, have, like all others, a symbolic value, which however obviously does not take away anything from its own reality as a fact, but that in addition confers such immediate reality a greater significance" René Guénon. *The King of the World* (1927) Conclusions.

factor, *opus* is synthetically sustained by the integration of the most diverse techniques of the cosmos, ranging from the more quantitative to the more qualitative.

# TECHNIQUE

*[The critique of technology] is an enterprise that is by nature controversial, that requires an interdisciplinary but rigorous approach, and that is often a lonely pursuit, both because its messages are almost always unwelcome by society and because its new insights only rarely seem plausible in light of conventional knowledge.*
T. L. Guidotti[391]

*If, then, we would attach a significance to technique, we must start from the soul, and that alone.*
Oswald Spengler[392]

*The more questioningly we ponder the essence of technology, the more mysterious the essence of art becomes.*
Martin Heidegger[393]

The concept of *technique* must be defined now, since there is whole spectrum of confusing concepts which surround this term. In the case of the Spanish, English, or French languages, the terms *technique* and *technology* are frequently used to refer to different concepts, since etymologically *technology* means a discourse on *technique*. On the other hand the concepts have had to vary over time due to sudden and vertiginous changes that have occurred in this technical/technological domain.

The term technique originates from the Greek *technikon* which is intrinsically linked to *techné*, a term embedded with artistic connotations referred to as *poiesis*. *Poiesis* constitutes a purely creative activity that reveals synthetic forms that are not initially

---

[391] Guidotti, T. L. 1994. "Critical Science and the Critique of Technology." *Public Health Reviews* 22:235–250.

[392] Oswald Spengler. *Man & Techniques*. Contribution to a Philosophy of Life. 1976 Greenwood Press.

[393] Martin Heidegger. *The Question Concerning Technology*. Garland Publishing, Inc. New York & London 1977, pg 35.

present, or are *hidden*, in correspondence with Heraclitus' thesis *physis kryptesthai philei* [nature (physis) loves to hide]. Heidegger writes:

> *"once there was a time when the bringing-forth of the true into the beautiful was called techne. And the poiesis of the fine arts also was called techne."*[394]

Whenever the phenomenon of *poiesis* is developed through a set of normative disciplines and teachings, we refer to the attainment as *techné*. The technical attainment of harmonic and integrated figures/symbols along a process of *poieses* shows in the case of *techné* the aim towards an overall figure of balance. "*Techné*," as a term that etymologically derives from the Greek word τέχνη, is defined as an "ancient form of art or craft." But it would be extremely impoverishing to confuse *techné* with our modern ideas on craft and art. The recovery of the primordial meaning of *techné* by resorting to philosophers such as Plato or Aristotle is still challenging due to the deep cultural gap existent between the Greek world-view and ours. As Spengler advises:

> *"The sum total of the Greek philosophy that we possess, actually and not merely superficially, is practically nil. Let us be honest and take the old philosophers at their word; not one proposition of Heraclitus or Democritus or Plato is true for us unless and until we have accommodated it to ourselves. And how much, after all, have we taken over of the methods, the concepts, the intentions, and the means of Greek science, let alone its basically incomprehensible terms?"*[395]

Today we can relate aesthetically to the arts of the Greek, but we can no longer be in the presence of their praxis, which in the case of approaching this issue is extremely important. However, if there is a term to define the normative and initiatory character of

---

[394] Martin Heidegger. *The Question Concerning Technology*. Garland Publishing, Inc. New York & London 1977, pg 34.

[395] Oswald Spengler. *The Decline of the West*. Alfred A Knopf. 1926. New York, pg 257.

specific operative and artistic procedures, the term techné is the best candidate.

One of the main difficulties that emerges in order to properly distil the term *techné* and extract all its powerful essences is that Greek civilization was highly focused on pure knowledge or *epistêmê*,[396] and consequently aspects such as technique, arts, craft, industry, and production were considered to be separate domains, and were rendered as "profane" in terms of knowledge. Whenever there is a fixed set of rules and disciplines that develop a technique on the part of the individual, in connection to the socio-political conditions of a given community or State, then we can consider that the creative process of *poiesis* is also *techné*, or a traditional mode of art. Due to the inevitable transformations that take place with culture and civilizations in combination with what Evola referred to as the inevitable *decline of the castes or superior races*[397]—which constitute the natural inheritors and depositories of psycho-biological operative predispositions—it can also happen that the rigorous following of a set of disciplines might not produce any substantial modification of the personality of the practitioner, hence impeding the emergence of the Magical "I." In this case, we can consider that there has been a "sterilization" of *techné*, as if one has managed to build an electric motor and yet the electricity is lacking. The capacity of *techné* to produce effective spiritual "fruits"—hence not becoming merely a mode of art that aims for arousing emotional feelings—depends on the specific "spirit of the times" (Zeitgeist), the

---

[396] Epistêmê is the Greek word most often translated as knowledge, while techné is translated as either craft or art.

[397] "When I criticized the racist world-view I mentioned that occult power when present, alive, and at work constitutes the principle of a superior generation that reacts on the world of quantity by bestowing upon it a form and quality. In this regard, one can say that the superior Western races have been agonizing for many centuries and that the increasing growth in world population has the same meaning as the swarming of worms on a decomposing organism or as the spreading of cancerous cells: cancer is an uncontrolled hypertrophy of a plasma that devours the normal, differentiated structures of an organism after subtracting itself from the organism's regulating laws. This is the scenario facing the modern world: the regression and the decline of fecundating (in the higher sense of the term) forces and the forces that bear forms parallels the unlimited proliferation of "matter," of what is formless, of the masses (...) a reversed selection and the ascent and the onslaught of inferior elements against which the "race" of the superior castes and people, now exhausted and defeated, can do very little as a spiritually dominating element." Julius Evola. *Revolt Against the Modern World*. Inner Traditions, pg 108-109.

character of the State. This is extremely crucial, because otherwise the processes of technical *poiesis* become impossible in practice, since the ultimate reference point for balance (the State) has mutated. In other words: the spiritual effects on the individual of, let's say, playing the violin in the 18th century were extremely more decisive than playing a violin in the 21st century, since the *resonance* with the character of the European States of those times is no longer possible. Ultimately, all instruments and devices that individuals handle as technical mediators correspond to elements of resonance aiming to "tune" into the mode of power that develops them, and in our times the psychological effects of, let's say, using a smartphone on a daily basis, are extremely more decisive than playing a violin, or even playing an electric guitar, since in our times all State-nations are characterized by cybernetic functioning, in operative terms.

## BRINGING FORTH TECHNIQUE

Heidegger employs the expression "bringing-forth" [Her-vor-bringen][398] in order to refer to *poiesis* as revealing the synthetic configurations or *essences*. One important feature that Heidegger reveals about the term *techné* is that because *techné* is connected to *poiesis, techné* itself is also a mode of revealing elements amidst *physis*. The term *physis* is used in Operative Traditions instead of the term "nature" as the domain where, as Heidegger, writes *"something arises from out of itself."*

By resorting to Evola's phenomenological views, it becomes clear that the term "nature" is epistemologically challenging to define unless it is referred to a given "I" that perceives and identifies specific "natural elements." This is because it constitutes an illusory privation or impotence to refer to anything existent outside the scope or focus the "I" can manage to configure in a specific domain. As Jacques Ellul requotes from the words of a modern scientist: *"Nature is what*

---

[398] Martin Heidegger. *The Question Concerning Technology*. Garland Publishing, Inc. New York & London 1977, pg 10.

*I produce in my laboratory."*[399] In effect, the domain of *physis* can be rendered or revealed as "nature" (in the modern sense) to the extent our instruments, devices, and operative procedures are capable of unveiling all the accessory elements present in *physis*, to discover the *essence* or core source of qualitative aspects (quality) present in a given domain. But it is worth recalling that the qualitative aspects are in truth produced by the "I."

For instance, *physis* is not the same universe for a human being who is developing activity in an office, as for a scientist in a laboratory, someone addicted to TV soap operas, a squirrel, or a lion. A perspectivist standpoint is thus inevitable to resort to when dealing with different states of being that illuminate reality through different "wave patterns," so to speak. An individual attached to a given "I" (in Evola's magical sense, of course) is like an individual moving amidst the darkness with the only assistance of a lantern (an operative "I") which illuminates reality based on its own projective characteristics. If the lantern emits pure white light, then all colours and essences shall be perceived easily; if the lantern emits a poor electromagnetic frequency, then the reality that is perceived shall be a function of this poorer vision. To speculate that beyond perception there is a domain of the "spiritual," the "gods," or even "conspiracy theories" does not actually reveal any essence in reality, but rather reveals alienation of the self and privation, that is, it reveals one of the biggest burdens when disciplining one's spirit according to an Operative Tradition. As Joachim Schumacher affirmed when referring to the alienation—even present in philosophical figures such as Immanuel Kant: *"Better our thickest darkness than your stinking light."*[400]

---

[399] Jacques Ellul: "Technology is both ahead of and behind science, and it is also at the very heart of science; the latter projects itself into technology and is absorbed into it, and technology is formulated in scientific theory. All science, having become experimental, depends on technology, which alone permits reproducing phenomena technologically. Now, technology abstractly reproduces nature to permit scientific experimenting. Hence, the temptation to make nature conform to theoretical models, to reduce nature to techno-scientific artificiality. "Nature is what I produce in my laboratory," says a modern physicist. Jacques Ellul. *The Technological System*. 1980. The Continuum Publishing Corporation, pg 114.

[400] Quoted by Anne Harrington, *Reenchanted Science: Holism in German Culture from Wilhel II to Hitler*. Princeton University Press 1966, pg 47 [Mocek, Reinhard. 1974. Wilhelm Roux-Hans Driesch.

The perspectivist standpoint was introduced into biology and ethology by German biologist and philosopher Jacob von Uexkiill (1864-1944) under the name of *Umwelt-Lehre* (environmental perception/study). Von Uexkiill and Georg Kriszat presented fascinating pictures of how *physis* is perceived by various animals, and this diversity of lower and higher perspectives on *physis* is not only applicable in the animal kingdom but can also applicable to the human condition in very specific cases. As we shall see, this extrapolation of the human condition only becomes structured and organised when a given culture is based on a caste organization and empirical correlations with *physis* depend on inner differences between members of society that determine the hierarchy. In a society divided into economic classes like today's, highly abstract links are established between individuals and the material domain (economic production) and the individual looses the capacity to produce in terms of *poiesis*, that is, in terms of extracting the qualitative aspects of *physis*.

Hence, the experience of men in relation to *physis* is always multi-dimensional and the *darshana* doctrine of the Sânkhya in Hindu cosmology, shows how every man experiences *physis* in a very particular *form*. According to this doctrine, Prakriti—the feminine pole—constitutes the "raw material" that, even though lacking formal definition, cherishes all potential multiplicities within. And it is by men's spiritual action *Purusha* (or more strictly speaking, by operating through *poiesis*) that *Prakriti* is manifested and revealed. *Prakriti* is passive, metaphysically limited to reflect the genuine quality of *Purusha* in the domain of manifestation. However, it is important to recall that when referring to *Purusha* we are not referring to any form of discursive/abstract thought (in the way Descartes conceived the spiritual) and *Prakriti* is not equivalent to matter, as considered quantitatively in classical physics.

The term *poiesis*—which is also the root of the term *poetry*—refers to a creative mode of activity that configures an undifferentiated and

---

Zur Geschichte der Entwicklungsphysiologie der Tiere ("Entwicklungsmechanik"). Jena: Gustav Fischer].

passive domain of *physis* through a process that implies *negative entropy*, that is, the production of ordered forms out of a universal chaos containing all potential possibilities of forms. Yet how can this be accomplished if according to thermodynamic laws all systems necessarily increase their entropy production, causing chaos to dissolve all forms? How can all elements subjected to specific laws that determine their mortality be superseded by powers that induce a counter movement towards immortality?

It is at this point where the term *techné* necessarily kicks in as a set of operative procedures aiming to discipline the spirit for selecting and "extracting," so to speak, diverse forms and figures out of the "sea" of *Prakriti*. Due to the intrinsic characteristics of any spiritual discipline based on operative premises, any laws—conceived in the sense of formal abstractions that define relations between a set of categories of thought—are in no case primordial, but rather derived from the specific character of *techné*, where the character of mobilization corresponds to the ultimate determining factor. This can be understood easily by realising that rendering a musical composition into exact mathematical functions and vibration propagation laws [*causa formalis*] in no case entails that our senses relate to the primordial musical configuration. And in the case of *techné*, the relation to whatever is primordial [*causa efficiens*] through the medium of our senses is accomplished through repetition of patterns and habit. Hence, *techné* entails a disciplining of the senses, allowing them to spontaneously perceive and adapt to specific configurations, according to a dynamics Heidegger referred to as *enframing [Ge-stell]*, which was what the German philosopher considered to be one of the key attributes, or essences of *techné*.

Not only *poiesis*, but also *techné*, do not derive from any formal laws, which constitute a clear mode of privation, as the source of production of the formal laws is beyond the laws themselves. Heidegger's concept of *enframing* as a feature of *techné* implies that all beings or elements that are susceptible to becoming mobilized, shall become mobilized whenever a higher principle of development is capable of integrating their qualitative attributes. This is what

Heidegger referred to as the "standing-reserve" [*Bestand*] of the elements exerted by the *enframing* dynamics that are typical of *techné*. The mobilization of resources and their further classification into specific categories or series at an industrial level can provide us with a good visualisation of the *enframing* character of *techné*, in the sense that a higher principle of development and production defines all elements based on their operative characteristics. A mere analytic or scientific description of the elements present in a modern factory, no matter how accurate the description is, it does not provide us any insight whatsoever into the principle of development of the configurations present in such a factory. As Heidegger stresses:

> *"Technique is therefore no mere means. Technique is a way of revealing. It is the realm of revealing, i.e., of truth."*[401]

A truth of a higher order escapes any purely analytical or instrumental view of the highly dynamic technical processes. For instance, in the case of many corporations and factories, the substitution of old equipment for more advanced equipment becomes a necessity in order to keep hold of the market, even if the inevitable choice for survival entails higher costs and less profit at a economic level, or if the substitution impels the corporation to increase diminishing returns in its typical processes.

As Dennis Gabor pointed out when referring to the overall power of technique: *"Anything that can be (technically) done will be done,"*[402] so not only economic laws or physical laws are unable to halt the mobilizing power of *techné*, but also human morals are as well powerless against the *enframing* character of *techné*, something that Ernst Jünger perceived extremely well in war situations, where all humanist conceptions on the universe were rendered totally

---

[401] Martin Heidegger. The Question Concerning Technology. Garland Publishing, Inc. New York & London 1977, pg 12.

[402] "It is its own acquired speed that makes technology progress. And there are two reasons for this. The first is that the traditional industries must be kept up . . . the second is nothing other than the fundamental law of technological civilization: 'Anything that can be done will be done.' This is how progress applies new technologies and creates new industries without seeking to find out whether or not they are desirable." Dennis Gabor, Survivre au future.

powerless when faced with an overall principle of development that was *making use* of military procedures and techniques in a way that resembled a game of chess. In effect, a higher principle of development *enframed* each piece in the war "chessboard" in a specific position, in a waiting position ("standing-reserve") or rendered mobile, depending on the specific attribute of each piece, in correlation with the overall forces of development, which Jünger himself considered to no longer constitute a conflict between nations, but *between ages*, that is, the principle development of the Zeitgeist, its synthetic figures, and its future impositions were humanly unstoppable and ultimately destined to prevail. The German war hero writes in regard to the *enframing* and *challenging* aspect of technique that:

> *"In all places where the human being falls under the jurisdiction of technique he is faced with an essential alternative. Either he accept the means peculiar to technique and speaks their language or he perishes. But when someone accepts these means, he then becomes, and this is very important, not only in the subject of the technical processes, but at the same time in its object."*[403]

The later irreversibility implies that through mobilization the process of revealing typical of *techné* activates operatively on factors and integrates them into a higher principle of development, that then finally consummates them. These dynamics of *techné* is what Heidegger referred to as a challenging *[Herausfordern]* which forces all elements to actively express their attributes. When using the term attributes, we refer to operative substantial attributes or *skills*, which define the character of mobilization and forms of action. For instance, independent of a given formal education in college or university, what determines the level of effective responsibility on the part of a technician in today's society are the specific skills

---

[403] "Überall, wo der Mensch in den Bannkreis der Technik gerät, sieht er sich vor ein unausweichbares Entweder-Oder gestellt. Es gilt für ihn, entweder die eigentümlichen Mittel zu akzeptieren und ihre Sprache zu sprechen oder unterzugehen. Wennman aber akzeptiert, und das ist sehr wichtig, macht man sich nicht nur zum Subjekt der technischen Vorgänge, sondern gleichzeitig zu ihrem Objekt" Ernst Jünger. *Der Arbeiter*. Ernst Klett, Stuttgart 1981. Klett-Cotta, pg 82.

and "know how." In many cases none of these skills can be taught through formal academic instruction, but rather have to be *trained,* that is, an individual has to be *enframed* in specific environments and power echelons where a series of automatisms and habits are induced in the worker in a "standing-reserve" state, ready to become mobilised for ulterior developments.

However, in spite of the fact that in the last paragraphs, we have resorted to a visualization of the idea of *techné* based on the aim of intending to capture the overall principle of development that takes place in modern industry, this industrial layer of visualisation is still not enough to capture the essence of *techné*, because in most modern industrial processes, operative efficiency corresponds to the main trait adopted in all decisions. Efficiency, as an operative procedure, necessarily entails developing configurations that minimise local dissipation, thus expelling all entropy outwards, towards the entire system[404] so this necessarily impedes any capacity for local prediction/foreseeing of the overall aim of the development, its *being* or what Heidegger referred to in German as its *Da-sein* ["openness-to-Being"]. The essence of *techné* therefore cannot be grasped when all operative procedures become extremely functional (like in cybernetic or modern techniques) and where chaos and dissipation remain "out of the equations." For instance in the case of the strict *techné* training that Eugen Herrigel had to go through based on the indications of Master Awa Kenzo, chaos and dissipation constituted aspects that were constantly present, and were absolutely required in order for Herrigel to perceive the aim (symbolically expressed in the arrow) of material development. In this spiritual discipline, Heidegger's conception of *enframing* was expressed in the specific modes of training that adapted the body's motion to symbolic patterns which, by being constantly *challenged* (Heidegger's *Herausfordern*) by that which Ernst Jünger referred to as the "elementary powers" of the irrational (emotion, fear, anger, hatred, passion, etc.) ultimately aimed to resonate with the Tradition incarnated by the Master.

---

[404] Jeremy Rifkin also perceives very well this process when writing: "Each technology always creates a temporary island of order at the expense of greater disorder in the surroundings." Rifkin, J. 1980. *Entropy — A New World View.* New York, NY: The Viking Press. p. 123.

But the case developed in the second part of this book involvng Eugen Herrigel's spiritual training in Japan before the 1940s, corresponds to a rather exceptional case of *techné* development, where Heidegger's notion of Da-sein is present, in this specific context *Mastery* corresponds to the function of incarnating and representing essences through technical dominion of a purely artistic kind.

So all the former introductions to the term *techné* can allow us to provide the most general definition of *technique* as, above all, the given exerting of a *praxis* (Greek: φύσις) by a specific operative "I," or potency that, *by partaking in an active organization, interacts with the environment and configures the territory according to a specific competence. This competence can be characterised by multiple modes of configuration that develop towards different aims.*

If there is a difference between the ancient and traditional conception of *techné* with the more general idea of *technique*, it is that while all states of being are characterised by a specific technique, an organization characterised by *techné* integrates and centralises diverse technical modes from the most functional techniques to the most magical, so in this sense, *techné*, as the highest artistic mode, refers to the presence of mastery of techniques and dominion of time, which is directly related to *opus*.

One of the greatest issues related to the former definition of technique, is that it corresponds to a concept related to *process*. Any attempt to establish a static or instrumental description of a given technique (techno-scientific discourse) shall always be a "shadow" of an intrinsically dynamic character that is typical of a technical process, and whenever a given technique is described systematically or through language/concepts, the representation can only constitute a partial projection of the technique upon space, time, and causality. As Heidegger points out with clarity: *"the essence of technique is by no means anything technological."*[405] For instance, we can read an accurate description of the movement of a dancer or martial artist,

---

[405] Martin Heidegger. *The Question Concerning Technology*. Garland Publishing, Inc. New York & London 1977, Introduction xix.

but the depiction tells nothing about the dynamic character of their techniques. The *dynamic character of technique* can be sensed from the impression that a given technical product or art produces in our senses, and yet such a character can only be described symbolically. In addition, the dynamic character of technique impedes it being grasped via speculation, laws, or scientific thought, which always correspond to intellectual shadows of the technical phenomena. A parallelism to the latter can be established in the domain of thermodynamics, where the heat exchange of an ideal gas from a given state A to a given state B, not only depends on the variables of each state but also on the type of process (isothermal, adiabatic, etc.). In non-ideal and real life processes, thermodynamics can only statistically address the dynamic process from one state to another (through Boltzmann's equations), and in the case of technique, exactly the same limitation takes place. No matter how well we address technique from a scientific viewpoint, we are always playing with "shadows" and "frozen pictures" of the technical phenomena in all its ultimate richness. This projection of technique on logos constitutes *technology*, which etymologically is a *discourse on technique*, causing the projection of socio-political, moral, ideological and scientific models on the "shadows," like when we address the "good" or "bad" effects caused by humans in the application of specific techniques. The term engineering (civil engineering, industrial engineering, electronics/computer engineering, social engineering, etc)—derived from the French *ingénieur*—corresponds to a mode of professional activity developed in those domains where a conflict arises amidst a series of technical configurations (an engineering problem), leading to the search for an integrating technical configuration that harmonizes all determinisms (an engineering solution). Engineering activity is also related to the idea of *invention*, where in many cases it is very challenging to encounter the cause-effect or clear scientific relation between a set of primary technical conditions and a final invention, or as Ernst Jünger remarks: *"often remains obscure the true source of very important technical and scientific inventions."*[406]

---

[406] "der wahre Ursprung der wichtigsten wissenschaftlichen und technischen Erfindungen häufig im

The correspondence between the aims and means required for a technical process to take place implies exact correspondences between the macrocosmos and the microcosmos. As formerly stated, the activation or not of given techniques at a local level is determined in human societies by the overall State that favors such modes of interaction. We can refer to any phenomena as *magical* when a given attractor is activated with a given technical process to go from a set of conditions in state A to a set of conditions in state B, which is qualitatively different from A. The attractor or figure symbolizes the teleological factor existent within the process, a factor that cannot be grasped by the methods of modern science.

These technical processes of a magical character or *poiesis* are thus responsible of favoring developments that induce what Norbert Wiener referred to as *"islands of neguentropy in the ocean of entropy."*[407] Let's provide a very simple example of this: From a classical or Newtonian perspective of physics, if a 70 kg person aimed to ascend the Everest mountain (8848 m) along a process where no aerodynamic or other kind of friction is considered, this person would require about 1450 Kcal of energy.[408] We are considering here a process with no friction and no dissipation of energy, and no matter how many times we repeat this process, 1450 Kcal corresponds to the minimum energetic requirement based on the first law of thermodynamics. If we ever design an aerial lift for such a process, 1450 Kcal would be the inevitable lower energetic limit. The second law of thermodynamics demonstrates that in a closed system, those 1450 Kcal can never be diminished. The more functional a given technique is that could allow us to elevate this 70kg man, the more closer we would get to that 1450 Kcal energy consumption, and thus, the more *efficient* our technique would be.

Yet, contrary to the aforementioned approach to *physis*, magical techniques or *poiesis* are characterized first of all by the maximization

---

Dunkeln" Ernst Jünger. *Der Arbeiter*. Ernst Klett, Stuttgart 1981. Klett-Cotta, pg 51.

[407] Wiener, Norber. (1950). *The Human Uses of Human Beings: Cybernetics and Society* (pg. 36). Boston: Houghton Mifflin Co.

[408] This is the theoretical energy requirement calculation: 70 x 8848 x 9.8 x 0.001 / 4.184 = 1450 Kcal.

of dissipation and friction, in an absolutely anti-economic way, transgressing all formal laws, and allowing the establishment of communion with all overall non-linear and irrational factors that compromise the thermodynamic isolation of the process within a closed and functional system. Magical techniques or *poiesis* require opening frontiers established by scientific reductionist paradigms in order to commune with long-range non-linear patterns, such as those studied by Ilya Prigogine.[409] This requirement of openness to the overall patterns introduces a concept of Jünger's: the Total Character of Work (*Der totale Arbeitscharakter*)[410] which as the German author writes *"breaks both collective and individual boundaries, and corresponds to the source to which are referred all productions of our time"*[411] and adds to the latter the core idea of the *poiesis* we are describing:

> *"in the workspace the only decisive aspect is the performance by which the entire space is expressed. Such performance is power, which institutes the reference point in a system whose situation can certainly be changed very significantly. Such performance is indisputable, as it is incarnated by objective and factual symbols. The virtue of the type is that of recognizing such symbols, in whatever place they might appear."*[412]

Robert M. Pirsig approaches this total character of work with that of the quest for *quality*, which is,

---

[409] Prigogine, Ilya; Nicolis, G. (1977). *Self-Organization in Non-Equilibrium Systems*. Wiley. ISBN 0-471-02401-5.

[410] "Der totale Arbeitscharakter aber ist die Art und Weise, in der die Gestalt des Arbeiters die Welt zu durchdringen beginnt." Ernst Jünger. *Der Arbeiter*. Ernst Klett, Stuttgart 1981. Klett-Cotta, pg 50.

[411] "Wichtiger ist es jedoch, zu sehen, daß der totale Arbeitscharakter ebensowohl die kollektiven wie die individuellen Grenzen durchbricht und daß es diese Quelle ist, auf die jeder produktive Gehalt unserer Zeit sich bezieht." Ernst Jünger. *Der Arbeiter*. Ernst Klett, Stuttgart 1981. Klett-Cotta, pg 51.

[412] "Im Arbeitsraume entscheidet nichts anderes als die Leistung, durch welche die Totalität dieses Raumes zum Ausdruck kommt. Dies ist Macht, und dies setzt den Bezugspunkt in einem System, dessen Lage sich sehr wohl und sehr bedeutend verändern kann. Diese Leistung ist insofern unbestreitbar, als sie durch objektive, durch sachliche Symbole verkörpert wird. Es gehört zur Tugend des Typus, daß er solche Symbole anerkennt, wo immer sie auch erscheinen mögen." Ernst Jünger. *Der Arbeiter*. Ernst Klett, Stuttgart 1981. Klett-Cotta, pg 75.

> *"neither a part of mind, nor is it a part of matter. It is a third entity which is independent of the two."*

The author of *Zen and the Art of Motorcycle Maintenance* writes:

> *"A real understanding of Quality doesn't just serve the System, or even beat it or even escape it. A real understanding of Quality captures the System, tames it, and puts it to work for one's own personal use, while leaving one completely free to fulfill his inner destiny (...) The sun of quality (...) that does not revolve around the subjects and objects of our existence. It does not just passively illuminate them. It is not subordinate to them in any way. It has created them. They are subordinate to it! (...) Quality is the Buddha. Quality is scientific reality. Quality is the goal of Art."*[413]

Thea idea of creative activity incarnated by Jünger's concept of the type—in contrast to the individual—corresponds to one of the only modes of activity capable of creating local order out of cosmic disorder, as a form of regal activity that aims towards the ultimate State of balance. And yet by considering that this process of negative entropy is contrary to the tendency in nature to maximize entropy production and chaos, this means in practice that Jünger's notion of type will have to play against the odds of nature. As we have presented in the second part of this book, an operative training enframed in a given Tradition, such as the discipline Eugen Herrigel had to cope with, aims to ultimately overcome and render subordinate all the so-called "laws of nature," causing laws to revolve around a core metaphysical power, draining all abstract or formal representations of physis towards a pure vacuum or emptiness that corresponds to what Buddhism referred to as Nirvana. There is no way the laws of nature can be overcome except by relating to a domain that transcends nature itself, and in order to accomplish this, it is precisely our notion of "nature" that must also be transcended. Ultimately, there is no such thing as "nature," but rather a highly

---

[413] Robert M. Pirsig. *Zen and the Art of Motorcycle Maintenance: An Inquiry into Values*. 1974 (William Morrow & Company).

multidimensional universe in which, as Nietzsche wrote, *"every center of force—and not only man—construes all the rest of the world from its own viewpoint, i.e., measures, feels, forms, according to its own force."*[414] The more we rely on a functional and mechanical perception of the universe, the more our senses also surrender to irreversible centrifugal processes when relating to this universe, and the more our minds demand complex representations of such a universe, to the point that today's physics and quantum mechanics is characterized by a complexity that is inversely proportional to its potential individual applications. An Operative Tradition it is intended to develop the senses of the individual in a centripetal and non-mediated way, counter to all laws and natural determinisms, and this necessarily implies for the student to develop his perceptive faculties beyond the common modes of representation, and to connect to a primordial domain that actually produces and develops all the categories of thought and actions that take place in the physical domain.

The idea of *poiesis* is extremely mind-boggling for modern science as well, since from a statistical viewpoint, the way life develops through *poiesis* or *auto-poiesis* is extremely improbable, and from a technical viewpoint, also very challenging. It was the mathematical physicist James Clerk Maxwell (1831-1879) whose work Einstein considered as the *"most profound and the most fruitful that physics has experienced since the time of Newton,"*[415] who perceived the technical challenges that present the production of negative entropy, and described this challenge through a thought experiment called Maxwell's demon or Maxwell's paradox.

This imaginary challenge discussed the actual possibility that, in a closed system constituted by a gas at a specific initial temperature, a hypothetic "demon" placed in a little central floodgate could selectively allow the entry of the quickest molecules into one of the compartments, and the entry of the slowest into the other, thus

---

[414] Friedrich Nietzsche, Walter Arnold Kaufmann, R. J. Hollingdale *The Will to Power* 1968, pg 339.

[415] McFall, Patrick (23 April 2006). "Brainy young James wasn't so daft after all." *The Sunday Post*. maxwellyear2006.org. Retrieved 29 March 2013.

creating theoretically a final state in which one of the compartments would eventually be at a higher temperature than it was initially, with the other compartment being at a lower temperature.

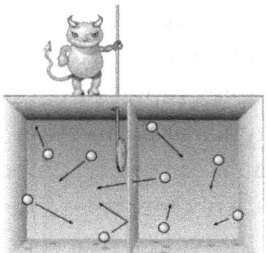

Illustration of the Maxwell's demon paradox

Due to the *process* put forth by this "demon" when facing a chaotic domain, the heat (the vibrational flow of molecules) would be flowing from a cold source to a hot one within a closed system, thus contradicting the second principle of thermodynamics. If in practice it could be technically feasible to create a "demon" with the characteristics, it would as well be possible to create thermal motors in ships that could hypothetically extract energy from sea water, freezing it along their way. However, the insurmountable technical difficulties that arise when trying to contradict the second principle were explained by Leon Brillouin (1889-1969), who solved this contradiction[416] by introducing the importance of the *information* concept, as a key aspect, apart from having introduced it in general physics. By briefly summarizing the contribution of Brioullin to this dilemma, we can realize that in practical terms the "demon" can not separate and filter the quickest molecules from the slow ones if such demon doesn't have specific information about the exact position and speed of those particles which are closer to him, and in relation to those which he actually has interactive capacities with; we are clearly not referring here to statistical information, but to *exact* and non-mediated information.

---

[416] Brillouin, L., 1956, *Science and Information Theory*, Nueva York, Academic Press; trad. fr., 1959, pg 20.

Leon Brioullin's solution to Maxwell's paradox shows us that whenever there exists precise information of the characteristics of the elements outside a system and their development, such information can permit developing processes of negative entropy, creating inner order inside such a system at the expense of the increase of disorder outside. All creative processes in nature require this process of negative entropy to take place, and the mode by which a given organism communes with the cosmos and obtains information is through processes of resonance, which require a common information/pattern shared between the specific creative organism and the whole.

In operative terms, all creative effort can only be accomplished by the embodiment of a principle of development and configuration—the Magical "I"—that is characterized by a *pattern, figure or chreode* that expresses a particular state of being which exists beyond time and space, just like in the case of a specific animal species.[417] When resonance is established between this principle of development and the cosmos, a creative process then takes place which does not produce matter but rather produces a specific form. This process has been called quite recently in biology as *auto-poiesis* (Greek: ποίησις) which can be translated as "the production of one's specific form/figure" and refers to the processes of creation of a form/figure that cannot be discursively explained nor deduced by merely analyzing the parts. Hence *auto-poiesis* corresponds to a synthetic process which is related to a Magical "I." The concept of *auto-poiesis* doesn't necessarily have to be restricted to the domains of biology or that of simple organisms, but can also be applied to the domain of high cultures and modes of civilization. For instance, in the past there were aristocracies and castes, and these groups were characterized by their natural development of qualitative forms at the level of politics, arts, religion, and culture; in this sense we can affirm that

---

[417] Titus Burckhardt. *The Theory of Evolution*: "a species is in itself an immutable "form"; it cannot evolve and be transformed into another species, although it may include variants, which are diverse "projections" of a unique essential form, from which they can never be detached, any more than the branches of a tree can be detached from the trunk." William Stoddart. *The Essential Titus Burckhardt: Reflections on Sacred Art, Faiths, and Civilizations* (Perennial Philosophy). World Wisdom (2003), pg 30.

they embodied operative "I"s of a free, magical character, allowing one to establish a natural and spontaneous territorial dominion. When the castes and aristocracies were overthrown by industrial and financial powers, individuals raised in modern societies no longer had the chance of embracing any operative and Magical "I" and compensated for this lack with a speculative and abstract sense of purpose and identity. In this latter case, *poiesis* necessarily became "outsourced" to external centers of development (like the surrendering of State-nations to international networks, agencies, and corporations) and the individual, to the extent it hasn't yet developed an operative "I," is inevitably forced to enhance economic production (activity or work conceived as *ponos*), at the expense of impoverishment in self-production of homeostatic forms (*poiesis*).

So for instance, if amidst a given fluid there are specific wave patterns (and I'm well aware of the amplitude and frequency of the patterns), it would be technically feasible to build a device characterized by a specific natural frequency (resonance frequency) that could partially absorb the highly dispersed kinetic energy present in the motion of the waves. This technique would induce negative entropy, just like a talented surfer aiming to make use of the waves for developing his own movement. In order for the surfer to actually be capable of accomplishing such movements, he has to be very well aware of the wave patterns present in the sea, and ultimately to be capable of adapting to their frequencies and amplitudes. So a surfer is actually a perfect example of the "demon" of Maxwell, because he induces order out of chaos and must at all time "keep up" his verticality amidst non-linear and highly turbulent conditions. By resorting to Evola's ideas on Magical Idealism, it has to be considered that the "I" can only recognize overall patterns to the extent such patterns are already present in the "I," and in this process of *poiesis*, *techné* constitutes the process of disciplining the surfer's senses in order for him to perceive the overall life present in the waves, beyond all formal interpretation arising from the mind. The Italian author writes that *"beyond the phenomenon, and the "passion" of the world, the "I" thus generates from its extension that which, with a mathematical image, we could say is the "first integral" of it."*[418] In order to understand what Evola means by this, let's suppose I have a rifle and I intend to shoot towards a given target. Any bullet trajectory will correlate very well to a parabolic curve. This curve would constitute the "integral" or long-range pattern I must perceive in order to decide the angle required to shoot. Ultimately, my perception of the whole determines my local actions. Is it the surfer who plays with the waves or is it the waves that play with the surfer?

In terms of local energy production, two other good examples of processes of negative entropy are tidal wave stations, which can only concentrate energy by resonating with the specific wave patterns present in the sea, and the classical Atmos clocks, which don't have to be wound, and obtain their energy via a mechanism of resonance with temperature oscillation. Oswald Spengler once wrote: *"he who knows the secret relationship between microcosm and macrocosm commands it also,"*[419] and in effect such a relationship is secret, not in the sense of "secrecy" or "occultism," but in the sense that our perceptive faculties and techniques are generally not developed enough in order to perceive the subtle connections, which correspond to the only way to create a synthetic *opus* that induces order out of chaos through a process of negative entropy, that is, through *poiesis*.

So in order for *poiesis* to be possible a process of *enframing* of the "I" into progressively more integrated patterns and symbols, causing to eventually mutate the senses and develop new techniques. Such a disciplined process of progressive *enframing* corresponds to *techné*, and it is clearly connected to Jünger's idea on the "total character of work," where *physis* constantly challenges the individual to relate to specific configurations, causing the individual to be forged into a given *type*, who then instinctively develops technical mastery through processes of negative entropy. However, the technician

---

[418] Di là dal fenomeno e dalla «passione» del mondo, l'Io dunque genera dal proprio interno ciò che, con una imagine matematica, potremmo dire il «primo integrale» di quello. Julius Evola. *Fenomenologia Dell'individuo Assoluto Iii* ed. corretta: Edizioni Mediterranee, Roma 2007.

[419] Oswald Spengler. *The Decline of the West*. Alfred A Knopf. 1926. New York., pg 271.

or technocrat profile who develops activity based on the operative requirement of efficiency can not be considered a *type*—in Jünger's sense—since such professional work is rather alien to the elementary powers present in chaos or dissipation, which allow a connection in order to attain *poiesis*, and in its most regal and free form: *opus*. It is precisely this direct contact with the "seas" and chaos, what allows the little "demon" presented by Jünger in his figure of the *Operator* to develop a synthetic mastery and justification of all long-range technical developments. He writes:

> *"there is no singular performance that is not related to everything else. In a word: the total character of work is made manifest in the sum of all special characters of work."*[420]

As Ilya Prigogine showed us, whenever matter is forced to attain a non-linear state out of equilibrium, it appears to gain an extremely mysterious form of life that transgresses any formal notion of separation between the "organic" world and the inert or "inorganic" domain; new subtle and extremely dynamic modes of communication emerge where the "I"—highly stressed by the situation—can face its weakest or strongest points. Ultimately this chaos constitutes the domain that most perfectly "mirrors" the core constitution of the "I." It's only through this contact with the elementary and irrational forces of physis, that it is possible to discern a harmonic array of patterns and figures that are progressively integrated and progressively synthetic, to the point that the "I" can discover its own typical figure of dominion when managing to finally "surf" this sea of chaos. Jünger writes: "The more we devote ourselves to movement the more firmly shall we be intimately convinced that underneath it there is a being at rest, and that any increase in speed is only the translation of an imperishable primordial language."[421]

---

[420] "daß es keine Einzelleistung gibt, die nicht zu allen anderen in Beziehung steht. Es schlägt, mit einem Worte, durch die Summe der speziellen Arbeitscharaktere der totale Arbeitscharakter hindurch" Ernst Jünger. *Der Arbeiter.* Ernst Klett, Stuttgart 1981. Klett-Cotta, pg 88.

[421] "Je mehr wir uns der Bewegung widmen, desto inniger müssen wir davon überzeugt sein, daß

So let's go back to the case of our 70 kg man aiming to climb the Everest. We showed that it was required 1450 Kcal of energy to be capable of accomplishing this, and that 1450 Kcal is the minimum energy required to develop the process. Yet let's suppose that our man developed his "I" through a specific Operative Tradition in order to perceive long-range patterns, attractors, configurations, or that which Ernst Jünger constantly referred to as *figures*. To accomplish this goal would be like developing the senses of an eagle when it comes to discerning the specific of air currents present in the atmosphere, or even the "sixth sense" that Rupert Sheldrake studied in the case of cats and dogs.[422] The development of the operative "antennas" is the ultimate goal of an Operative Tradition, like a lightning-rod that itself attracts or rather *produces/places* particular phenomena that have extreme power for determining the direction of processes that take place in the material domain. If there is any power that resembles Obi Kenobi's "force" in the Star Wars movies, it is precisely in these subtle domains where the "force" has to be found. As an authentic Master of the 19th century, Julius Evola, wrote in this regard: *"the supersensible reality is the true synthetic integration of phenomena."*[423]

If he Magical "I" our 70 kg man acts like Luke Skywalker at the critical moment and rejects any assistance; he will produce a new reality, and a new set of phenomena that determine his gesture and motion. He will not allow conditions to perturb his ascent, because he is deeply aware of the patterns present amidst chaos that are primordial. He will let butterfly effects intervene with his ascent, and open his plight to a long-range system characterized by chaos. Risk, danger, and the elementary forces of nature shall show their decisive power. This contact with a higher form of life goes hand in

---

ein ruhendes Sein sich unter ihr verbirgt und daß jede Steigerung der Geschwindigkeit nur die Übersetzung einer unvergänglichen Ursprache ist." Ernst Jünger. *Der Arbeiter.* Ernst Klett, Stuttgart 1981. Klett-Cotta, pg 17.

[422] Rupert Sheldrake. *Dogs That Know When Their Owners Are Coming Home.* Fully Updated and Revised. Three Rivers Press (2011).

[423] La realtà sovrasensibile è una vera e propria integrazione sintetica del fenomeno. Julius Evola. *Fenomenologia Dell'individuo Assoluto Iii ed corretta:* Edizioni Mediterranee, Roma 2007.

hand with the possibility of death. The divine forces present in chaos can both push him or pull him, to heaven or to hell. As the poet Friedrich Hölderlin wrote: *"where the danger is, also grows the saving power,"* and it is at this point that the heroic capacity for self-sacrifice serves as the best way to enlighten the body and let it be exposed to determinations that open a "portal" or "window of opportunity," where theoretical assumptions fall, and where he can attain his goal by reducing the amount of physical energy required. Evola writes:

> *"Active faith and creative spiritual energy is essentially ecstasy, transcendence. Fearlessness when launching beyond oneself as the generative organ of the dynamic world, as well as all other supersensible worlds, and who can not become self-sufficient, who fears loss, remains in an abstract and empty consistency, or plane of privation. Here it becomes literally true that those who want to save their lives will lose it, and that the only ones who give it shall become really alive."*[424]

This process has been described symbolically in the Grail legends when the challenge of the "invisible bridge" emerges for the Grail hero.[425]

Not just mountain climbing, but also many other activities provide a chance for the individual to be released from individual modes of identification, and to reconnect with the greater forces present in the cosmos which impel action and gesture in a very specific and symbolic way. The ascent to a mountain represents symbolically this process of centripetal convergence of efforts into one single point, going against the odds, and opening one's efforts to the divinities of the mountain, where at the top is Zeus, inaccessible to all titanic attempts that aim to dominate the mountain.

---

[424] "La fede attiva, l'energia creatrice spirituale è, essenzialmente trascendersi. L'intrepidezza di gettarsi di là da sé è l'organo generatore del mondo dinamico, così come degli altri mondi soprasensibili, e chi ad essa non sa farsi sufficiente, chi teme di perdersi, resta in un'astratta e vuota consistenza o nel piano deprivazione. Qui è letteralmente vero che chi vorrà salvare la propria vita la perderà e che soltanto chi la donerà la renderà realmente vivente." Julius Evola. *Saggi Sull'Idealismo Magico*. Edizioni Mediterranee. 2006, pg 97.

[425] *The Romance of the Grail*. Fanny Bogdanow. Manchester University Press. Barnes & Noble. New York, pg 29.

By resorting to the philosophical perspective of Evola present in Magical Idealism, we might agree that mount "Everest" exists because we've seen it in the books, in pictures, or in video. And yet such type of knowledge constitutes a mode of privation, or alienation of the "I," compared to the experiential knowledge gained by those who have actually ascended it, and in opposition to those who boldly aimed—mostly in the past—to ascend the mountain with a minimum amount of technical equipment. Due to its functional characteristics, technical equipment seldom allows reconnecting with the elementary powers present in the cosmos, and protects the individual from such powers, which are extremely decisive as the subtle "wind" which allows life to run counter-current to the inevitable tendency in nature towards material chaos and low energy states.

Yet the attainment of highly concentrated levels of local energy, which corresponds to the spiritual "flash/thunderbolt" mentioned in the second part of this book, is the ultimate goal of an Operative Tradition, symbolizing a process of resonance with the highest State of being which allows the emergence and channeling of a physical phenomenon ("qi" concentration in Eastern traditions) that appears to be a miraculous concentration of power in one single point. In this process of *poiesis*, *physis* reveals its qualitative aspects through the mediation of hand movements, which appear as "magnetized" by a flow of forces that aim to pilot the material development. It is this dominion and mastery of technique (opus) which is capable of revealing the creative-destructive processes taking place, and the mission of Ernst Jünger's main figure: the *Operator*.

# TECHNIQUE AS A CIVILIZATION FORCE
## THAT TRANSCENDS SCIENTIFIC AND ECONOMIC DISCOURSE

"But to tear down a factory or to revolt against a government or to avoid repair of a motorcycle because it is a system is to attack effects rather than causes; and as long as the attack is upon effects only, no change is possible. The true system, the real system, is our present construction of systematic thought itself, rationality itself, and if a factory is torn down but the rationality which produced it is left standing, then that rationality will simply produce another factory. If a revolution destroys a systematic government, but the systematic patterns of thought that produced that government are left intact, then those patterns will repeat themselves in the succeeding government. There's so much talk about the system. And so little understanding."
Robert M. Pirsig, *Zen and the Art of Motorcycle Maintenance: An Inquiry Into Values*
"What is addressed here is no longer a new way of thinking or a new system, but rather a new reality"
Ernst Jünger. *Der Arbeiter*.[426]
"The relation of techniques to the course of civilization, Russia, Science? Goethe would have understood all this and loved it, but there is not one living philosopher capable of taking it in."
Oswald Spengler.[427]

Technique corresponds to the most primordial creative and operative mode of any being. By identifying the active techniques of a given being we can then identify the territorial ties of this being. For instance, in the case of an animal the specific development of a given function or set of senses, once it becomes a factor of territorial dominion is afterwards "sealed" forever as a new technique by natural selection. Jünger writes on technique that: *"every life has the technique that suits it, that is congenital to it."*[428] Technique corresponds in all beings to the most

---

[426] "Da es sich hier weniger um neue Gedanken oder ein neues System handelt als um eine neue Wirklichkeit" Ernst Jünger. *Der Arbeiter*. Ernst Klett, Stuttgart 1981. Klett-Cotta, pg 4.

[427] Oswald Spengler. *The Decline of the West*. Alfred A Knopf. 1926. New York, pg 43.

[428] "Jedes Leben hat die Technik, die ihm angemessen, die ihm angeboren ist" Ernst Jünger. *Der*

important determination and operational habit; it is a "way of doing things" that, as formerly pointed out transcends the categories of thought, especially due to its dynamic and subtle character, which can be perceived as mysterious and unexplainable, especially in the case of magical techniques, *poiesis* or *techné*, where the individual, through developing mastery and dominion becomes an Individual, thus correlating to a Magical "I."

Therefore, the "elusive" character of technique in general impedes it being addressed by scientific thought. However, there is a technical mode that can be approached scientifically quite well: *functional techniques*. These are characterized by the aim of minimizing energy dissipation, which itself entails higher performances and higher wields in most domains of activity. The core idea of functional techniques is expressed by Jacques Ellul when considering that the technique is *"the translation of the concerns of men to dominate things by reason, to disregard the subconscious, make quantitative what is qualitative, make clear and precise the contours of nature, hold the chaos and provide it form."* [429] A cyclist, for instance, can have a predictable outcome in a given competition once his main biomechanical faculties are accurately described and measured. Scientific experts in many domains are prone to cope with areas of activity where the main aim is to develop highly functional techniques. Therefore, within highly functional domains, techniques have very little margin for improvisation, and scientific predictability corresponds to a feasible task.

Yet functional techniques, though more comfortable for the individual who develops them, entail individual risks that can eventually become irreversible. One of the main challenges experienced by any individual who receives training in highly functional environments is that there is the emergence of a progressive gap between the technique developed by the individual and the capability of the functional technique itself to provide self-awareness, since consciousness itself is not of a functional nature.

---

*Arbeiter*. Ernst Klett, Stuttgart 1981. Klett-Cotta, pg 37.

[429] Jacques Ellul, *The Technological Society*. ISBN 13: 9780394703909, pg 43.

This risk takes place because the knowledge is outsourced to scientific experts, who are not trained to embrace the qualitative aspects of the cyclist's personality, which transcend the rationalized conscious mind. And there is always a price to pay for the outsourcing. When referring to this specific mode of technique, Ernst Jünger writes: *"Each of the victories of technique is here a victory of comfort, and who determines the access of the elements is the economy."*[430] Or as Barry Commoner states in his fourth law of ecology, *"There is no such thing as a free lunch."*[431]

In all Operative Traditions, the progressive development of technical activities aims to concentrate responsibility into one single point, the Magical "I," and this can only be achieved by training techniques that are progressively less and less functional and more and more integral, in the sense that in such dissipation of energy/efforts, and the progressive immersion into risky situations that are not controlled—like in a laboratory—is encouraged, since in these situations the specific technique forges the individual's personality in a substantial way than through any scientific understanding of the phenomena, which only acts on the surface of the individual.

While scientific knowledge can only act on the visible "tip of the iceberg" of one's personality, we have learnt in this book that an Operative Tradition forges the invisible but more determining parts of the unconscious individual drives, and the specific character of body movements, which ultimately make all conscious and scientific notions "float" on the visible surface of life. In this sense, it is not what one does but rather *the way one does what one does* which ultimately provides the most crucial insights. The active mode of relation with given means always expresses the place of the individual in the operative scheme of things. Ernst Jünger pointed this out when saying that *"the gesture that the singular person shows when opening his newspaper, and the way his eyes move in such action*

---

[430] "Jeder Sieg der Technik ist hier ein Sieg der Bequemlichkeit, und der Zutritt der Elemente wird durch die Ökonomie bestimmt" Ernst Jünger. *Der Arbeiter*. Ernst Klett, Stuttgart 1981. Klett-Cotta, pg 23.

[431] Commoner, B. 1971. *The Closing Circle: Nature, Man, and Technology*. New York, NY: Alfred A. Knopf, pg 41-45.

*is more instructive than all the journalistic articles in the world."*[432] This standpoint on the part of the German author is related to the heroic dismissal of the cultural assumptions inherited from the world-views of the 19th and 20th century, and the reemergence of operative aims on the part of his main figure, the *Operator*, who absorbs into a single body all the techniques that develop the material productions of the entire world, from the most functional to the more synthetic, integrated, and magical.

The *Operator* doesn't aim for more knowledge, but to *be* and *incarnate* in his own flesh a more integral and synthetic power which justifies technical developments at a worldwide scale. Hence, technique is considered a crucial field for the development and awakening of the Magical "I" than scientific discourse or modern culture in general, which are both considered as secondary and highly dependent; mere "varnishes" on the crucial *opus* substance. Jünger reminds us with a radical inquiry: *"But doesn't anybody realise that all our so-called "culture" is unable to prevent even the smallest of states bordering us to violate our territory?"*[433] For the *Operator*, the field of economics is also considered secondary; such a field can only become the main intellectual concern for individuals who have not yet recognized the powerlessness of modern cultural inheritances, and who are determined by technical drives and habits they haven't yet identified, due to the highly functional and scientific character of most work and academic environments. Consequently, through the eyes of the *Operator*, economic problems are irrelevant problems, since what provides actual solutions and outcomes to the economical developments are factors that transcend the economic domain itself, and these are always technical factors developed by military-industrial complexes. The modern classical idea that politics determines the economic realm (political-economy) is also considered by the *Operator* as a mummified world-view for one

---

[432] "Die Geste, mit der der Einzelne seine Zeitung aufschlägt und überfliegt, ist aufschlußreicher als alle Leitartikel der Welt" Ernst Jünger. *Der Arbeiter*. Ernst Klett, Stuttgart 1981. Klett-Cotta, pg 67.

[433] "Weiß man denn nicht, daß unsere ganze sogenannte Kultur selbst den kleinsten Grenzstaat nicht an einer Gebietsverletzung zu verhindern vermag." Ernst Jünger. *Der Arbeiter*. Ernst Klett, Stuttgart 1981. Klett-Cotta, pg 103.

simple reason: such ideas have no effect in the way the *Operator* handles his body, his movements, his reactions, his relations to technique nor his attitude or style, his *ethos*; all these operative aspects are the main ingredients of responsibility, and the more the individual discards the crucial importance of these empowering elements the more he shall have to assign responsibility/credibility to political leaders, celebrities, scientific experts, etc.

So, as we shall discuss more deeply in the following sections, the "liberation or redemption through work" promised by Karl Marx and most socialists—which aimed to dignify work at a socio-political level—dismissed completely the key importance of understanding that work activities ultimately grant an individual the chance of relating to the effective powers in act in the world (industry, techniques, etc.). When approached with an impersonal sense of devotion an almost military/warrior need to relate to power structures without expecting anything in return, has a deeply transmuting effect on the individual, a transmuting power capable of releasing the activity of the individual from the dangerously functional domains where individual responsibility can be dispersed and lost forever. Yet this progression and aim for liberation is not a comfortable path at all; it implies a progressive release from the self and from all forms of egotistic self-interest; it implies not being concerned about the economic results of an activity or to even think that money is the ultimate reward of any activity. As Nietzsche's Zarathustra writes:

> *"At you, ye virtuous ones, laughed my beauty today. And thus came its voice unto me: "They want—to be paid besides!" Ye want to be paid besides, ye virtuous ones! Ye want reward for virtue."*

The emergence of economic concerns reveals that the ascetic capacity of work conceived as opus has ultimately diminished its presence, and consequently that both need and desire have emerged to consciousness, as a mode of privation. However, the latter does not mean to discard economic production or economic profit, but rather to realize that if financial gain emerges during a state of individual privation, it corresponds to a sign of dependency, which instead of liberating the worker, instead traps him more in complex economic determinisms (debt mechanisms, etc.). Furthermore, the modern conception of work as the goal of economic production was also projected into the natural domain, where it is still considered by most Darwinists that there is a struggle and competition for material appropriation among species. This corresponds to a myopic view. First of all, what a strong and healthy species aims for is territorial dominion through activation of the highest technical modes capable of confronting the lower and impoverished techniques of other species. Once the territory is established, the material relations with the territory (economy) are stabilized, and they become homeostatic. Marxists and liberals—who are rarely capable of distinguishing the economic domain from the technical—got the whole idea of "liberation through economics" wrong, since economic production on its own never liberates the individual unless it goes hand in hand with technical dominion, as is the case with aristocracies where material dominion goes hand in hand with freedom. Due to this crucial confusion, it can't be a coincidence that most of the wealthy people of our era, scarcely resemble the rich and powerful of other ages, and have no time for anything. Their operative predispositions, molded by the modern technosphere determine their actions, beyond liberal or socialist ideas, and because as Friedrich George Jünger pointed out: "capitalist mechanization and Marxist mechanization are brothers."[434] At this point it has to be remarked that neither capitalism nor socialism are economic systems, but rather different modes of justifying economic development on the part of specific political and economic authorities. Since the Industrial Revolution, the systemic aspect of economic production was defined by the progressively systemic aspect of the techniques applied (infrastructures, communications, etc.) and the further assignation of the production within societies that were no longer

---

[434] "Der Kommunist hat von seinem kapitalistischen Bruder viel gelernt und ist nicht umsonst in dessen Schule gegangen" *Maschine und Eigentum* (Klostermann, Frankfurt, 1949).

divided in castes but into classes, required that the societies could accept and find justification of the higher or lesser economic discrimination between the elite and the workers. Such justification was always ideological, and the greater or lesser acceptance of capitalist/liberal or socialist/Marxist ideologies on the part of the diverse populations highly depended on the cultural roots of the society.[435] No ideology can modify reality; ideologies only justify a given reality and configuration of means by relating the means to given aims through a very simple narrative.

What are these narratives? Capitalist or liberal ideologies assume that individual competition for economic appropriation is beneficial for the overall regulation of economic assignation through the markets, while socialist/Marxist ideologies assume that State regulation of all economic production and work activities is beneficial for the State, and beneficial for the proletarian.[436] The latter descriptions show the highly sociological and economic aspect of ideologies such as capitalism and socialism which still don't address the key issue:

> "What to do if eventually all our economic needs become satisfied? What to do if we are finally released from economic determinisms and constraints?"

---

[435] Those peoples who are more prone to think in collective terms and who feel a deep skepticism regarding any particularity or genuine expression of the individual shall be specifically those that shall be more suggestive by Marxist ideology, which was a very efficient propagandistic factor in the case of the Slavs, and those peoples conforming to the Soviet orbit, where there already existed forms of exalted hypersensitivity and collective/messianic sentimentalities.

However, in the case of Europe and the U.S., Protestantism and its Calvinist/puritan branch served to fortify the strictly bourgeois character and traits, which on one hand are characterized by the promotion of organizational/technical skills, and on the other hand by the deep need of self-justification before the masses, mainly through individual interpretations of religious and moral values. This predisposition was practically absent in the soviet/communist countries and however not only did it start to be the "minimum common denominator" in the Western soul, but even began to constitute the "minimum common denominator" in the ambitions of the working classes of the West.

[436] Neil Postman is extremely synthetic when defining Marxist core narrative. He writes: "All forms institutional misery and oppression are a result of class conflict, since the consciousness of all people is formed by their material situation. God has no interest in this, because there is no God. But there is a plan, which is both knowable and beneficent. The plan unfolds in the movement of history itself, which shows unmistakably that the working class, in the end, must triumph. When it does, with or without the help of revolutionary movements, class itself will have disappeared. All will share equally in the bounties of nature and creative production, and no one will exploit the labors of another"Neil Postman. *Technopoly*.

The answer to these tough questions would necessarily imply the addressing of otium, hence implying the reconnection to the Political and to the State as determining factors of human existential integration in the cosmos. But ultimately, as Jünger remarks: *"The real problem is rather that a large majority does not want freedom and are even afraid of it."*[437] So whenever virtue is no longer conceived as an unconditional lifestyle, but rather as a conditional individual and moralistic sign that rewards a given economic appropriation, then otium can become the worst of nightmares, since an individual deprived of needs has to eventually face the question of meaning, not only that of the meaning of the surrounding economic means, but that of his own life. Spreading ideologies through propaganda is always a clear sign that the power of the State by political elites has been broken, and that they are no longer capable of embodying the pontifex between political power and the people. It is then that proselytizing and social agitation become the inevitable practice of resort to by those who Nietzsche would have referred to as the "preachers of death," those who,

> *"If they believed more in life, then they would devote themselves less to the momentary. But for waiting, they have not enough of capacity—nor even for idling!"*

As described in the first part of this book, the decline of the West can be visualized as a progressive downfall taking place already over centuries—especially since the constitution of the modern State-nations—in regard to the capacity of the States to *master* the material developments that transcend the limitations imposed by their national frontiers. This loss of operative faculties on the part of the modern States was compensated for with networked infrastructures and organizations (multinationals, agencies, etc.) which, based on a cybernetic operative character, allowed the functional integration of political, economic, and social elements that were becoming fragmented and complex. Yet we should emphasize that this type of

---

[437] "Das eigentliche Problem liegt eher daran, daß eine große Mehrzahl die Freiheit nicht will, ja daß sie Furcht vor ihr hat."Ernst Jünger. *Der Waldgang*. Klett-Cotta. Stuttgart 1980, pg 85.

integration doesn't correspond to mastery in operative terms, since new organizations lack the stable and *homeostatic* character that is typical of any State in the Traditional sense. This is exactly what Jünger referred to when expressing such provisional trait as a *forge workshop* [*Schmiedewerkstätte*]. He writes:

> *"our monstrous space resembles a forge workshop. To our eyes can not escape that in our space nothing is created in view of duration, towards the sort of duration we admire, for example, in old buildings, nor does it create anything in the sense of an art attempting to produce valid forms of language."*[438]

This transitional character can be observed in the case of the production of today's technology—where quick obsolescence is the norm—or it can also be observed in the case of the configuration of international corporations, which are constantly subjected to managerial and technical restructuring, in order to adapt and prevail in conditions of constant change.

This massive rate of change and the transitional character of developments necessarily compromises the perception of any metaphysical or transcendent stability. The political representatives of the States also lack the capacity to see any synthesis, and the lack of meaning is compensated for with an over-emphasis and focus on the means, that is, an inevitable hypertrophy of the economic domain. When this takes place, the diverse ideologies (liberalism, Marxism, etc.) that justify the actions of the *homo economicus* are absolutely incapable of capturing any synthetic mastery of means, since the ideologies correspond to speculative "shadows" that derive from bourgeoisie or proletarian views of the world that arose in the 19th century, a time when those who embodied mastery and imperial dominion in the West—the former aristocratic castes—

---

[438] "So kommt es, daß unser Raum einer ungeheuren Schmiedewerkstätte gleicht. Es kann dem Auge nicht entgehen, daß hier nichts im Hinblick auf dauernden Bestand geschaffen wird, wie wir es etwa an den Bauten der Alten schätzten, oder auch in dem Sinne, in dem die Kunst eine gültige Formensprache hervorzubringen sucht. Jedes Mittel trägt vielmehr provisorischen, trägt Werkstättencharakter, ist zu befristeter Anwendung bestimmt." Ernst Jünger. *Der Arbeiter*. Ernst Klett, Stuttgart 1981. Klett-Cotta, pg 85.

were overthrown by the new structures of industrial and economic power.

Hence, in terms of mastery of the means in our times, the subjection of the individual to any liberal or Marxist ideology is already a clear sign of the individual surrendering to the dynamic character of the economic means. In these cases of ideological attachment, the individual is forced to act in terms of constant *competition* with other individuals, since the amount of economic means available is finite, and what's more, becoming obsolete at a rapid pace. No matter how transitional or unstable might appear to be the economic means of a given age, this urge on the part of the individual for attaching to the economic realm is deeply related to the territorial traits implicit in any living being.

* * *

Technique is clearly linked to territory, and we are not referring to economic magnitudes. The configuration of a given natural environment into a territory is accomplished by the successful projection of a set of techniques. Hence, dominion is ultimately established through the *efficacy* of techniques. If we consider a machine as a specific territorial configuration determined by technique (a very specific handling of objects), we can resort to Gilbert Simondon when he states: *"The machine is a human gesture that is deposited, fixed, converted, stereotyped and with restarting power."*[439] Karl Marx, on the other hand, referred to the machine as *"crystallized human work."* Hence, a weapon can be considered as a static materialization of a given technique, yet the *handling* of the weapon on the part of the warrior is what actually reveals the dynamic aspect of technique and its state of activation, that is, if the technique is still *alive*, dynamic. The dynamic aspect of a technique is revealed in the character of the processes it determines, that is, in the *efficacy* of its projection on territorial conditions. Territorial

---

[439] « Ce qui réside dans les machines, c'est de la réalité humaine, du geste humain fixé et cristallisé en structures qui fonctionnent » Gilbert Simondon. *Du mode d'existence des objets techniques.* Aubier 2001 pg 12. Introduction.

conquest ultimately depends on the capacity of a technique to integrate all other techniques. Hence, we can realise here that whenever addressing technique we have to resort to the idea of hierarchy and homeostasis.

The latter implies that there is not "one technique" but countless technical modes depending on the character of the active organization that integrates the techniques. In this regard, animals are also characterised by diverse techniques which allow them to establish territorial dominion. In a organization, the "usefulness" of these techniques at a political, economic, and social level, favours their activation. In the case of the animal kingdom organization corresponds to the ecosystem, which selects techniques based on natural selection processes. In the case of human societies this selection corresponds to the specific character of a given State. In the case of human techniques, there are also different levels of integration, ranging from *magical techniques/techné* (upper castes) to *functional techniques* (lower castes) and the application of a specific technique within the range of techniques also determines the belonging or not to a given caste. The importance of naturally predisposed actions as the ultimate sign of belonging or not to a caste was already exposed in the Buddhist Pali Canon, where it is stated: *"Not by caste is one a pariah, not by caste is one a brāhmana; by actions is one a pariah, by actions is one a brāhmāna."*[440] Frithjof Schuon, relates the natural and spontaneous development of an activity or profession as determined by caste factors. He writes: *"One does not belong to a given caste because one exerts a given profession or one has specific parents, but rather—at least in normal conditions—one exerts a given profession because one belongs to a given caste."*[441]

---

[440] Suttanipata, I, vii, 21 (Uragavagga).

[441] Fritjof Schuon. *Castes and Races*. Sophia Perennis (June 1982).

## CASTE AS AN OPERATIVE ATTRIBUTE

In all integral Traditions that succeeded in establishing immanent transcendence or a homeostatic relation between men and the cosmos, the idea of *caste* was also of primordial importance. In modern societies that are technologically and industrially developed, the idea of *class* has clearly supplanted the idea of *caste*, and today in many sectors of our society, *caste* is conceived as a reactionary and outdated form of human organization which provides social privileges or discriminations based on the belonging or not to a specific lineage. If we understand the meaning of caste as referring to the assignation of political, social, or economic privileges provided by blood, then it isn't hard to agree that the concept of caste can easily be embedded with a reactionary and even sectarian undertone in the case of a society, such as the modern, where the circulation of elites has been much more dynamic than in other moments of History.

The issue that appears here is that the term *caste*, as many other words in our times, has also gone through a considerable impoverishment of meaning. It is precisely whenever applying the framework presented by Julius Evola in TAI-PAI, that the core implications of the traditional idea of caste emerge to the foreground. But it must be clear that Evola's model of application is diametrically different to the urban/industrial activity in our times, where the operative aspects that characterize any caste have been repressed and in many cases extinguished or marginalized, due to the imposition of new configurations that have succeeded in prevailing upon the atavistic caste tendencies.

The latter is so because caste is overall an *operative attribute*. It corresponds to a qualitative factor that is expressed in terms of action and technique. It is a term that refers to very specific territorial attachments that are determined at psycho-biological levels and transcend those of the conscious mind. Fritjof Schuon points the highly empirical character of the caste attribute when writing:

> *"What are the fundamental tendencies of human nature to which castes are more or less directly related? They could be defined as so many different ways of envisaging an empirical "reality": in other words the fundamental tendency in a man is connected with his "feeling" or "consciousness" of what is "real."*[442]

Belonging to a caste is expressed by instinctive reactions that are not irrational but culturally configured and forged; in this sense we can not equate caste to any primitivism or naturalism where irrational human drives are not configured nor channeled adequately in political, economic, and social terms. In Traditional societies, caste was a highly determining factor in terms of the attachment of the individual to given *techniques*.

Economic factors are considered secondary and derived when intending to define a caste, where qualitative attributes are defined by operative/technical factors. Hence, it is not material possessions that define an individual as a member of a caste, but rather the *configuring mode of matter*, that is, the intrinsic technique or set of techniques a given individual is capable of applying, naturally and spontaneously. As shown by caste organization in Medieval Europe, inherited biological factors such as racial or ethnic traits can predispose some individuals to have a more natural adaptation to a given activity, thus allowing them to belong to a given caste, yet without going through an ascetic "forging" process of the spirit towards a very specific form of activity. Biological (racial, ethnic, and physical) characters are not sufficient to determine if a given individual belongs to a caste. It should be kept in mind that we

---

[442] Frithjof Schuon. Seyyed Hossein Nasr. *The Essential Frithjof Schuon* (Library of Perennial Philosophy)-World Wisdom (2005), pg 201.

are referring to *active* castes, castes that had an active dominion of the territorial conditions in their era. For instance, the Jesuits, like other caste organizations of the past still present in our times (in the military, monarchies, etc.), are no longer active castes, but rather declining and inactive castes, since in terms of territorial dominion they have been overthrown by technicians, bankers, and the media. These castes have inherited cultural modes that are no longer applicable in our times, and if there is an applicable value for the world-views it corresponds to the propagandistic contents they provide to reinforce an already outdated sense of individuality.

Therefore, in correspondence to the most diverse and integrated modes of technique a appropriate equivalent classification of castes can be also established. In traditional societies, castes have been mostly divided into four qualitatively different groups or functions.

|  | Ancient India | Ancient Persia | Medieval Europe | Plato's Classification |
|---|---|---|---|---|
| *Servants/Workers* | Shudra | Huti | Servants/Proletarians | Derniourí |
| *Merchants* | Vaishya | Vastriya | Bourgeoisie |  |
| *Warriors* | Kshatriya | Rathaestra | Nobility | Psúlakes |
| *Regal Authority* | Brahmana | Athreva | Regality | Árkontes |

This is obviously an ideal categorization not intended to express that Traditional societies were strictly divided by rigid frontiers among castes, but rather intended to establish the idea of hierarchy based on different *operative levels*. It is not preposterous to imagine that under specific conditions the requirement of a classification that is defined by three groups or even five or more groups, and that the number of differentiated castes depends on the higher or lesser need of specialization of functions. What is also important to recall, is that hierarchy is determined by qualitative operative aspects, which distinguish the structure from the concept of *class*, which is dependent on quantitative and speculative/abstract aspects. Another key differentiation is that a caste organization relies on an understanding of the cosmos as *physis*, where any sense of the "individual" separated from greater forces is inconceivable. Contrarily to the Individuals who have related to a Magical "I" or *poiesis* within an active caste system, the individuals who compose a class organisation of society are much more prone to conceive of the cosmos in a dualist form, that is, divided into two separate domains that are conceived at best through scientific discourse: the sciences of man and society (psychology, sociology), and the sciences of nature (natural sciences). In the case of the individual who belongs to a class structure, economic sciences constitute the "bridge" between natural sciences and human sciences. Let's recall that it was in times of Jean Rousseau that the separation between nature and civilization (or natural sciences and human sciences) became a cultural trait for the spirit of the West, where the progressive urban development favored a higher sense of separation between nature and civilization with societies progressively divided into classes, and where all former caste organizations were becoming progressively inoperative in arts, culture and politics. The more the separation is emphasized *culturally*, the more the economic "bridge" has to be developed in order to *materially* fill such gap. In the West this "bridging" or "wielding" of nature and society was favoured by the Industrial Revolution, fostering the extraction, processing, transportation, distribution and commercialization of natural resources into consumable goods for society. Despite the cultural separation between nature and society that is typical of a class organization, it is obvious that nature and civilization are in strict opposition: "*Urban areas need nature in order to subsist. Nature doesn't need urban areas in order to subsist...,*" Spengler writes: "*The gigantic city sucks the blood of the countryside, insatiably, demanding more and more men, swallowing them, until it finally dies amidst the unpopulated prairies.*"[443] So one of the limitations in the cultural traits of a society divided in economic classes, occurs as Ernst Jünger says, when it "*has rank in the same extent one has an individuality.*"[444] It

---

[443] Oswald Spengler. *The Decline of the West*. Alfred A Knopf. 1926. New York, pg 287.

[444] "in diesem Raum in demselben Maße Rang besitzt, in dem man über Individualität verfügt" Ernst

fosters the predominance of the sociological and economic aspects over any aim intending to create homeostatic and harmonic links between man and the cosmos, that is, *it doesn't encourage any convergence of both domains*. In this context, it is obvious that no matter how developed all the diverse branches of economic science, such balance/homeostasis between man and nature cannot be achieved exclusively through sciences, because on one hand the reason for the ever-widening gap is of a cultural nature, and secondly because the extraction, processing, transportation, distribution and commercialization of natural goods are not strictly economic processes, but overall *technical* processes.

It was Lionel Robbins who first pointed out the latter. He writes: *"It is not an exaggeration to say that, at the present day, one of the main dangers of civilization arises from the inability of minds trained in the natural sciences to perceive the difference between the economic and the technical."*[445] Technique constitutes an operative factor, a mode of domination of the cosmos determined by the character of a civilization or being. In the case of a civilization that is characterized by a class system, technique gains a more functional character and progressively the higher techniques (techné, crafts) become eroded as the individuals who embrace the operative treasures extinguish due to the power gained by mass production and industry. This process inevitably causes technique to lack "upper" homeostatic factors, to become autonomous from any anthropomorphic standpoint, thereby challenging all other autonomies (ecosystems, native cultures, etc.), and consequently technique becomes a factor in the exploitation of traditional custom and cultures by causing the urban-industrial form of society to prevail and impose itself on the configuration of the entire cosmos.[446] In this context,

---

Jünger. *Der Arbeiter*. Ernst Klett, Stuttgart 1981. Klett-Cotta, pg 70.

[445] C. Wright Mills quoting Lionel Robbins. *The Sociological Imagination*, pg 80.

[446] David S. Landes, *The Unbound Prometheus:* "A competitive industrial system-whether the competition takes place internally, between productive units, or externally, with rival systems, or both-will therefore place a premium on easy movement of labour power, technical skills, and managerial talent. It will encourage geographical mobility, separating men and women from their ancestral homes and families, to work in strange places; and it will increase social mobility, raising the gifted, ambitious, and lucky, and lowering the inept, lazy, and ill-fortuned" *The Unbound Prometheus: Technological*

economic sciences serve as an intellectual framework allowing the individual to relate to the production that is "drained" into society from the territorial conquest on the part of the urban class-defined civilization over nature. Economic sciences afterwards address issues regarding the social distribution of industrial goods, by considering the most effective modes of social and individual assignation. In the context of a class-system it becomes a progressive disconnection between the products that the individual consumes and their specific technical process of production. In other words, *on one hand the individual is progressively alien to the configuring forces that developed the goods he consumes, and on the other progressively alien to the infrastructures that shelter the society he lives in*. And yet, as a residual existential urge passed down to traditional societies divided into active castes (where there was still existent a correspondence between the member of a given caste and the surrounding objects as an expression of the symbolic aspects of the caste), even today the individual who belongs to a economic class still feels the need of affirming his self or ego through a projection of property on the external domain, but in this occasion with objects as *finished products*, as *images of class consciousness*. This is what Thorstein Veblen refers to as *conspicuous consumption*, and in his *Theory of the Leisure Class* writes:

> *"In order to gain and to hold the esteem of men it is not sufficient merely to possess wealth or power. The wealth or power must be put in evidence, for esteem is awarded only on evidence."*[447]

Consequently, no longer having the chance of relating to the processes that transform the materials of nature into the products consumed, the individual who is inserted into a class society becomes progressively alien to the technical cosmos, and alien to the operative magnitudes that determine the character of individual actions, thoughts, and paradigms in connection to the overall

---

*Change and Industrial Development in Western Europe from 1750 to the Present.* Cambridge University Press.

[447] Thorstein Veblen, *The Theory of the Leisure Class* (1934 ed.), pg 36.

determinations of a given civilization. And the opposite is also true: in the case of a society divided into classes, technique becomes an alien factor, not because it doesn't exist, but because apart from the technician's profile who applies technique in a rather unconscious way, technique is no longer recognized as an *operative* factor, not even by scientists. Hence, in a civilization divided into economic classes, there is not much difference between the scientific mode of thinking about reality and the economic mode of thinking about reality, since both modes of thinking tend to objectively describe the *products* of technical/territorial conquest, both in the case of intangible and tangible goods. In effect, on one hand modern reductionist science requires a controlled environment (laboratories) in order to obtain data from experiments, and this control is determined by technical equipment; and on the other hand modern economics requires a socially stable environment in order to obtain economic magnitudes (GDP, etc.), and the stable environment is determined by the technical power of the industry and war.

The specifically *titanic* and *unbound* character of the technique which is acquired by a civilization divided in classes—already pointed out by David S. Landes[448]—is very well symbolized by the tragedy of the Titanic transatlantic, where the decorations, imagery, and comfort easily persuaded most of the first-class passengers to believe that they were not actually traveling in a ship. Due to the replica of the societal conditions of those times, most of the first-class passengers had their perception conditioned to not take account of the technical development required in order to transport this class-system "microcosmos" from England to United States across the Atlantic. In effect, between the passengers (social domain) and the sea (natural domain) there existed an intermediate transitional infrastructure determined by technical processes and a specific technical configuration, allows it to "float" and subsist upon natural resources. If there is a tragedy that can be demonstrated by modern Western culture when starting to be impelled by technique

---

[448] Landes, David S. (1969). *The Unbound Prometheus: Technological Change and Industrial Development in Western Europe from 1750 to the Present*. Cambridge, New York: Press Syndicate of the University of Cambridge pp 324-340.

as a determining factor,[449] it is that *the more the individual is alien to the power of technique, the more the individual is determined by it, and the more the need emerges to justify ideologically the doses of manufactured freedom provided by technical determination.* Eventually this alienation in regard to the operative powers that dominate the cosmos is compensated for with ideological, scientific, or religious justifications on the part of the individuals who compose a class-divided society.

In the case of an active caste organization based on operative relations to the cosmos, alienation with the territorial conditions configured by technique doesn't exist: the member of a given caste can recognize what in the Hindu tradition was referred to as *dharma,* the "law of blood" where typical modes of dominion over specific means and different levels of technical/operative integration permit different levels of self-recognition. In this context, each mode of dominion is determined by the development of a specific mode of technique, and the power provided by a given technical mode is so determining that it even establishes firm engagements to scientific paradigms on the part of the person who exerts the activity. When the homeostatic character of a caste organization is active, strong and balanced, its members do not need to ideologically justify their

---

[449] Jacques Ellul. *The Technological Society*, 1980: "Saying that technology is the determining factor of this society does not mean it is the only factor! Above all, society is made up of people, and the system, in its abstraction, seems to ignore that (...) During the past half century, the state has taken over education, welfare, economic life, transportation, technological growth, scientific research, artistic development, health, and population. And it is now moving toward a function of sociological structuring (national development) and psychological structuring (public relations). This simple enumeration points out that the present-day state has nothing in common with the state of the eighteenth or nineteenth centuries. The organism of the state has augmented along with its functions and areas of intervention. (...) I believe that the reason for the system is technological growth. On the one hand, if the state is expanding its jurisdiction, then this is not the result of doctrines (interventionist, socialist, etc.), but rather of a kind of necessity deriving from technology itself. All areas of life are becoming more and more technicized. In proportion, actions are becoming more complex, more intervolved (precisely because of extreme specialization), and more efficient. This means that their effects are vaster and more remote while their realization implies the use of costlier apparatuses and a sort of mobilization of all forces. In all the technicized activities, a programming is now necessary. And this programming must have a national, often an international, framework. Hence, only the state organism is able to carry out this coordinating and programming, just as it alone is capable of mobilizing all the resources of a nation to apply one or several technologies; just as it alone is in a position to measure, and take upon itself, the long-term effects of such a technology. We could go into detail here and cite countless examples to show that in modern society it is always because of technology that the jurisdiction of the state keeps expanding."

lives, and are less susceptible to being "brainwashed" by ideological propaganda, since a deeper power is at the center of their actions and engagements. The *dharma* of such a member requires the affirmation of the center, figure, or *attractor* in the material conditions available, and the character of the center or *axis* is determined by the character of the State the caste organization is linked to.

* * *

The former introduction to technique has allowed us to perceive the idea of caste as an operative factor based on *dominion*, in contrast to the idea of an economic class, which is more in consonance to Evola's concept of *privation*. By considering the economic projection of the territorial idea of dominion, Karl Marx was the first thinker in the West to refer to the possession of economic means of production as the core factor for the establishment of classes and class consciousness, and though this view is of a materialistic nature, it however *points to* the idea that consciousness and territorial dominion are both linked, not only for humans, but for all beings. In operative terms, what one produces and how one produces it, is an expression of one's consciousness, and this understanding is present in all Operative Traditions. And yet, as we've shown, there are also many different levels of production and there are also different technical levels.

In modern times, and especially in the urban-industrial domain, the emergence of hierarchies based on the criteria of economic production constituted societies divided into economic classes. When faced with this change, the old castes, nobilities, and aristocracies risked losing their socio-political privileges unless they established adequate dealings with the new economic powers. The progressive loss of power on the part of these castes is due to the combination of two factors; the progressive intergenerational loss of contact with the spiritual traits of their ancestors, and the significant changes taking place within their specific territorial conditions. As formerly pointed out, an active caste constitutes an operative factor that relates to specific territorial conditions, so when the territorial conditions change it is practically like the situation of a "fish out of the water."

When a caste organization disappears, all operative faculties in terms of craft and production also vanish, as well as all integrated forms of technique. The full integration and hierarchy of techniques and castes is always supported by the embodiment of virtuous lifestyles, ritual, and discipline, to the point that the given qualitative attribute of a product or artistic creation is a reflection and "sign" that indicates that the magical power of virtue is still present. In these Operative Traditions, the process of inner development is in parallel to the development of gesture, handling, and technique of the creative process, so that both subject and object converge at the same point that has materialized in the highest symbols attained by the process. But when the tension of *virtus* diminishes, then inevitably the "negative feedback" or negative entropic conditions that are expected to emerge at the highest levels of technical development can no longer produce a "stable balance" of metaphysical figures (*poiesis*) only an unstable set. It is in these situations that if an individual is released from technical determination, and approaches a state of *otium*, a feeling emptiness can easily emerge, which is incapable of concentrating all emotions centripetally. This "erosion at the top," once commenced can only become a progressive avalanche of artistic decline, and there is no way of stopping it. The Masters of the craft are those who incarnate the pinnacle, they incarnate *techné*, and are those who directly relate to the State or Empire through their capacity to live in conditions of *otium* that are free from the lower levels of technical determination and activity.

The disappearance of *techné*, by constituting a factor that expresses the disappearance of the power of a caste and its incapacity to further establish dominion over territorial conditions, can be justified through a socio-economic or religious denial of *otium*; denied as the higher aim of a life devoted to virtue and art. In the West contemplative predispositions that are consubstantial to *otium* and *techné* began to be ridiculed as "lazy and idle" individuals dominated by vice and having no public sense of service.

In effect, the West devised sophisticated modes of disguising this incapacity for *otium*, an incapacity that reveals the lack of a virtuous life, that is, a dignified life that is not conceived of in petty moralistic, ideological, or conformist terms. It is obvious that deep feelings of guilt and resentment can cause contemplative *otium* to become a hellish situation for most individuals, yet all guilt complexes—already studied by psychoanalysis—can be directly or indirectly related to self-repression, a trait that was very much reinforced through the Puritan lifestyle. This dominated the Anglo-Saxon countries in the last two centuries, and advocated fear and repression of subconscious traits, treating them as "evil." In addition to the influence of Calvinism it began to be conceived that the redemption of the "diabolic" human traits could be accomplished and sanctified by the economic "fruits" arising from a strong and mechanistic work ethic. Robert K. Merton even shows in *Science, Technology and Society in Seventeenth Century England* that the scientific revolution that was consubstantial to the Industrial Revolution wouldn't have taken place without the Puritan morals and lifestyle.

These strong religious predispositions are at the cultural roots of the North American people, and it was precisely in this nation that the bourgeoisie classes managed to mutate virtue into the pursuit of happiness, the "happy end" reinforced by cinema and literature, to the point that happiness itself became ideological and political.[450]

---

[450] "And if a new ideology existed that would come forward very powerfully during the 19th century, and that would allow to a great extent the bracing of this typical entrepreneurship optimism of the bourgeoisie -no matter if the bourgeois was a politician, scientist, philosopher, worker or artist- it would be the ideology of happiness, which constituted one of the most perfect justifications for the hypertrophy of productive development and the enslavement of human activity in general. It is precisely in the 19th century that there a began a proliferation of abundant moral, economic and philosophical treatises that were justified to a great extent by this new derivation of the concept of happiness. Yet this concept had already lost the connotation of felicitas that in Rome used to be intimately linked to fatum and fortuna, which implied the positive and "happy initiative" of those actions that are in accordance with cosmic forces of the time or the Zeitgeist. It must be recalled for a while that in Ancient Rome, this concept of felicitas was very much related to that of bonum augurium, that is, related to the accurate prediction of the "currents" of a given age and place in order for the initiative and cause to become effected, not in the technical sense of efficient, but in the sense of not dissipating one's efforts fighting powers very superior to one's own. This concept was used continuously in Rome during the military initiatives and was associated with pietas or the subordination of the lives of men on a daily basis -vita buona- to forces that were beyond oneself." *The Solar Warrior*. Origins, Rise & Decline. Miguel A. Fernandez

Hence, happiness in the new continent became ideologically conceived as the levels of comfort gained by the individual, that is, the higher or lower level of attachment to economic products (property, luxury, etc.) which on one hand could constitute the infrastructure to progressively release and gain "free time" from the tension caused by work, and on the other hand contributed to the promotion of an economically-based self-image that caused psychological satisfaction. Yet in the case of a life devoted to virtue, *otium* constitutes the "cherry at the top" that beyond the realm of necessity, allows one to gain stability, balance, and reflection of the self in connection to the whole, such as takes place in a negative feedback reaction in a servo-system that favors a "stable balance" situation. This was clearly the case of Ernst Jünger, who after a life of adventure, virtue, and heroism, devoted most of his years to develop deep political and philosophical insights that embraced the conditions of his time.

Yet the case of Ernst Jünger, who existentialist Jean-Paule Sartre affirmed *"I hate him, not as a German, but as an aristocrat,"*[451] was a very unique case of attainment of *otium* in the conditions of modern times. It was his brother, Friedrich Georg, who wrote:

> *"Leisure is the prerequisite of every free thought, every free activity. And this is why only the very few are capable of it, since the many, when they have gained time, only kill it. Not everyone is born for free activity, or else the world would not be what it is."*[452]

It is not a coincidence that both brothers, who devoted their lives to artistic, philosophical, and political reflection could withstand freedom and *otium* because they were also one of the very few at that time who understood the implications of the technological phenomena, to the point that Ernst Jünger wrote *"Technology is the real metaphysics of the 20th century."*[453]

---

[451] « Je hais Ernst Jünger, pas en tant qu'Allemand, mais en tant qu'aristocrate. »
Jean-Paul Sartre, quote recalled by Jean-Pierre Péroncel-Hugoz to Domminique Venner.

[452] Friedrich Georg Jünger. *The perfection of Technology*.

[453] "Technology is the real metaphysics of the 20th century" *Ernst Jünger Spretnak*, 1997, pg 128. Also

Aristocratic castes that are capable of relating to the new territorial configurations through instinctive mastery of the overall operative aspects have always corresponded to the "negative feedback" or "cherry on the top" of the development taking place at the political, social, and economic level, where justice is not defined in terms of economic capacity, but in terms of mastering the territory through dominion of the diverse techniques. This kind of mastery is an imperial kind of mastery that is substantially different from an imperialistic dominion that is unable to centripetally configure forces towards the same aims. But these Aristocracies, who were not only characterized by having dominion over the conditions of space but also the conditions of time (*otium*) were progressively overthrown by new political and economic configurations, and the West eventually became orphaned from these "pilots," who were substituted by State politicians and governors who were focused on dominating the minds of their voters through ideological imposition, instead of grasping the whole meaning. Under these conditions, the lack of dominion over free time (*otium*) has to be necessarily compensated for with dominion over free space; free time thus makes no sense, and progressively had to be destroyed as an *inefficient* attribute. Deprived of political or State-related meaning, the appropriation of material wealth lacks a higher political meaning, causing the means themselves to subversively take over the configurations and developments. A pure warrior caste, which is naturally devoted to establishing a territorial dominion that transcends individual economic interest, becomes a useless and disturbing figure, especially for the merchant spirit. In this regard, Jünger writes: *"he [the bourgeois] exterminates who made and committed the acts and attacks that opened to him by violence the doors of dominion as soon as they finish their task."*[454] And it is at this point that the absurdity and the irrational start to take over societies that can only be handled by resorting to the lower modes of technical

---

quoted by Jacques Ellul, in "*The Technological Society*" Vintage Books 1958.

[454] "Kein besseres Beispiel ist dafür zu nennen, als daß er den eigentlichen Täter und Attentäter, der ihm die Tore der Herrschaft erst sprengte, vernichtet, sowie dessen Aufgabe beendet ist" Ernst Jünger. *Der Arbeiter*. Ernst Klett, Stuttgart 1981. Klett-Cotta, pg 9.

application. As Julius Evola points out: *"Thus we often see, especially in America, powerful capitalists who enjoy their wealth less than the last of their employees; rather than owning riches and being free from them and thus employing them to fund forms of magnificence, quality, and sensibility for various precious and privileged spectacles (as was the case in ancient aristocracies), these people appear to be merely the managers of their fortunes."*[455]

*Otium* doesn't entail inactivity, but rather a form of activity that on one hand is free from economic necessity of production, and that on the other hand corresponds to the synthetic affirmation of technical determinations that condition its existence. *Otium* blooms in states beyond necessity and desire. Hence, it would be misleading to think that a flower blooms due to an economic necessity; the blooming process corresponds to the expression of the plant that integrates biological laws, it is the expression of the capability of given organisms to direct scientific determination and scientific laws towards a purpose that is symbolic and homeostatic. Such a blooming always expresses the victory of cosmic destiny over natural law, the materialization of a destiny that is highly improbable from a statistical viewpoint, and therefore the phenomena can be considered as miraculous. It is pure economic dissipation. *Otium* is the maximum expression of the accomplishment of *freedom* in the highest sense; it implies dominion and mastery over the material conditions, not the material conditions considered quantitatively, but in terms of appropriation and integration of the *essences*[456] of objects. *Otium* also implies *responsibility*, that is, the ability of a being to respond to external stimuli and conditions to the point that all the diverse aspects converge, are affirmed, and conflicts are pacified into a higher symbol; it expresses the heart and sun around which all material laws revolve. As *otium* corresponds to an anti-economic activity only accomplished by the higher caste of responsible, free, and sovereign beings, it is impossible in practice to quantify the

---

[455] Julius Evola. *Revolt Against the Modern World.* Inner Traditions.

[456] See René Guénon, *Symbolism of the Cross* (Ghent, NY: Sophia Perennis et Universalis, 1996), chapter 14, "The Symbolism of Weaving."

economic value of the artistic symbols developed by this caste that relate to the State. Like the flames caused by the nuclear fusion in the sun, *otium* can only be possible through material dissipation, material sacrifice, and by release from sociological constraints and attachments; it is related to the periodical festivities, where dancing, singing, and love served as manifestations of the integrating force capable of taking over members when free from economic necessity. Ernst Jünger writes: *"The deepest human happiness is to be sacrificed, and the supreme art of command is to identify goals that are worthy of sacrifice."*[457] *Otium* was only the privilege of noble and regal individuals who lived a life of virtue, whose commitments and loyalties were unconditional and never based on self-interest, who were willing to lose it all in order to gain it all.

Amongst several other factors, it was the Industrial Revolution that altered the territorial conditions (*physis*) in the West, and later in the East: new infrastructures, new techniques, new political configurations, and new economic structures. Only those individuals or organizations who could establish strong territorial and juridical links with these new techno-industrial and economic conditions of existence had a chance of firstly acquiring juridical power, and secondly profiting economically from this power. The more complex and systematic these infrastructures began, the more they required international corporate and governmental planning, and the more the only profile capable of relating properly to the new conditions of existence were the technocratic profiles (engineers, scientists) in combination with the liberal business profile (capitalists), or proletarians (selling work/*ponos*[458]) as a more profit-driven type aiming to assign capital to the new techno-industrial developments.

With the advent of industry, and the development of modern science, especially in a society where every single thing began to be considered in utilitarian and economic terms, the cultivation of

---

[457] "Das tiefste Glück des Menschen besteht darin, daß er geopfert wird, und die höchste Befehlskunst darin, Ziele zu zeigen, die des Opfers würdig sind." Ernst Jünger. *Der Arbeiter.* Ernst Klett, Stuttgart 1981. Klett-Cotta, pg 36.

[458] On the Greek term *ponos* and its crucial implications we shall deal with in the third part of this book.

virtue was being discarded as useless. This is because a virtuous life entails the establishment of unconditional loyalties that transcend individual interest, and under the liberal conditions that prevailed in the West, individual interest was embedded by utilitarian attributes. *Virtue, otium, techné,* and *poieis* correspond to operative attributes that address the finalities of means, actions, and developments, something totally alien to modern science. So whenever we consider economics from the viewpoint of modern economic schools, by their own nature, *virtue, otium, techné* and *poiesis* are strictly anti-economical attributes. In effect, in modern times economic sciences were affected by the reductionist paradigms of science itself, where any teleological instance could not be grasped through the limits imposed by materialistic frameworks. It must be recalled that the concept of economic work addressed by Adam Smith in *The Wealth of Nations* and Karl Marx in *The Capital* are very much equated with the concept of work conceived in classical or Newtonian physics, yet if we ever intended to calculate the amount of theoretical economic work required in order to develop an object that was produced through *techné* assisted by means, resources, and tools we would soon realize that in strictly physical terms work can be actually calculated, like in the case of the theoretical amount of work required to set up a pyramid. When it comes to the actual process, we can easily realize that today's engineering means are no longer capable of addressing productive ends, inevitably leading us to conclude that the economic value of any cultural development that entails *poiesis* tends to be infinite.

As it was pointed out in the beginning technique is a magnitude that involves *process*: it's character not only depends on the variables that define a given state A towards a state B, but also the character of the process of transformation from A to B. So this framing of technique allows us to introduce the *magical* character of *techné*, and its relation to *ritual*.

# A BRIEF INTRODUCTION TO ERNST JÜNGER'S NOTION OF TYPE & INDIVIDUAL

*"To renounce to a sense of individuality is presented as a process of impoverishment only for the individual, who sees death in such renunciation. For the type, however, such renunciation is the key to enter a different world, which is not subject to a critique determined by individual criteria."*
Ernst Jünger[459]

As humans, our adaptation and behavior are no longer related to the conditions of nature, if by nature we refer to what is generally conceived of as "natural conditions" determined by flora and fauna. For about a century, rural domains that were integrated with local natural conditions have lost their prevalence, and consequently their habitants, in the urban-industrial domains which have produced the selection of types of individuals who are better adapted to primary industry sectors. Hence, the transformation of the peasant into the industrial proletarian—as occurred in England about two centuries ago—is not determined by nature, but by a completely new and artificial environment. This causes intergenerational "breeding" of human types which adapt better under the new conditions. Under these conditions exerted by the adaptation to a new environment, in the family lineage the

---

[459] "Der Verzicht auf Individualität stellt sich als ein Vorgang der Verarmung nur dem Individuum dar, das in ihm den Tod erkennt. Für den Typus bedeutet er den Schlüssel zu einer anderen Welt, die der Kritik durch überlieferte Maßstäbe nicht untersteht." Ernst Jünger. *Der Arbeiter*. Ernst Klett, Stuttgart 1981. Klett-Cotta, pg 114.

physiological, psycho-somatic, and psychological predispositions can mutate considerably in a matter of just two generations. The crucial influence of technique in the deepest recesses of the human being who relates to it is pointed out by Heidegger when writing:

> *"The threat to man does not come in the first instance from the potentially lethal machines and apparatus of technology. The actual threat has already affected man in his essence."*[460]

However, a shift in adaptation, such as that of the peasant into the proletarian doesn't take place without important shortcomings. In effect, the peasant corresponds to an almost extinct type where dominion and mastery manifested in terms of integrating all the conditions of flora and fauna. Such integration constitutes the *country*, where tradition, ritual, and custom still serve as guiding principles in order to internalize the correspondences required between man and nature, so that a "piloting" and mastery of means can be feasible. In the case of the peasant, crafts are still a very important activity which depends on the capacity for material manipulation. Under these circumstances, the economic function becomes stabilized, sustainable, and integrated, to the extent that the inexistence of any outsourcing of power doesn't even require money, since no financial circulation or debt is necessarily required. A Master peasant owes it all to his tradition, and to nobody else. Peasants who attain this level of mastery are, however, different to the *farmer*, who has adapted to more mechanistic forms of relating to the earth, and who conceives of the traditions of the peasant as cultural "superstructures" that he cannot relate to for the simple reason that he belongs to another *type* or *caste*. Nietzsche's Zarathustra stated:

> *"Best and dearest to me today is a healthy peasant, coarse, cunning, stubborn, enduring; that is the noblest species today. The peasant is the best type today, and the peasant type should be king."*[461]

---

[460] Martin Heidegger. *The Question Concerning Technology.* Garland Publishing, Inc. New York & London 1977, pg 28.

[461] Friedrich Nietzsche. *Thus Spoke Zarathustra.* Conversations with the Kings. Friedrich Nietzsche,

What actually took place during modern times was the progressive extinction of the peasant, who could not compete against the predispositions of the farmer when adapting to the conditions of industry, which required the individual to adapt to functional physical movements. The adaptation on the part of the farmer was much easier, since in terms of instinctive reactions and operative gestures, the farmer is already "molded" for physically functional activities, yet if there is an optimal adaptation to any functional activity, any mode of mastery is inevitably lost. Peasants or Martial Arts Masters who were forged in the climate of their traditions could not adapt to functional activities, and whenever they were forced to adapt the result was an intergenerational impoverishing of their *type*, and finally its extinction. Related to this issue concerning the crucial effect of adapting to new technical means, Jünger affirms that: *"when someone accepts these [technical] means, then he becomes, and this is very important, not only in the subject of technical processes, but at the same time in its object."*[462]

Let's recall that Mastery is expressed exclusively in terms of homeostasis, that is, whenever unpredictable and chaotic situations take place in a situation, there emerges a very specific set of instinctive reactions that reestablish dominion and order, and the reactions are not based on a tyrannical mode of control, but rather on a regal mode of power that spontaneously integrates and provides purpose to the qualitative elements. Tyranny, control, or any other form of totalitarian character emerges where the homeostatic character has weakened, not only in the symbols of a civilization when attempting to relate to the developments in the material domain, but also in the instinctive response on the part of the civilization's elite or rulers. The human species is the inheritor of a vast set of genetic traits that contributed to affirm dominion over natural environments and animal species during the evolutionary process, and yet the activation or deactivation of the traits also depends on the cultural

---

Walter Kaufmann. *The Portable Nietzsche*. Penguin Books, pg 357.

[462] "Wenn man aber akzeptiert, und das ist sehr wichtig, macht man sich nicht nur zum Subjekt der technischen Vorgänge, sondern gleichzeitig zu ihrem Objekt." Ernst Jünger. *Der Arbeiter*. Ernst Klett, Stuttgart 1981. Klett-Cotta, pg 82.

formation or Heidegger's *enframing* on the part of the civilizational forces. In the *Fourth Political Theory*, Alexander Dugin writes that civilization *"does not merely remove 'savagery' and 'barbarism', entirely overcoming them, but itself is built precisely on 'savage' and 'barbaric' grounds, which transfer to the sphere of the unconscious, but there is not only nowhere to escape from this, but, on the contrary, they acquire unlimited power over man, to a large extent precisely because they are thought to be overcome, and even non-existent."*[463] In our time, we live in a transitional phase characterized by the swift mutation of the State and its mode of its power, and this mode of power affects the activation and deactivation of specific human instincts, actions, and reactions. No intellectual, moral, or economic justification can protect men from the intrusiveness of the powers, which as Jünger points out, is: *"Approaching us with their questions, what they expect from us is not a contribution to the objective truth (...) Those powers do not grant value to our solution, but to our response to the questions asked."*[464] The encroaching, networked, and cybernetic infrastructure of power in our times, propitiates and favors the activation of human instincts that have a functional character, whereas all qualitative attributes of dominion that transcend the territorial configurations determined by cybernetic infrastructures are marginalized, "silenced," destroyed or virtualized, where they are rendered into spectacle and image. When introducing the idea of technique we resorted to Heidegger's notion of *enframing* and *standing-reserve* as two modes in which a technique reveals the operative attributes amidst a domain. Referring to the key importance of these concepts, the German philosopher points out that:

> *"everything is ordered to stand by, to be immediately at hand, indeed to stand there just so that it may be on call for a further ordering. We call it the standing-reserve [Bestand]! The word*

---

[463] Alexander Dugin. *The Fourth Political Theory*. Arktos 2012, pg 106.

[464] "Wir leben in Zeiten, in denen ununterbrochen fragenstellende Mächte an uns herantreten. (...) Indem sie sich mit ihren Fragen nähern, erwarten sie von uns nicht, daß wir einen Beitrag zur objektiven Wahrheit liefern, ja nicht einmal, daß wir zur Lösung von Problemen beitragen. Sie legen nicht auf unsere Lösung, sie legen auf unsere Antwort Wert." Ernst Jünger. *Der Waldgang*. Klett-Cotta. Stuttgart 1980, pg 5-6.

*expresses here something more, and something more essential, than mere "stock."*[465]

Furthermore, Jünger points out: *"such enemy cannot be faced with mere concepts."*[466] This process can be compared to the transition that takes place in the case of the life of a lion that is forced to leave its natural territory and live in a zoo. "Silencing" or marginalizing the former territorial instincts is not necessarily a conscious decision made by individuals who actively participate in the cybernetic infrastructure (businessmen, technicians, technocrats, etc.), but a subconscious one, determined by their operative links with the structure they develop. The more the operative procedures become fixed between an individual and an organization, the more these modes of action determine the predispositions and commitments of the individual, and the character of his interests. It was Upton Sinclair who perceived the power of this organizational conditioning when writing: *"You'll never succeed in making someone understand something if their salary depends on them not understanding it."* The restriction of personal freedom of choice whenever induced by individual commitment to a organization, was considered by Cicero in *De Officiis* as a sign of slavery, and thus defining salaried work as the activity of slaves. "I*psa merces est auctoramentum servitutis*" (who takes a salary becomes a slave).[467] Karl Marx also understood that the individual relation with technical developments creates the way people perceive reality as a whole, and that penetration into methods of perception are the key to understanding diverse forms of social and mental life. In *The German Ideology*, the German author writes: *"As individuals express their life, so they are."*

In the context of cybernetic power, the individuals who adapt to functional activities accumulate more or less economic resources to

---

[465] Martin Heidegger. *The Question Concerning Technology*. Garland Publishing, Inc. New York & London 1977, pg 17.

[466] "Dem Gegner kommt er nicht mit bloßen Begriffen bei." Ernst Jünger. *Der Waldgang*. Klett-Cotta. Stuttgart 1980, pg 66.

[467] Manual de la sociología del trabajo y de las relaciones laborales. Antonio Martín Artiles. Delta Publicaciones.

the extent that the activities more or less develop with the particular infrastructure mode. This territorial and economic "rooting" with the structure favors the possibility of constituting a family and having progeny, thus transmitting genetically to the next generation those instincts and predispositions that formerly conquered the specific territorial configurations. Yet whenever a structure of power propitiates the reproduction of individuals based on the criteria of their higher or lesser adaptation to functional activities which are filtered by cybernetic processes, it has also propitiated transitional and unstable family "cellules" in correspondence with the transitional character of the dominion over the means and their development by the social, political, and corporate institutions. Due to the impossibility of mastery over cybernetic infrastructures, the States in decline inevitably had to go through a crisis of dominion, and this crisis is transferred from the top to the bottom towards the family units, since marriage—even the modern marriage based on "love"[468]—corresponds to a political constitution based on the modes of power that develop a specific socio-economic context. Under the determinations imposed by a vastly expanded cybernetic infrastructure of power, all morals, ethics, religious ideals, gender, or principles become factors that have less influence in the constitution of marriages, and what becomes more relevant in these family arrangements is the adequate adaptation of the couple to the determinations of the system. So if a marriage is successful under such circumstances and finally constitutes a stable family unit, the stability is mainly due to its capacity for adaptation to constant change. This implies that the offspring of successful marriages within the context of these power modes, shall also be instinctively predisposed to adapt very well and with no trauma.

---

[468] Author Stephanie Coontz, in her extensive work regarding the History of marriage, restricts exclusively in our times the decision of marriage according to economic, utilitarian and practical criteria: "Men must grapple with new questions: "Do I really want children if I have to do half the work of raising them?"; "Am I willing to stay involved with my children if I'm not still with the mother and getting the benefits of having a wife?" Women, on the other hand, grapple with different questions: "What does marriage really offer?"; "What are its costs compared to its benefits?"; "How would marriage be a help to me in raising any children I choose to have?" Stephanie Coontz. *Marriage, a History*.

The "domestication" of the human animal by the cybernetic configurations of civilization, and the intergenerational reinforcement of passive traits, also imply the fortification of *competitive* traits in the individual. Robert M. Pirsig writes that,

*"It's not the technology that's scary. It's what it does to the relations between people, like callers and operators, that's scary."*[469]

Contrary to what adepts of the Darwinist ideologies believe, competition among individuals is not typical of all civilizations, and not even a trait that is typical in the animal kingdom. Competition is rather a particular case, and an inevitable consequence of the activation of operative modes of power that favor and promote functional interactions. When interactions are functional or mechanistic, no qualitative criteria is then available in order to establish a homeostatic hierarchy or organization. This inevitably causes struggle among individuals for economic resources that are under the dominion of the new territorial configurations. The competitive notion of "survival of the fittest" can be also conceived here as the greater or lesser selection of those individuals adapt to the character of the State. So, if the State promotes functional modes of power—which is exactly the case in modern States—then inevitably the "winners" shall be those individuals who better adapt to functional activities, having a better performance when it comes to competing against other individuals when attaining functional mastery over the means. The more the cybernetic modes of power encroach on a State, the more they are promoted in activities (sport in particular) to obtain a better performance, in order to outrun other individuals that are restricted to highly functional activities. Sports are considered as an activity of leisure, but they are much more than that. Not only Ernst Jünger,[470]

---

[469] Robert M. Pirsig. *Zen and the Art of Motorcycle Maintenance: An Inquiry into Values.* 1974 (William Morrow & Company).

[470] [The opposite of work is it not rest or leisure; from this point of view, there is no situation that can not be conceived as work] "Das Gegenteil der Arbeit ist nicht etwa Ruhe oder Muße, sondern es gibt unter diesem Gesichtswinkel keinen Zustand, der nicht als Arbeit begriffen wird" Ernst Jünger. *Der Arbeiter.* Ernst Klett, Stuttgart 1981. Klett-Cotta, pg 44.

but also Seneca,[471] considered leisure time as the best opportunity to relate to the power of a State, and in effect, modern sports represents the diverse nations best modes for assimilating the rules of the State, by experiencing participation and communion with its power. These rules favor functionalism, performance, and competition among individuals or teams, which are in practice are the same rules that favor a liberal economy.

Today's modes of power favor adaptation of the individual at all levels, and the process of adaptation is characterized by the legitimizing of the entire cultural "atmosphere" existent in public opinion, where in sports, show-business, and politics, the core operative ideas of individual performance and competition must never be publicly questioned. For instance, no professional commentator can affirm in the media that political parties, when competing against each other for government privileges—and regardless of their political ideals—the means, techniques, and strategies they resort to in order to gain the acceptance of the voter are in practice always the same. Hence, competition of the individual itself cannot be questioned. In order for the individual to gain predominance, competition is assumed as an inevitable requirement for our societies. But if we intend to overcome competition (or its economic counterpart: *liberalism*) as a particular case of power appropriation, the only way possible is by first overcoming the *individual* itself. Not only in the case of State or Imperial affairs, but also in the case of many ritualized physical activities, competition was not the primary mode to relate to power. For instance, the concept of "sports" is alien to any Tradition based on operative premises. Even in most modern competitive sports, the individual is ultimately encouraged to adapt to a specific type, and the greater the adaptation, the greater the chances of success and appropriation of power in order to compete with other individuals. By discussing

---

[471] Seneca, in his treatise De Otio conceives a distinction between the general state of affairs of a person integrated in a given society and political structure, and then another higher State that integrates the former which can be accessed and served through otium. It is precisely this latter State the one that is consubstantial to Jünger's Operator. Erskine, Andrew (1990). *The Hellenistic Stoa: Political Thought and Action*. London: Duckworth. ISBN 0-7156-2326-5, pg 68.

modern sports, we can start to understand here Ernst Jünger's concept of *type*, where in spite of the high focus in performance, functionalism, and competition, there are many diverse types of sportsmen characterized by significant differences in terms of genetic predispositions, nervous reactions, and psychosomatic traits. For instance, it can be easily distinguished that the proto-*type* of the gymnast, the cyclist, the formula 1 pilot, or the basketball player, etc., are all characterized by physical and psychological attributes that are not exchangeable. The traits that favor the constitution of given *types* are highly hereditary, thus corresponding at a physical level to the idea of *caste* which is characteristic of Operative Traditions. In modern sports, the better the individual adapts to an optimal type defined by the sport scientists, the better he can compete in terms of performance and functionalism. This sophistication leads to high levels of professionalism in this domain. And yet a tragic dead-end appears once the affirmation of the individual in competition is no longer possible. The prospect of retirement appears at an early age for most sport professionals, when their physical attributes no longer match the optimal constitution defined by the *type*. It is at this point that the anxiety of being forced to live outside of the rules of competition emerges, in which the individual previously found their self-justification. This crisis expresses the limitations of the modes of power defined by competition and the individual. In the case of many corporations, the predominance and power the organizations gain by competing with other corporations eventually has a price to pay, as new competitors can easily pop up in the industry gaining a temporary mastery over technical development. The crisis of adaptation is then transferred from the corporation to its employees, who face unemployment and the anxiety that the situation causes to their existential sense of individuality and to the justification of their lives.

In a liberal economy, competition conceived of as a mode for the appropriation of power is inevitable when there is a mutable mastery over technological and industrial means on the part of diverse organizations. The extinction of the old crafts, which in order to gain mastery required years of dedication and self-discipline, has been overthrown by activities and professions where the individual is forced to enter into a race of constant being *up-to-date*. Hence the instability, contingency, and ephemeral character of the individual is equated with the same instability, contingency, and ephemeral dominion over the means the individual has at reach. This lack of stability is then compensated for with highly subjective, abstract, and stereotypical constructs that have been expressed in the tendencies of modern art, where the massive outflow of novel techniques create a kaleidoscopic universe of images where any sense of creative meaning cannot be found, it is also nihilistically denied by the same artists.[472] Ultimately, technique is the power that drives the process. This was recognized by one of the most popular artists of the 20th century, Andy Warhol, when stating: *"What prompts me to paint this way is that I want to be a machine, and I have the impression that whatever I do like a machine, that will be what I want to do."*[473]

From an evolutionary viewpoint, it appears that the individual is incapable of establishing firm territorial connections with an environment composed of an ensemble of means that are in constant and accelerated change. As Ernst Jünger points out: *"dominating means is very different from just using them."*[474] Because of this constant need for adaptation and competition on the part of the individual within the mutable conditions, the individual is forced to disconnect from his economic property, and to disconnect from the product of his production activity. This disconnection is what Marx referred to as the alienation of the worker, a phenomenon that he conceived of from an economic viewpoint, by assuming that the capitalist appropriates the surplus value from the productive activity

---

[472] Un irrazionalista coerente dovrebbe limitarsi al semplice vivere e ad un agire secondo impulsi ed ispirazioni momentanee, al massimo darsi ad effusioni liriche o a qualche specie di sponta-neità «creativa,» libera il più possibile da una forma. [A coherent irrationalist should merely be bound to living and acting based on momentary impulses and inspirations, bound to lyrical effusions or whatever kind of "creative" spontaneity] Julius Evola. *Teoria Dell'individuo Assoluto*. Edizioni Mediterranee, 1988, pg 139.

[473] Andy Warhol. *Beaux-Arts Magazine*, May 1986, pg 70.

[474] "Diese Meisterung ist sehr unterschieden vom bloßen Gebrauch." Ernst Jünger. *Der Arbeiter*. Ernst Klett, Stuttgart 1981. Klett-Cotta, pg 39.

of the workers, depriving the workers of relating to the profit of their activity. From a strictly economic viewpoint applied to an industrial economy, Marx's analysis is convincing, yet History has shown that in developed countries where workers have had the chance of gaining access to middle-class material privileges and who have also had the chance to participate in capitalist activities, the separation between the individual and the economic properties has not been reduced, but has instead widened. Today, the individual's economic possessions (which to a great extent are technical objects and devices such as automobiles, gadgets, etc.) have less relation to the productive activity of the same individual. The more the individual participates in productive activities of high value that develop the global territorial cybernetic configurations, the more capacity for consumption the individual will have for the products that the global infrastructure manufactures. Yet, in the case of the individual who is determined by a functional and technocratic environment such as those of the *technopolis*,[475] his "will to power" corresponds to that of adapting to the professional types promoted by the cybernetic modes of power in order to compete with other individuals, and at this level one of the most prevailing types is the *business man*.

The businessman corresponds to a well defined *type* whose activity is that of facilitating the connections, networks, and progressive centralization of the cybernetic infrastructures of power at a corporate transnational level. Because of the developments taking place during the last decades, the most effective businessman is he who applies functional thinking to all domains, hence succeeding in appropriating domains through the techniques. These domains are mostly social (R.R.P.P), economic, and political. The businessman

---

[475] In *Technopoly*, Neil Postman refers to the key importance of the technocratic profiles when developing such highly functional and almost utopian environments. The American author, educator, media theorist and cultural critic writes: "Technocracy gave us the idea of progress, and of necessity loosened our bonds with tradition—whether political or spiritual. Technocracy filled the air with the promise of new freedoms and new forms of social organization. Technocracy also speeded up the world. We could get places faster, do things faster, accomplish more in a shorter time. Time, in fact, became an adversary over which technology could triumph. And this meant that there was no time to look back or to contemplate what was being lost. There were empires to build, opportunities to exploit, exciting freedoms to enjoy, especially in America" Neil Postman. *Technopoly. The Surrender of Culture to Technology*. Vintage Books A Division of Random House, Inc. pg 45.

has to be capable of perceiving an *interest* in all activity, that is, the capability of centralizing capital and resources into the functional determinations of a vast network of relations, where financial profit arises as a by-product of the rationalization of processes, in a similar way to how the reduction of entropy in a given organism provides information. The businessman corresponds to a very well defined *type*, in the sense that his functional operative capacity prevails, in spite of being immersed in environments that are not necessarily characterized by functional traits. Hence, the businessman has an instinctive capacity to perceive opportunities for business in practically any social context, like parties, weddings, celebrations, etc. and other ceremonies that are not based on functional or mechanistic interactions.

The historical "grandfather" of the businessman is the bourgeoisie, but the businessman is a more consistent and well-defined *type* than the historical bourgeoisie of the West, which by corresponding to an economic class, was still burdened by irreconcilable traits, arising from the influence of former aristocracies, and the productive traits assimilated from the lower classes. Historically the bourgeoisie was always unable to attain operative mastery over territorial means as the aristocracies did, however the class was responsible for rendering the concept of property into something no longer related to territorial dominion of means and developments—which characterized the aristocratic castes—but related overall to the capacity to *concentrate capital, labor* and *resources*.[476] Hence, the Industrial Revolution would have been impossible if the bourgeoisie classes hadn't managed to transfer and concentrate the rural and agricultural capitals into the development of new urban-industrial domains. Yet these dynamics of developed by the bourgeoisie were projected exclusively to the domain of capitals, workforce, and resources. The purely economic factors became more centralized, in a highly *functional imitation* of what had formerly been the character of the State or Sacred Empires incarnated in aristocratic castes. But the centralization that the

---

[476] See Julius Evola. *Revolt Against the Modern World.* Inner Traditions, pg 103.

bourgeoisie implemented was that of the economic means, not a centralization of political aims. This is clearly related to the fact that the bourgeoisie has always been alien to the idea of *caste*, by equating it with "rusty" political and juridical privileges granted to individuals by a title of nobility. This conception was vastly spread as Operative Traditions were progressively marginalized in the West, since *caste* is overall an *operative* attribute related to the development of an instinctive mastery of economic functions, and based on mastery of specific territorial configurations. Hence, a *caste* corresponds to the fortification of a given lineage of instinctive or genetic capacities capable of responding to external territorial influences, where the economic configurations become stabilized and sustainable. The implementation of a debt-based economy makes no sense in these cases. The term "type" employed by Ernst Jünger, and the term "caste" are thus partially equivalent: they refer to operative traits intending to establish territorial links. While the bourgeoisie class bases its class-consciousness—as Karl Marx pointed out—in its specific relation to the economic means of production. The *caste* or *type* is linked to a State of territorial dominion that transcends the economic sphere. The historical incapacity on the part of the bourgeoisie to relate to the State beyond any economic consideration impedes their constitution as castes, and instead they became classes. Now the question arises here: can the businessman be considered a caste in our society?

Participating in the political and socio-economic roles of a given caste is expressed by adopting given attires or uniforms, which symbolizes the links of the individual to external territorial configuration. In all cultures of the past the wearing of certain clothing expressed the link between a given personality and the production techniques of a given territory. There are countless examples of this in the past, and even today there is still specific attire for some types (e.g. sportswear), and in the military uniforms are still worn. Yet in the civil domains, a very specific suit has prevailed which is the *international business* or *western business* attire, exemplified by a suit and necktie. This suit was "sealed" in

19th century English society, during times when international corporate domains were developing, and the need for expanding networks was very starting to become dynamic. This suit is worn in the workplace, and has always expressed the surrendering of the individual to activities aiming to foster business, which as we've formerly explained, consists in a very specific operative trait. The prevalence of this suit in most domains, shows that *a very specific way of doing things* has prevailed in our times, and this is because it has established dominion over the old modes of power. As the configuration of global powers demands functional adaptation on the part of the individual, it is inevitable that a process of selection shall take place by the power structure which favors those individuals who adapt to this, and who also contribute to developing their networked power configuration. A caste is proliferating here, and business has encroached on all domains of modern life. The irrational feelings of respect granted to anyone who wears a business suit is related to the fact that it symbolizes the prevailing mode of operative power in our times.

Julius Evola and René Guénon referred to the idea of the *regression of the castes*, which they considered an irreversible process according to the cyclical forces in Traditions. They also mentioned the idea of a take-over by the *fourth caste*, where work was deprived of any meaning. When interpreting Evola's and Guénon's approach, Russian politologist Alexander Dugin[477] suggests that the ideas presented by the traditionalist thinkers in terms of the prevailing castes were wrong, yet Dugin also associates the *fourth caste* to the social levels determined by the proletarian, and the *third caste* to the capitalists/liberals. Like Karl Marx, it appears that Dugin is unable to release himself from a mode of thinking based on socio-economic and production terms, since the concept of the proletarian and the capitalist/liberal individual are defined by their relation to the means of production, and yet the concept of *caste* or *type* transcends that of economic production by that of

---

[477] "both (Evola & Guénon) anticipated the victory of 'the fourth caste', in other words, the proletariat (as represented by the Soviet Union) over 'the third caste' (the capitalist camp), which proved incorrect" Alexander Dugin. *The Fourth Political Theory*.

*territorial dominion*. When addressing this difference between class consciousness and the Operator's consciousness forged by the idea of technical mastery, Ernst Jünger wrote: *"what matters is not to create a scheme of the world, in order to adapt to the mold of this or that claims; what matters is to digest it."*[478] Caste refers to the mode of operating on a given territory, or in other words, it refers to the set of techniques that allows the establishment of territorial dominion. In this sense, if an individual is conceived of in socio-economic terms as a proletarian, and another individual is conceived of in socio-economic terms as a capitalist, both become equally determined and well-adapted to this modern mode of power configurations that favors functional, technical, and business operations. The "natural selection" taking place on the part of the territorial configuration of powers determines their belonging to the *fourth caste*, and it does not matter if one is poor and the other is a multi-millionaire. The predominance in today's society of rich and poor individuals who belong to the fourth caste would have been impossible without the promotion of science *as a faith*. For instance, Denison Olmsted wrote in the 19th century that *"science, in its very nature, tends to promote political equality; to elevate the masses; to break down the spirit of aristocracy; and to abolish all those artificial distinctions in society which depend on differences of dress, equipage, style of living, and manners; to raise the industrial classes to a level with the professional; and to bring the country, in social rank and respectability, to a level with the city."*[479]

Based on the classification of Evola and Guénon in terms of the historical decline of the castes, the castes are overall classified depending on the type of State the caste serves. Hence, the *first caste* serves a Holy or Sacred form of State, the *second caste* serves the State conceived of in terms of temporary powers, the *third caste* serves the State conceived of economically, and the *fourth caste* serves the State

---

[478] "nicht darauf kommt es an, die Welt zu schematisieren, sie über die Leisten irgendwelcher Spezialansprüche zu schlagen, sondern darauf, sie zu verdauen." Ernst Jünger. *Der Arbeiter*. Ernst Klett, Stuttgart 1981. Klett-Cotta, pg 45.

[479] Olmsted, D., "On the Democratic Tendencies of Science," *Barnard's Journal of Education* (1855-1856), cited in Hughes (1975), pg 144.

conceived of as a mechanistic system. In this classification we can see a transition from highly homeostatic, vertical, and integrated States towards non-homeostatic, horizontal, and progressively centralized networks of power.

There is the common notion that the task of the businessman is to "make money." This is quite a superficial approach that confuses causes with effects, since in operative terms, the businessman is devoted to establishing new networks among diverse economic, financial, and productive agents. These liaisons become formalized in social, juridical, and corporate terms, and in this context profit arises as consequence of the added value gained by all encompassed processes. The higher or lesser value gained by a business is a function of the higher or lesser capacity to centralize activities, and this principle is what allows corporations to be founded in the first place. Yet it would be misleading to think that the activity of the businessman is of an economic nature, no matter how spectacular his salary is. The character of the businessman is that of developing functional relations and connections at all levels. This favors the constitution of formal and informal modes of relation that adapt well to systems of production, yet the characteristics inherent to this kind of configuration eventually take over economic structures regulated by the State. In economic terms, the businessman is not interested in relating to the services he participates in developing, and not interested in resource depletion rates, diminishing returns, or debt dependencies, since the businessman is prone to trust reports on these issues provided by international agencies. Hence, becoming integrated into the determinations of a cybernetic configuration necessarily fosters short-term thinking and the incapacity for forecast. Let's recall that the capacity for forecast can only be accurate in the long-term when provided by organizations that have politically mastered domains, because these organizations are characterized by a notion of destiny, as all States during their blossoming are. This capacity for forecast facilitates State planning and regulation based on the predictions of growth and resource availability, among other factors. When a new configuration of power emerges that is based on functional operations, then the synthesis and political mission of the former States are transformed into a massively complex array of networks which are not only more difficult to forecast in terms of their development, but even more challenging to register in terms of their real-time variables.

Hence, even though the businessman has the capacity for vision typical of the classical Statesman, it corresponds to a type that has taken over political activities in the modern era. The businessman has gained predominance in all fields, yet this predominance must not be confused with *dominion* or *mastery*, since the businessman corresponds to the "outsourcer" *par excellance*. This necessarily implies the hypertrophy of socio-economic liaisons and the proliferation of debt mechanisms both at a economic level, which is configured operatively through network structures based on high levels of connectivity, yet lacking any purpose beyond the mere dynamism and mobilization of factors. The Promethean tragedy of the businessman is that the more he gains financially, the less he can master and he cannot transcend any ideological discourse, and here we use "ideology," the way Jacques Ellul referred to the term when stating that an ideology *"veils the actual situation by transposing it towards an ideal domain, focusing all attention towards the idealistic, the noble, the virtuous, and in addition, to justify a given situation by coloring it with what is 'good' and what is 'common sense'."*[480]

---

[480] « voiler la situation réelle en la transposant dans un domaine idéal, en attirant toute l'attention sur l'idéal, l'ennobli, le vertueux, d'autre part, justifier cette même situation en la colorant des couleurs du bien et du sens » Jacques Ellul, *Pour Qui, Pourquoi Travaillons-Nous ?* Editions de La Table Ronde, 2013.

# THE OPERATOR:
## A FIGURE THAT TRANSCENDS NATIONS

Even though most of today's economic means are highly dynamic in terms of their constant change and development (especially in the case of technological configurations and infrastructures) there is still a territory which is linked to an individual sense of attachment: the *nation*.

It is natural for peoples and individuals to have a sense of identity, the psychological attachment or deep respect to the custom, habits, and traditions of a cultural heritage serves as an important disciplining for the individual in order to experience socio-cultural and historical entities that transcend his own individuality, and that the capacity to *serve* the forces forms and moulds the personality of the individual and how he relates to the cosmos. In regions of the world where there has not been any significant irruption of modern technique or technology. Respect for one's cultural heritage contributes to maintaining the homeostatic relations between people, political organization, and local territorial conditions. These autochthonous relations have been highly compromised during the last century by the mutation of infrastructure and technique. At this stage, respect for one's regional or national heritage is challenged by the use of technical means that are alien to the culture. The culture can not produce on its own, as it doesn't have the operative capacity to do so.

All these territorial mutations are still not completely capable of uprooting the cultural *sap* of a national heritage, since many peoples around the world cherish emotional and romantic/folk valuations of their past and their national cultural heritage. By themselves, the atavistic feelings are incapable to develop an operative mastery over the conditions of existence. In effect, if the mastery was actually intended, it would first demand knowledge of the aims that new technical or economic means serve, and the governors of the nation would have to consider if the nation is capable of operatively assimilating these productions. The implicit impotence of the romantic traits when aiming for dominion is expressed by Ernst Jünger with these words:

> *"the romantic domain appears as the past and also appears as a past colored by the reflection sentiment (re-sentment) against the specific situation of each moment (...) One of the loopholes of the vanquished is whatever is wonderful, understood in the sense of who knows how to magic and lovingly evoke peals of medieval bells or perfumes of exotic flowers. (...) The romantic man seeks to establish his own assessments of an elementary life that he does no longer participate, but whose validity grasps; hence deception or disappointment do not lack here."*[481]

However, during the last decades, conservative positions on the part of the State-nations had to be discarded, since modern technological configurations require the exploitation of raw materials that exist beyond most regional or national frontiers. This compromises any regional or national mastery of means and territory, so the only option remaining for maintaining operative mastery is by radically rejecting the use of alien tools and equipment. The adoption of the latter is highly exceptional, but still takes place today within groups such as the Amish in Pennsylvania and Ohio (US) who are reluctant to adopt modern techniques. Ernst Jünger points out that a radical conservative standpoint in regard to technique follows an instinctive cultural preservation when stating that:

---

[481] Ernst Jünger. *Der Arbeiter*. Ernst Klett, Stuttgart 1981. Klett-Cotta, pg 26.

*"In no way, therefore, is technique a neutral power, a stock of effective and comfortable means which the traditional forces can freely resort to. What is hidden behind this appearance of neutrality is, rather, the mysterious and seductive logic with which technique knows how to offer itself to human beings, a logic that is becoming more and more evident and irresistible to the extent it is encroaching the entire domain of work. And to the same extent it also weakens the instinct of those affected by it. The Church possessed instinct when it tried to destroy a kind of knowledge that who saw the Earth as a satellite of the Sun; instinct was possessed by the cavalry soldier who despised firearms and the weaver who broke the machines, and the Chinese who prohibited importing machines to their country."*[482]

Hence, technique is impregnated with a set of values that cannot be discarded, and has decisive cultural influences. The embracing of a radical alternative, such as practiced by the Amish, follows an instinct of preservation and conservation of heritage, which even Mahatma Ghandi embraced when advocating the use of a traditionally-based developed spinning wheel. As Neil Postman wrote in this regard: *"in every tool we create, an idea is embedded that goes beyond the function of the thing itself."*[483] However, from an operative perspective that aims to re-establish an integral Tradition, the adoption of conservationist practices lack the imperial character of actual mastery of the cosmos. For instance, groups like the Amish might be capable of establishing relative dominion over their local territory based on their customs and spiritual understandings, but

---

[482] "Die Technik ist also keineswegs eine neutrale Macht, kein Reservoir von wirksamen oder bequemen Mitteln, aus dem jede beliebige der überkommenen Kräfte nach Gutdünken zu schöpfen vermag. Gerade hinter dem Anschein dieser Neutralität versteckt sich vielmehr die geheimnisvolle und verführerische Logik, mit der die Technik sich den Menschen anzubieten versteht Diese Logik wird immer einleuchtender und unwiderstehlicher in demselben Maße, in dem der Arbeitsraum an Totalität gewinnt. In demselben Maße auch schwächt sich der Instinkt der Betroffenen.
Instinkt besaß die Kirche, als sie ein Wissen zerstören wollte, das die Erde als einen Trabanten der Sonne sah; Instinkt besaß der Ritter, der die Gewehre verachtete, der Weber, der die Maschinen zerbrach, der Chinese, der ihre Einfuhr verbot" Ernst Jünger. *Der Arbeiter*. Ernst Klett, Stuttgart 1981. Klett-Cotta, pg 82.

[483] Neil Postman. *Amusing Ourselves to Death*. Penguin Books. 1986 pg 12.

they are no longer free from the effects of the developments taking place worldwide, such as climate change, resource depletion, or industrial pollution.

Yet as History has shown, the seductive new industrial means of power (industry, commerce, financial alternatives, etc.) were spontaneously adopted by a vast percentage of nations and peoples to the detriment of more traditional forms of operative dominion. Once the automobile was adopted, it became practically impossible to make a living by resorting to a horse and carriage. When this kind of technical transition takes place, the atavistic national feelings of the individual have no actual outcome when being forced to compete against new technical powers and, repressed by the new conditions of power, they are finally *transposed* or *sublimated*—in psychoanalytical terms—in the ideology of *nationalism*, which provides a sense of collective identity to an individual by identifying with a given collective that speaks the same language, or that shares the same *signs* of cultural heritage. These *signs* can also correspond to what are collectively considered to be the specific *signs* of belonging to a given nation, like having a certain skin color, or other specific physical attributes that are often conceived as "racial" attributes. Yet whenever these particular signs become the main factors of discernment, then the ideological character of nationalism becomes very obvious, since all ideologies intrinsically foster a quantitative and mechanistic view of the individual in relation to the whole, as Friedrich Georg Jünger points out when writing: *"all ideology already presupposes mechanization, a machinelike uniformity in the thinking of its followers."*[484] Like all forms of propaganda, an ideology triggers inner drives that are present in the individual, and then provides these irrational tendencies with a socio-political outcome, so all atavistic tendencies or instinctive predispositions—mostly derived from belonging to a specific lineage—can easily find in a nationalistic ideology a pretext for affirming the individual. At this point we are no longer referring in any case to the Individual presented by Evola, or the *type* presented by Ernst Jünger, but rather we are dealing with

---

[484] Friedrich Georg Jünger: *Perfektion der Technik* (1946).

the individual as ephemeral and abstract conceptions derived the time of the Enlightenment. Concepts of the "individual" as a social atom exposed to external determinisms is included in Evola's first and second stage of TAI-PAI; thus the individual is in a state of privation, where he is alien to the configuring force of his actions. In the state of need, desire, and privation, an individual who demands a nationalistic ideology as an existential justification, shall easily surrender to propagandistic bombardment and be the main target of merchants and business types who aim to gain power in modern State-nations by gaining social acceptance, conceiving of society as a mass-aggregate of individuals.

An individual attached to a nationalist ideology is deep down incapable to foresee any other goal for his life other than that of gaining individual predominance over other individuals. Yet ultimately, the only way this can be accomplished is by enhancing economic capacity, fame, and popularity, which entails competition with countless other individuals in a struggle for success in the "rat race" confirms many Neo-Darwinist postulates, such as those presented by the author of "*The Selfish Gene*," Richard Dawkins.

It could be a coincidence that the term *nationalism* appeared rather recently in modernity, corresponding with the cultural takeover provoked by the foundation of Speculative Masonry in London in 1717. Formerly, *nation* was not a term related to a population belonging to a given geographical area, but rather entailed the idea of an organic structure capable of establishing operative links with a given territory, which could be local, regional level or even continental. The idea present in the term *fatherland* expresses very well this subtle and decisive difference. As Fritjof Schuon affirmed that *"caste is superior to race because spirit is superior to form; race is form, caste is spirit,"*[485] here we can say that *the fatherland is superior to the nation because spirit is superior to form; nation is form, fatherland is spirit*. Fatherland implies political authority over the nation or territory, established by a minority of individuals of the higher caste, individuals who have overcome the

---

[485] Fritjof Schuon. *Castes and Races*, London, Perennial Books. 1982. The Meaning of Race.

merchant spirit and whom embrace warrior or priestly caste traits. This perception of the nation is what caused Ernst Jünger to coin in 1926 the term "Neo-nationalism" in opposition to the extremely ideological view of the nation which he referred to as "nationalism of the antecedents" [Altväternationalismus]. Yet whenever the operative mastery implicit in an active caste is extinguished and the decline of the States or dynasties is expressed in their incapacity to master national and territorial configurations, it is precisely at that point when the term *nation*, no longer referring to operative dominion, adopts a highly speculative and abstract character. Ultimately those individuals who, disconnected from an integral Tradition, lack self-mastery and find the ideology of nationalism appealing, as it justifies the ethical and moral surrender to all external forces This therefore expresses that nationalism becomes an ideological force.

Recent History has shown that whenever any modern State-nation is threatened by the irruption of transnational powers that are characterized by a temporary capacity to control the dynamic economic and technical means, the propagation of nationalism on the part of the State has always been a rather efficient option to resort to, in order for the State-nation to *compete* against the dominion of the economic factors that are in risk of being controlled by international networks and corporations.

Hence, *as an ideology,* nationalism favors a sense of identity in the individual by attachment to a romantic memory of the former cultural structures of the nation, that is, to those structures that had previously mastered the social and economic conditions of the region. Whenever the State becomes powerless to master and integrate external forces of a military, industrial, technological or economic nature, the alternative of resorting to nationalistic propaganda reinforces the political commitment of the individuals to accept demagoguery. This occurs in State-nations that are already decadent in terms of mastery and integration of territorial conditions and are *"deceptively making the masses believe that individual traits are something that are felt to be needed,"*[486] as Jünger

---

[486] "durch die der individuelle Charakter der Masse als Bedürfnis vorgespiegelt wird." Ernst Jünger.

remarks. Decadence and corruption always go hand in hand, as in this case of modern State-nations which is expressed by the fact that State-nations have to constantly resort to the propagation of a highly deceptive *lie*; the lie that the political representatives of the State-nation are succeeding in mastering the social,cultural, and economic conditions of their nation. No matter how this affirmation is propagated in the masses, this affirmation is still an illusion, since as we've shown previously, mastery implies the power that is capable of "piloting" a new configuration of means and territorial structures towards an aim determined by the State-nations. The new territorial configurations transcended national frontiers (the so-called "globalization" process), by resorting to transnational planning on the part of corporations or international agencies, who had the temporary capacity to master the extremely dynamic means of power. Under this context, the nations appear to Ernst Jünger as one operative magnitude [*der Arbeitsgröße erscheint*][487] among many others, and as an inevitable consequence of this: "*the power of all those negotiators, diplomats, those lawyers, those businessmen is an apparent power, a power that each passing day loses ground.*"[488]

The ideological character of liberalism, Marxism, or nationalism is therefore expressed in the fact that they justify the attachment of the individual to economic factors, yet do not assist the individual to relate to the operative and technical modes of power that supersede such factors. For instance, liberalism justifies the activities of amassing capital, resources, and industrial goods by considering the market as the "regulator" instead of the State. Marxism justifies the activities of work and industrial production, and nationalism justifies the individual attachment to national capitals, workforce, and resources. All of these ideologies refer exclusively to the economic domain, but do not address the fact that after they were spread, national territories have been progressively invaded by

---

*Der Arbeiter*. Ernst Klett, Stuttgart 1981. Klett-Cotta, pg 65.

[487] Ernst Jünger. *Der Arbeiter*. Ernst Klett, Stuttgart 1981. Klett-Cotta, pg 75.

[488] "Die Herrschaft dieser Verhandler, Diplomaten, Advokaten und Geschäftemacher ist eine Scheinherrschaft, die Tag für Tag an
 Boden verliert" Ernst Jünger. *Der Arbeiter*. Ernst Klett, Stuttgart 1981. Klett-Cotta, pg 81.

a technological and industrial infrastructure that, by having a cybernetic character, is alien to the national artistic and productive heritages. In other words, the territorial conditions developed by urban-industrial modes of society constitute a mutation of the national operative traits that compromise any national mastery of technical or economic means. These ideologies are alien to the fact that before economic factors can develop, a territorial power has to be established and developed through given *techniques*. In our times economic thinking has completely invaded everything, but should not impede us observing that if no stable links exist with a given territory, no economic production is possible. The establishment of these links do not initially have an economic character, but rather a technical and operative one. Hence, first comes the conquest of a territory through the operative application of given techniques (warfare in most cases) and then the economy emerges as the specific transformation of the economic factors into products that ultimately symbolize the technical conquest. Economic factors are therefore configured through specific technical development and then *flow* into material production that symbolizes conquest. Today, all operative factors supersede the ideological and theoretical. Even Oswald Spengler, more than a century ago, realized this when he wrote:

> *"For us, too—let there be no mistake about it—the age of theory is drawing to its end. The great systems of Liberalism and Socialism all arose between about 1750 and 1850."*[489]

For instance, currently an Iphone corresponds to a material production that symbolizes temporary mastery on the part of Apple over diverse global technical and industrial conglomerates. And the fact that a corporation such as Apple has achieved spectacular economic performance in the stock markets, is not due primarily to its financial focus on profit, but its aim for technological and technical mastery. Apple, Samsung, and Microsoft, like many other

---

[489] Oswald Spengler. *The Decline of the West*. Alfred A Knopf. 1926. New York, pg 454.

corporations, correspond to *technocratic* configurations where the structures of power have an operative character, which transcends any speculative or ideological way of thinking. What do liberals *make use of* today in order to spread their ideological world-view? They resort to technological gadgets such as Iphones. What do cultural Marxists *make use of* today in order to criticize liberals? They resort to technological gadgets such as smartphones. What do State-nations *make use of* in order to propagate nationalism? They *make use of* propaganda via mass-media control and Internet based social media. This panorama allows us to recall Jünger's words:

> *"all of them have accepted their peace with technique, the kind of peace that reveals the vanquished. The consequences [of such defeat] are then presented with an increasingly obvious and thoughtless way, at higher paces"*[490] *and the specific language of technique then becomes "accepted, willingly or unwillingly, by those who surrendered to it,"*[491]

But, ultimately, there is a price to pay for using the technical means as *"there are certain powers that can not be accepted in terms of their legality without becoming accomplices of them, just as one can not accept gifts from a scammer without being his accomplice."*[492]

"Making use" of an object corresponds to a technical and operative trait that is characteristic of human beings, due to the special capacities provided by our hands, which were molded by the entire evolutionary process. Yet today we mostly "push buttons," and a mediation takes place between the movements of our hands and the consequences of the movements on the world. An Operative Tradition aims to restore, the power of the hand to relate artistically

---

[490] "Sie alle aber haben ihren Frieden geschlossen, jene Art von Frieden, die den Unterlegenen verrät. Die Konsequenzen stellen sich mit immer größerer Beschleunigung, mit immer rücksichtsloserer Selbstverständlichkeit ein." Ernst Jünger. *Der Arbeiter.* Ernst Klett, Stuttgart 1981. Klett-Cotta, pg 82.

[491] "Dieser Akt deutet sich an, indem der Unterworfene, sei es freiwillig oder unfreiwillig, die neue Sprache akzeptiert." Ernst Jünger. *Der Arbeiter.* Ernst Klett, Stuttgart 1981. Klett-Cotta, pg 84.

[492] "Es gibt Mächte, von denen man ebensowenig Legalität wie von einem Hochstapler Geschenke annehmen kann, ohne daß man sich zum Mitschuldigen macht." Ernst Jünger. *Der Arbeiter.* Ernst Klett, Stuttgart 1981. Klett-Cotta, pg 123.

to the world by minimizing mediations, and this constitutes the empiric proof of Mastery, in the strictest sense. As Jünger writes: *"it is expected that the service that now hands refuse to take, will emerge again in those places where humans appear as Masters, where they appear linked to their means without contradictions."*[493] Hence, the hand, the *"tool of tools"*[494] from the viewpoint of an Integral Tradition —fuses the operative and speculative spiritual traits—considering that the *use* of a given object is always embedded with a ritual and symbolic significance, to the extent that through repetition and habit, it forges the individual's spirit and predispositions in connection to the cosmos. These predispositions can be perceived in the adoption of bodily gestures determined by the subconscious. In this technical context the propagation of ideologies on the part of political classes has absolutely no effect at an operative level, where the actual interplay of powers develop. Recovering the symbolic importance of technique and movement, Jünger asserts that *"the movement of the hand by which the collector of a tram sounds the bell is much more important than all those [ideological] attacks."*[495] In effect, Jünger had confidence that the magic implicit in hands and *handling*, considered as raw operative faculties, would eventually return to men. He writes:

> *"Also is worth waiting for, last but not least, to see eliminated the stupid arrogance that has led to see in manual work a pitiful situation. Such arrogance is the natural consequence of a purely abstract and economic concept of work; in correspondence with the unfortunate character of the "educated man" who never was lucky enough, in any area, to start serving from the lowest step and start climbing all steps of rank one by one. All manual activities, including that of cleaning excrement in the stables, to the extent*

---

[493] "Um dieses Verhältnis zu veranschaulichen, sei noch einmal die eben erwähnte Rolle der Hand als des Werkzeugs der Werkzeuge gestreift: es ist vorauszusehen, daß dort, wo der Mensch als der Herr und in widerspruchsloser Verbindung mit seinen Mitteln erscheint, auch die Hand den Dienst wieder aufnehmen wird." Ernst Jünger. *Der Arbeiter*. Ernst Klett, Stuttgart 1981. Klett-Cotta, pg 119.

[494] Idem.

[495] "Uns ist die Handbewegung wichtiger mit der ein Straßenbahnschaffner seine Klingel bedient" Ernst Jünger. *Der Arbeiter*. Ernst Klett. Stuttgart 1981. Klett-Cotta, pg 120.

*they are not felt as an abstract work, are rather executed within a larger order full of sense."*⁴⁹⁶

Completely independent of any liberal, Marxist, or nationalist standpoint, an individual who feels deeply attached to the constant use of an Iphone is actually symbolizing a profound sense of service towards the temporary operative powers (corporations such as Apple or Google, etc.) that, by having established strong technical and territorial dominions, determine the material and economical developments taking place globally. These power configurations also integrate and render virtual any ideological standpoint inherited from the last decades. However, this virtual character does not imply that the ideologies or political standpoints have become less speculative; quite the contrary. Jünger had hoped that operative disciplining on the part of the individual in order to *"deal with romantic or traditionalist influences and to develop an attitude which will not be possible to convince with mere words"* would help them face a scenario where *"soon there will be no political magnitude that will not try to act by invoking socialism and nationalism, and we must see that this phraseology is accessible to anyone who is aware of the twenty-eight letters of the alphabet."*⁴⁹⁷

As an inevitable consequence that was formerly developed in TAI-PAI, the more operative mastery and dominion diminishes, the more speculative aspects hypertrophy as forms of privation, and the more they appear as *a posteriori* justifications of human

---

496 "Nicht zuletzt ist auch zu erwarten, daß eine alberne Überheblichkeit, die dazu geführt hat, in der Handarbeit einen bemitleidenswerten Zustand zu sehen, ausgetrieben wird. Diese Überheblichkeit ist die natürliche Folge eines abstrakten, etwa rein wirtschaftlichen, Arbeitsbegriffes; ihr entspricht die unglückliche Figurdes »Gebildeten," der niemals das Glück hatte, auf irgendeinem Gebiete von der Pike auf im Dienst gewesen zu sein. Jeder Handgriff, selbst das Ausmisten von Pferdeställen, besitzt Rang, insofern er nicht als abstrakte Arbeit empfunden, sondern innerhalb einer großen und sinnvollen Ordnung geleistet wird" Ernst Jünger. *Der Arbeiter*. Ernst Klett, Stuttgart 1981. Klett-Cotta, pg 145.

497 "Freilich muß hier alles verschwinden, was romantischen oder traditionalistischen Einflüssen nicht gewachsen ist, und es muß eine Haltung stattgreifen, die durch bloße Worte nicht zu überzeugen ist. Es wird binnen kurzem keine politische Größe mehr geben, die nicht durch den Appell an den Sozialismus und an den Nationalismus zu wirken sucht, und es muß gesehen werden, daß diese Phraseologie jedem offensteht, der den Gebrauch der vierundzwanzig Buchstaben beherrscht" Ernst Jünger. *Der Arbeiter*. Ernst Klett, Stuttgart 1981. Klett-Cotta, pg 124.

actions in order to maintain the illusion of the individual, which as Ernst Jünger pointed out, corresponds to the most *"prodigious and abstract form of human being, the most precious discovery of bourgeois sentimentality."*[498]

In opposition to the decadence of the *individual* in modern times, by developing an activity that aims for integration and meaning, the *type* can overcome decadence and create new modes of art that constitute a redeeming element in the core personality, to the point that gesture is substantially modified naturally, as a symbolic expression. Jacques Ellul writes that,

*"Motion is the spontaneous expression of life, its visible form. Everything alive chooses of itself its attitudes, orientations, gestures, and rhythms. There is, perhaps, nothing more personal to a living being —as far as the observer is concerned—than its movements. In reality there is no such thing as movement in general; there are only the movements of individual things."*[499]

The power of *gesture* has to be discovered again as the most revealing attribute of a personality; a set of instinctive movements and reflexes that arise in moments when the conscious mind and all its subjective contents "let go." Gesture is ultimately a character that defines caste; it is forged by *technique*, and can be perceived in the way individuals walk, talk, play, and write. As Evola writes in *Orientations* (1950): *"A point must be reached where the human type we are referring to, must constitute the cellular substance of our troops, easily recognizable, impossible to confuse, differentiated, who he can say about him: Here is someone who acts as a man of the movement."*

The more an individual is attached to liberal, Marxist, or nationalist ideologies, the more the individual becomes threatened by new territorial and technical configurations of power, and the more the standardizing of gesture is imposed by the use and

---

[498] "Im engsten Verhältnis zur Gesellschaft steht endlich der Einzelne, jene wunderliche und abstrakte Figur des Menschen, die kostbarste Entdeckung der bürgerlichen Empfindsamkeit und zugleich der unerschöpfliche Gegenstand ihrer künstlerischen Bildungskraft" Ernst Jünger. *Der Arbeiter*. Ernst Klett, Stuttgart 1981. Klett-Cotta, pg 10.

[499] Jacques Ellul. *The Technological Society*. Vintage Books, pg 482.

attachment to standardized means (such as technological gadgets) is a revelation of the standardizing of subconscious traits that determine psychological reactions or thought processes on the part of the individual. A good artistic exposition of the standardizing caused by modern modes of highly functional technique is exposed in Godfrey Reggio's film *"Visitors"* (2014), where the American filmmaker exposes how the technological cosmos perceives the individual and how it molds its reactions.

Not only are today's State-nations threatened and disempowered by the new configuration of territorial operative powers, the *individual* is also threatened... and the more threatened the individual is, the more they rely on self-security in the use of products derived from a cybernetic structure, which demolishes their sense of individuality even further, thus fostering neurotic and narcissistic traits. *Is there any way out of this vicious cycle?*

Most people do not even consider this vicious cycle a problem. Like the case of Spiral Dynamics, the technical and cybernetic paradigms have also integrated spiritual traditions and rendered them into ideological spiritualism. However, in the West and in the East, there were times when the aristocratic castes who embraced real (royal) mastery and dominion, were surrounded by artistic and technical creations that expressed their beings: these creations were the honest expression of their origins and destinies. Today, however, the aristocratic traits have been completely extinguished. The old aristocracies who still live in luxurious palaces are more related to the cybernetic infrastructure of technologies than to the artistic creations they've inherited from the past. If there is any chance of an "awakening" from the *vicious cycle* fostered by cybernetic configurations of power, some former aristocracies and noble castes should not be excluded from recovering dominion and operative mastery, since these individuals still cherish atavistic memories in regard to dominion over territorial means through an artistic self-discipline that focuses on the importance of cultivating *virtue, opus,* and *art (techné)*. Even though these old castes present a rather romantic and conservationist perspective on the importance of the State and its relation to the activities of the individual, the romanticism and conservative traits—which are sometimes labeled as "reactionary"—are still valuable in order to "protect" individuals from the *counter-initiatory* effects caused by ideological attachment to speculative views on the cosmos. This "protection" of the individual facilitated by respect for aristocratic ancestors has a "prophylactic" effect that should not be discarded. It should be emphasized, however, that these old castes are favored only in terms of an "awakening" or realization of the absurdities caused by human attachment to the vicious cycle triggered by the global cybernetic infrastructure of power. In truth, at the end of the day, what shall count for advancement through the stages presented in this book, is not only an "awakening" but the whole development of practice and discipline, which can be extremely demanding for these old castes, just as it is for anybody else. When contemplating this issue, Jünger wrote: *"More important than the comparison with copies of past times and missing spaces is for us whether we ourselves maintain a new and unique primordial relationship."*[500]

Let's present at this point of the exposition that the whole aim of theory and practice exposed in this book is to put forth stages for the recovery of operative mastery and dominion, thus intending to link one's spirit to a State characterized by regal and homeostatic traits. This State or kingdom doesn't yet exist anywhere, and it can't be "planned," nor "projected" into the future. This kingdom is not determined by time considered chronologically, but by time conceived ontologically, and can only be visible with *privileged* eyes that are capable of perceiving the actual structure of power in our times, and capable of relating to the ultimate symbol or *figure* of its development. Secondly, one also has to be capable of disciplining one's spirit in service to another specific civilizational development which, as with all kingdoms, ultimately symbolizes freedom and sovereignty over the conditionings and determinisms present in modern power configurations. As formerly stated, these

---

[500] "Wichtiger als der Vergleich mit den Abbildern dahingeschwundener Zeiten und Räume ist für uns die Frage, ob wir nicht in einem neuen und eigentümlichen Urverhältnisse stehen." Ernst Jünger. *Der Arbeiter*. Ernst Klett, Stuttgart 1931. Klett-Cotta, pg 102.

configurations are characterized by a transitional "workshop" character, mainly because of their technocratic and cybernetic structures.

\* \* \*

After 1717, the eyes of modern men could no longer be present at the miraculous foundation of kingdoms or States, as occurred in former times where myth emerged in order to meta-historically express this potential, even if the Zeitgeist was not favorable. So when intending to put into practice any endeavor that plans to reconnect with an *Absolute State*, one of the most important dangers is that a passive surrender to the technical determinations of our times can easily foster a sense of self-satisfaction. This danger can't be circumscribed to any specific socio-economic sector, since the traits are extremely widespread in all domains, to the same extent as the cybernetic and global infrastructure of power.

The *Absolute State*, as with all Sacred Imperial States that historically expressed through culture integrated stages of metaphysical mastery over temporary powers, required men to be free from any sense of individualism and to define their sense of being in terms of operative mastery over the technical, economic, and political domains of the State. The establishment of territorial and organic links implies the constitution of *caste* structures, where what counts in terms of mastery and dominion are not the abstract and subjective attachments that foster modern individuality, but rather a whole set of disciplines and practices aiming to *cultivate instinct*. In general, the modern individual represses instinctual drives at work, and is urged to liberate them during the weekends with alcohol, drugs, or through the entertainment industry. In the case of caste structures, the subconscious and instinctive traits become permanently "bridged" with the homeostatic and metaphysical character of the State, and such a man in service of the State is characterized by an individuation expressed by the symbolic character of his productions. By *Immanent Transcendance*.

\*\*\*

As Nietzsche wrote when recovering the golden core of European traditions:

> *"O my brethren, I consecrate you and point you to a new nobility: ye shall become procreators and cultivators and sowers of the future;—Verily, not to a nobility which ye could purchase like traders with traders' gold; for little worth is all that hath its price. Let it not be your honour henceforth whence ye come, but whither ye go! Your Will and your feet which seek to surpass you—let these be your new honour!"*[501]

---

[501] Friedrich Nietzsche. *Thus Spoke Zarathustra*. LVI. Old and New Tables.

www.ingramcontent.com/pod-product-compliance
Lightning Source LLC
Chambersburg PA
CBHW080237170426
43192CB00014BA/2477